RADIO DETECTION OF HIGH ENERGY PARTICLES

Related Titles from AIP Conference Proceedings

To learn more about these titles, or the AIP Conference Proceedings Series, please visit the webpage **http://www.aip.org/catalog/aboutconf.html**

RADIO DETECTION OF HIGH ENERGY PARTICLES

First International Workshop
RADHEP 2000

Los Angeles, California   16–18 November 2000

EDITORS
David Saltzberg
University of California, Los Angeles, California

Peter Gorham
Jet Propulsion Laboratory, Pasadena, California

AMERICAN INSTITUTE OF PHYSICS

Melville, New York, 2001
AIP CONFERENCE PROCEEDINGS ■ VOLUME 579

SEP/AE
PHYS

Editors:

David Saltzberg
Department of Physics and Astronomy
University of California, Los Angeles
Box 951547
Los Angeles, CA 90095-1547
USA

E-mail: saltzberg@physics.ucla.edu

Peter Gorham
Jet Propulsion Laboratory
California Institute of Technology
Mailcode 238-600
4800 Oak Grove Drive
Pasadena, CA 91109
USA

E-mail: peter.gorham@jpl.nasa.gov

L.C. Catalog Card No. 2001091910
ISBN 0-7354-0018-0
ISSN 0094-243X
Printed in the United States of America

CONTENTS

CURRENT AND FUTURE EXPERIMENTS USING RADIO EMISSION IN DENSE MEDIA

ACCELERATOR MEASUREMENTS

RADAR AND OTHER TECHNIQUES

APPENDIX

Preface

As the study of cosmic rays has turned to rarer and rarer events, astrophysicists have been driven to build larger and larger detectors. Radio detection has long offered the possibility of instrumenting the largest volumes of material. After forty years of research, researchers in the field met at the University of California, Los Angeles for the "First International Workshop on the Radio Detection of High Energy Particles" (RADHEP-2000). From November 16–18, 2000 we discussed theoretical and experimental aspects of the detection of high-energy cosmic rays and neutrinos using radio waves and related techniques. We gratefully acknowledge the U.S. Department of Energy for supporting the publication of the proceedings.

This workshop was in many ways indebted to two outstanding twentieth century physicists, Gurgen Askaryan the remarkable theorist, and John Jelley the consummate experimentalist, both of whom have recently passed away. Their ideas and methods have now resonated with us for nearly forty years, sometimes with great strength, and other times without much notice. But from the enthusiasm and creativity evident at this meeting, it appears that the seeds planted by these physicists a generation ago are now bearing fruit.

David Saltzberg
Peter Gorham

Los Angeles, California

HISTORICAL OVERVIEW

Radio Pulses from Cosmic Ray Air Showers

Trevor C. Weekes

Whipple Observatory, Harvard-Smithsonian Center for Astrophysics,
P.O. Box 97, Amado, AZ 85645-0097.
e-mail: tweekes@cfa.harvard.edu

Abstract. The first experiment in which radio emission was detected from high energy particles is described. An array of dipoles was operated by a team of British and Irish physicists in 1964-5 at the Jodrell Bank Radio Observatory in conjunction with a simple air shower trigger. The array operated at 44 MHz with 2.75 MHz bandwidth. Out of 4,500 triggers a clear bandwidth-limited radio pulse was seen in 11 events. This corresponded to a cosmic ray trigger threshold of 5×10^{16} eV and was of intensity close to that predicted. The early experiments which followed this discovery and their interpretation is described.

I BACKGROUND

It is a great honor to be asked to give the opening talk at this, the first workshop on the radio emission from high energy particles. I should stress that my involvement with the field was ephemeral and that although I was the first one to actually see evidence of radio emission from high energy particles, I have made no contribution to the field for 34 years. This invitation motivated me to open my Ph.D. dissertation, probably for the first time since its completion [1], and to realize how lucky I was to be involved, albeit in a junior capacity, with some outstanding physicists in a classic experiment in radiation physics. Since I am now struck by how different the methods used in this simple experiment are from those used by the contemporary student I will belabor the experimental method and stress the historical background as best as I remember it.

Experimental physics in the fifties was still very much conducted in the shadow of post-war politics, economics and experience. The positive contribution of physicists to the war effort ensured that physics research received some support even in those austere times. Still, ingenuity and creative solutions were necessary to accomplish anything meaningful. As a graduate student in the Experimental Physics Department of University College, Dublin, I was fortunate to join Neil Porter's Cosmic Ray group (in 1962) which was building a very high energy gamma-ray astronomy experiment. In those days graduate students really did build experiments, perhaps

CP579, *Radio Detection of High Energy Particles,* edited by D. Saltzberg and P. Gorham
2001 American Institute of Physics 0-7354-0018-0

because little time was spent at computer terminals. Porter had recently returned to Dublin after a stint at the Atomic Energy Research Establishment (AERE), Harwell in England and continued to collaborate with John Jelley at AERE on these cosmic ray studies. It was a measure of the times that our gamma-ray telescope consisted of ex-World War II searchlight mirrors mounted on an old British-naval gun mounting and that major support for the experiment came via a grant from the U.S. Air Force. These were exciting times in high energy astrophysics (as the field came to be called) with the first X-ray sources yet to be discovered and pulsars and quasars still over the horizon. The post-Sputnik explosion in physics research was still in its infancy and its effects had not yet really reached the British Isles.

The discovery of cosmic ray particle cascades, the so-called Extensive Air Showers (EAS) [2] opened a new era of cosmic ray studies with most interest centering on using them to explore the highest energy cosmic rays. The sharply falling spectrum meant that the events of greatest interest were detected infrequently by conventional means (arrays of spaced particle detectors). In the wake of World War II and the availability of radio equipment and expertise, Blackett and Lovell [3] were moved to attempt to detect EAS using radar techniques to detect the ionizing trail left in the wake of the shower. These experiments were unsuccessful (the lifetime of the ions was overestimated) and the technique was not pursued. The feasibility of using simple optical radiation telescopes to detect EAS was demonstrated by Galbraith and Jelley [4] following the suggestion by Blackett [5] that the combined Cherenkov emission from all secondary cosmic rays in the atmosphere might constitute 0.01% of the night-sky light background. Since the Cherenkov radiation was concentrated in the forward direction with a lateral spread similar to that of secondary particles, this technique did not offer too much for the detection of very large EAS, particularly since it was limited to clear dark nights giving duty cycles of $< 10\%$; it did however lead to the development of an effective technique for very high energy gamma-ray astronomy [6]. Jelley [7] proposed that Cherenkov radio emission at radio wavelengths might be detectable but at the microwave frequencies necessary to prevent destructive interference; the predicted signal was small although potentially detectable [8]. The possibility that EAS might be detectable by the fluorescent emission that they caused in the atmosphere was explored, without success, by Greisen [9]; it was later to be revived with great success by the University of Utah group [10].

In 1962 G. Askaryan published a short paper in Russian in which he suggested that the particle cascade resulting from the interaction of a high energy particle in a dense medium would not be electrically neutral since the resulting positrons could decay in flight; also the cascade would accumulate delta-rays and Compton scattered electrons [11]. Cherenkov radio emission could then occur at longer wavelengths where there would be coherent emission from the net negative charge. Askaryan was primarily concerned with the emission of radiation from the particle cascade that would result from the interaction of a cosmic particle in a dense medium like rock, e.g., on the moon. Since this dielectric material is essentially transparent to radio waves, it can provide a large target mass for such elusive particles as neutrinos. Askaryan also pointed out that geomagnetic effects might also contribute to separation of the charged components in the shower and this dipole might provide an additional emission mechanism. He estimated the negative excess, ϵ in rock to be about 10%. If N = the number of particles in the shower, then instead of incoherent emission from N particles, we must consider coherent radiation from $(\epsilon \cdot N)^2$. If N=10^6 and ϵ=0.1, then the coherence factor is 10^4, a huge gain. The coherence condition would require that the dimensions of the shower-emitting region should be comparable with the wavelength. The Cherenkov emission is proportional to $\nu.d\nu$, so that the reduced Cherenkov emission at, say, 50 MHz compared with 5 GHz would be more than compensated for by the increased coherence factor.

The Askaryan paper appears to have gone unnoticed by experimentalists until a follow-up paper by Alikanyan [12], on the emission of radio emission by high energy particles, was noted by Neil Porter who was asked to write an abstract for its publication in English. The paper referenced the earlier Askaryan paper but it was not available in Ireland. Knowing that Jelley had been interested in such phenomena, Porter sent a copy of the Alikanyan paper to Jelley with a note to the effect that it did not seem a very feasible technique (Porter, private communication). Jelley acquired the English version of the Askaryan paper and realized that the negative excess offered new possibilities for the radio detection of air showers. After an exchange of letters Jelley and Porter agreed to attempt a simple experiment. Since neither of them had access to radio telescopes they decided to enlist the help of the extensive post-war British radio astronomy community. F. Graham Smith, then at the Cambridge Radio Observatory, offered to assist; as he was about to take up a position of Deputy Director at the Jodrell Bank Radio Observatory, it was agreed that the experiment should be done there.

II   THE EXPERIMENT

The experiment was to consist of a large area, broad band, medium wavelength, wide angle radio telescope coupled to an adjacent air shower array which would trigger the recording of the predicted radio analog signal from high energy showers at a reasonable rate. A radio telescope with these parameters is best matched by a simple dipole array. It was proposed to use the frequency reserved for the new BBC video signal at 44 MHz where a relatively noise-free bandwidth of 2.75 MHz could be achieved when the BBC transmitter was turned off. This was approximately from 00.00 to 9.00 AM (no late night talk shows!) so this was to be a night experiment with limited running time. I do not recall whose idea it was to use this band but, in retrospect, it was key to the successful outcome of the experiment. With the increased use of the radio bands for communication it would not be possible to do this today.

As the junior graduate student in the Dublin group I was assigned the task of designing and building the EAS trigger that would be used to signal the arrival of a large EAS. The design was based on calculations using analytical models of shower particle distributions. I believe these simple calculations (on our state- of-the-art IBM 1620 computer) constituted the only involvement of a computer in the experiment; computers were not involved in data taking or data analysis.

The construction of the air shower array entailed a number of trips to AERE where the array would be built and tested. AERE, as a large British government research laboratory with seemingly limitless resources, was a wonderland to a young Irish student. It was possible to put together a somewhat primitive but functional array of Geiger counters in a matter of weeks. Geiger counters were chosen because they were cheap and easily available. The simple design called for three trays of Geiger counters at the corners of an equilateral triangle of side 50 m; a trigger required a signal from each tray within 0.5 microseconds. In each of the three trays we had counters of various sizes so we could get a rough determination of the particle density and hence the shower size and its proximity. Geiger counters are slow and noisy and precautions had to be taken to ensure that the radio system was not susceptible to pickup from the shower trigger. To achieve this we planned to run the complete shower array from one battery supply, the radio detectors from another. In June 1964 the array was operated for a week at AERE (beside the Tandem Generator Building) where the trigger rate was about 2.5 events per hour, roughly corresponding to an EAS threshold of 10^{16} eV.

In July, 1964, the array was moved to the pastoral setting of Jodrell Bank, some 20 miles from Manchester. There, under Graham Smith's direction, a dipole array was being assembled, primarily by Bob Porter, a University of Manchester graduate student.

With John Jelley and senior Harwell technician, John Fruin, I assembled the small air shower array adjacent to the dipole array and in the shadow of a 15 m radio telescope whose instrument shack we commandeered. As with any new experiment in the field it was an exciting time. Within a week we had the array up and running and the Harwell pair left me to operate it for a couple of months.

The putative shower radio signal was delayed by passing it through one kilometer of coaxial cable in a rack enclosed in a Faraday cage with amplifiers distributed along the line to compensate for cable attenuation. The output of the radio detector was to be displayed on the 20 microsecond time-base of one channel of a two channel oscilloscope; a time signal was displayed on the other. The scope was triggered by the air shower array with the radio pulse expected between 5.0 and 6.0 microseconds from the beginning of the trace (there was considerable jitter in the array trigger) (Figure 1). This long time-base gave a display of the radio background noise both before and after the shower and was important in establishing that any detected bandwidth-limited pulses were really associated with the shower. The oscilloscope display was recorded on 35 mm film by a Shackman camera; a small lens also brought into focus a clock and hodoscope indicating which of the smaller Geiger counters had been struck.

We quickly learned that radio pulse investigations operate in a quite different environment from that seen by conventional radio astronomy. The long integrations

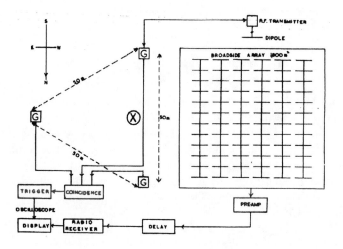

FIGURE 1. The Geiger counter array was set to one side of the dipole array which had a total area of 1,700 m². The rf transmitter was used in calibration experiments to measure the delay in the analog signal [1].

used by radio astronomers smoothed out the man-made pulsed radio background which we soon discovered to be dominated by such things as the nearby electric trains, the typewriters in the administration building and the erratic ignition systems of old cars driven by senior faculty. These problems were to be rediscovered by the radio astronomers who within a few years would discover the radio pulsars with very similar equipment.

Sadly I find I have no photographs of the dipole array which did not lend itself to photography (no more than the air shower array of Geiger counters did!). In contrast I have a detailed written log of those days; I was scheduled to get married in September and out of guilt at leaving all the marriage arrangements to my future wife, I wrote to her every day; romantic soul that I was, I filled my letters with day-to-day accounts of the progress of the experiment! The Geiger array ran pretty much without problems (apart from the inevitable eating of the high voltage cables by rabbits). The trays and cables were 0.5 m above ground level and kept air tight; even when the grassy area in which they were located was partially flooded they continued to operate!

The success of the experiment was in no small way due to the choice of frequency and bandwidth and relative radio quietness at that time. A chart recorder that recorded the integrated radio brightness demonstrated the nightly passage of the Milky Way and confirmed that we were indeed galactic noise-limited; the sky-brightness temperature varied from 6,000° to 20,000°. The receiver noise temperature was 450°. No form of automatic gain control was used. By phasing the dipole array we were able to observe the transit of Cassiopeia A and thus determine our beamwidth (10° FWHM). This was my first experience of real astronomy using non-visible wavelengths!

III   THE RESULTS

On August 19, 1964 all was ready for the first night of data taking. John Jelley was a meticulous experimentalist and had left a check list for the night operator to complete. Radio observatories are often lonely places at night and it was a little eerie setting up an experiment in a remote location to begin operation after midnight. All was now ready to test the hypothesis that air showers produced detectable radio emission. I confess to some anticipation and excitement as I set the system to operate that first night. These emotions were balanced by my discovery the next morning that after the first event the camera had jammed and no data was taken! The next night I was more careful: I double checked the Jelley list and waited for an hour to be certain that data was being recorded.

It was Sunday morning when I unloaded the camera, checked that the various housekeeping chart recorders showed nothing unusual and developed the long roll of film in the observatory dark room. In the dark room lights I made a quick check on the film and was delighted to see on the fifth image recorded exactly what we had been looking for, evidence for a very large radio pulse at precisely the right point on the time-base. Later that day I examined the full roll of film and noted no other evidence for emission but no background events either.

This was my first experience of a scientific discovery and young that I was, I assumed this was the norm of scientific research and that I should expect such happenings on a regular basis! Little did I realize that (a) I would wait another twenty years for a comparable happening (b) that this was the largest pulse we would record and (c) there would be no other evidence for emission in the remainder of the eight days of the run. Wisely I decided I should celebrate immediately; I did not have access to a telephone on site, there was no e-mail or other way to communicate with my supervisors, so I mounted my trusty government-issue bicycle and repaired to the nearest village some miles away where I downed some good English bitter.

The next week was a disappointment. All systems seemed to run perfectly but no radio pulses were evident. We were running with just one quarter of the dipole array completed. At the end of the week we were faced with the big decision; did our preliminary run justify the effort and expense of completing the array? I opted for completing the array which was an easy decision on my part since I was leaving for a few months and the cost, which was minimal, would be borne by Jodrell Bank.

In a month the array was completed and the experiment was run on almost every night until March 1965 (I did another stint of operation in December, 1964). To my immense relief more pulses appeared but only one off-scale pulse. In some 4,500 air shower triggered events, there were eleven with clearly discernible pulses in the right delay window. There were no comparable pulses anywhere else on the event times-bases or on time-bases triggered by a clock every half-hour. Some of these

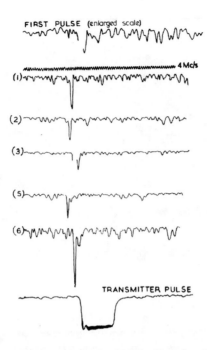

FIGURE 2. Some of the oscilloscope traces showing the "large" radio pulses. The first (offscale) pulse was recorded on a different timebase than the others. The arrival time of the calibration transmitter pulse must be corrected for the time delay in the Geiger circuitry [1].

events are reproduced in Figure 2.

We also performed an analysis on the traces that did not show an obvious large pulse. This analysis was done in Dublin and consisted of recording the position of the maximum "noise" fluctuation on each oscilloscope trace. A clear peak was seen in the distribution in the anticipated interval (and none in the clock triggered events) providing independent evidence for the detection of radio emission. This data was folded with the distribution of measured pulse sizes in the large events to give the size spectrum shown in Figure 3.

IV   INTERPRETATION

The observed rate of detected radio pulses was consistent with emission from cosmic ray air showers initiated by primaries of energy 5×10^{16} eV. The energy in the received radio pulse was about 1 eV. The air shower array did not give any information about the shower arrival direction and the small counter triggers did not allow a precise definition of either the size of the shower or the position of its axis. A series of follow-up experiments involved (a) use of an optical air shower to trigger on smaller showers with known arrival directions (b) operation in

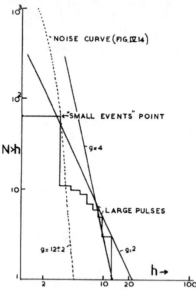

FIGURE 3. The distribution of pulse sizes measured in the delay window where radio air shower pulses are expected is plotted with a single point for the observed small events. The normal receiver noise spectrum pulse height distribution is also shown [1].

coincidence with the 15 m radio telescope fitted with a radio receiver at 150 MHz (c) a short run with one section of the dipole array rotated in orientation to favor North-South polarization. After 1966 (when I left UCD) the Dublin group extended the observed radio pulse frequencies into the UHF band and detected radio pulses that were probably radiated incoherently [13]. None of these experiments gave conclusive results on the nature of the emission but all were consistent with the hypothesis that the radio emission originated in large air showers.

Alternative radiation hypotheses were also considered including: (a) nuclear field bremsstrahlung; (b) transition radiation; (c) induction effects; (d) cosmic rays striking receivers; (e) reflection of TV signal; (f) molecular emission. Preliminary estimates eliminated all these as possible explanations of the observed radio emission [14] [1].

In the original Askaryan paper the possibility that radio emission might also result from geomagnetic effects on the shower particles were mentioned. This concept was developed by Porter in Dublin as we built the experiment. Two mechanisms were recognized in addition to the Cherenkov radiation from the negative excess: i) the separation of the positive and negative charges as they traversed the earth's magnetic field would create a dipole and give rise to radio emission; ii) as shower develops in the atmosphere, electrons are being moved to one side, positrons to the other by the magnetic field; the result is that the shower constitutes a current element which will also radiate radio waves. The full treatment of these processes was done eventually by Kahn and Lerche at Manchester University [15] and Colgate [16] at Los Alamos. A good summary of all aspects of these early experiments can be found in a review article by Allen [17]. Remarkably all three processes seemed to give signals of comparable magnitude (but with the detection of polarization and the measured distribution of radio signal with frequency, the consensus now seems to be that the current element is the dominant emission mechanism).

V CONCLUSIONS

The results from the initial experiment were published in a Letter to Nature [14] and in a Nuovo Cimento paper [18]. They were also reported at the 9th International Conference on Cosmic Rays (ICRC) in London in 1965 [19]. By that time the phenomenon had been independently confirmed by an experiment performed by the UCD group in the Dublin mountains, using a helical antenna with a receiver at 70 MHz with bandwidth 20 MHz and a plastic Cherenkov detector shower trigger [20]. These early results were given some prominence in the report of the ICRC EAS Highlight speaker, K. Greisen [21], from whose paper the following is quoted: "The technique is barely in its infancy, and too little is known about it to justify elaborate predictions. However, we feel confident that this achievement is a significant breakthrough, and that further study will reveal ways of obtaining types of information about the showers that were not available by other means. The signal will not, of course, be sensitive to fine details like the structure of the shower core; nor does the method seem adaptable to surveying at the same time all directions in which showers may arrive. However, it appears that the method may offer new orders of angular resolution; and it may complement particle detectors by being sensitive to the conditions of showers far above ground level. Also it will be able to detect showers in steeply inclined directions, in which the particles are absorbed before reaching the ground. Time will probably reveal other possibilities that are not apparent at this early stage." Sir Bernard Lovell was to note that this experiment fulfilled the original purpose of the first telescope constructed at Jodrell Bank in 1941 [22].

This optimistic assessment led to a spate of experimental activity which was mostly aimed at an exploration of the radiation mechanisms. Interest in the field can be assessed in the numbers of papers presented on the technique in subsequent ICRCs (Figure 4). Early interest in the potential of the technique seems to have waned as it was recognized that the emission was directed and thus not suited to distant shower detection and that the man-made pulse radio background was noisy and becoming more so. Thus the early hope that radio detectors might be used in stand alone systems to detect distant very large air showers was not realized. The recent interest in the detection of showers of energy $> 10^{20}$ eV has renewed interest in the radio detection technique.

I am grateful to Peter Gorham and David Salzberg for allowing me to indulge in this reminiscence of the early work and to Neil Porter and David Fegan for jogging

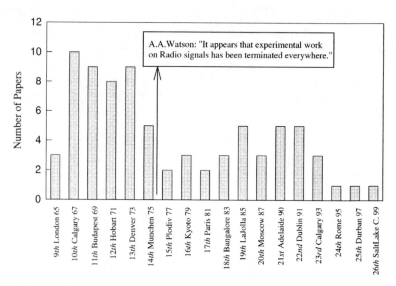

FIGURE 4. The number of papers on the radio emission of air showers presented at International Cosmic Ray Conferences from 1965 to the present day (compilation by H. Badran).

my memory on important details.

REFERENCES

1. Weekes, T.C.. Ph.D. Thesis, National University of Ireland, "Radio and Optical Methods of Detecting Cosmic Rays" unpublished (1966)
2. Auger, P.R. et al., Acad. des Sciences, Paris **206**, 410 (1938)
3. Blackett, P.M.S., Lovell, A.C.B., Proc. Roy. Soc. **177**, 183 (1941)
4. Galbraith, W., Jelley, J.V., Nature, **171**, 349 (1953)
5. Blackett, P.M.S., Phys. Abst., **52**, 4347
6. Weekes, T.C., Proc. Symposium on GeV-TeV Gamma Ray Astronomy (Heidelberg, June 2000), AIP (in press) (2000)
7. Jelley, J.V. "Cherenkov Radiation and its Applications", Pergamon Press (1958)
8. Jelley, J.V., Suppl. Nuovo Cimento **8**, 578 (1958)
9. Greisen, K., Ann. Rev. Nucl. Sci. **10**, 63 (1960)
10. Baltrusaitas,R.M et al. Nucl. Inst. Meth. **A240**, 410 (1985)
11. Askaryan, G.A., Soviet Physics, J.E.T.P., **14**, (2) 441 (1962)
12. Alikanyan, A.I., Laziev, E.U., Tumanyan, W.A., N.I.M., **20**, 276 (1963)
13. Fegan, D.J, Slevin, P.J., Nature, **217**, 440 (1968)
14. Jelley, J.V., Fruin, J.H., Porter, N.A. et al., Nature, **205**, 327 (1965)
15. Kahn, F.D., Lerche, I., Proc. Roy. Soc. A, **289**, 206 (1966)

16. Colgate, S.A. J. Geophys. Res., **72**, 4869 (1967)
17. Allen, H., Prog. in Elem. Part. and Cos. Ray Phys. (N.Holland Publ. Co.) **10**, 171 (1971)
18. Jelley, J.V., Fruin, J.H., Charman, W.N. et al., Nuovo Cimento, **46**, 649 (1966)
19. Smith, F.G., Porter, N.A., Jelley, J.V., Proc. 9th I.C.R.C. (London), **2**, 701 (1965)
20. Porter, N.A., Long, C.D., McBreen, B et al., Phys. Lett., **19**, (5), 415 (1965)
21. Greisen, K., Proc. 9th I.C.R.C. (London), **2**, 609 (1965)
22. Lovell, A.C.B., "The Story of Jodrell Bank", Oxford University Press (1968)

GURGEN A. ASKARYAN (1928 - 1997)

B. M. Bolotovskii

*P.N.Lebedev Physical Institute, Russian Academy of Sciences,
Leninsky Prospect 53, 117924, Moscow, Russia*

All his life Gurgen Ashotovich Askaryan lived and worked in Moscow. He spoke about himself: "I am an Armenian of the Moscow bottling". He very seldomly left Moscow, and if he left, always short term — several days, no more than a week. He never visited Armenia, despite numerous invitations of his friends and colleagues — physicists from Yerevan, the Armenian capital, where he was highly appreciated for his outstanding scientific achievements and for his numerous jokes.

Gurgen Askaryan was born in Moscow on December 14, 1928. His father, Ashot Askaryan, was a physician, a general practitioner. He was well known in Moscow. Gurgen's mother, Astrik Askaryan, was also a doctor and dentist. The parents were widely educated people, connoisseurs and judges of painting, music and fiction, and G. Askaryan inherited these qualities. Gurgen was the second child in the family. He had a sister, Goar, who was two years older. From early childhood Gurgen was taught to play the violin. He learned music willingly, and the parents expected that music would be his occupation in mature years.

In 1940, when Gurgen was 12, the head of the family, Ashot Askaryan, passed away. The family went into a grave condition. But in spite of all difficulties, Astrik alone raised the children although it was at the breakpoint of her forces and the family sometimes underwent severe hardships.

In 1946 Gurgen Askaryan completed his school education with a gold medal. In the school his favorite subjects were music and natural sciences, physics and mathematics. One event helped him to choose between his preferences. Not long before G. Askaryan left the school, a great violinist Yehudi Menukhin came to Moscow to give a concert. As a promising young violinist, Askaryan was allowed to attend Y. Menukhin's rehearsal. During that rehearsal G. Askaryan realized that he never would attain so high a level of performance. Therefore he decided to break his musical education and entered the Physical Faculty of Moscow State University.

Among the students, Gurgen Askaryan occupied his own place. He had sharp wit and sharp tongue. His knowledge in physics went beyond the frame of university programs. Usually he expressed his opinions in bright and aphoristic manner, and

CP579, *Radio Detection of High Energy Particles,* edited by D. Saltzberg and P. Gorham
© 2001 American Institute of Physics 0-7354-0018-0/01/$18.00

sometimes (not seldom) his remarks were caustic in form, though right in essence. Therefore, some students (and even lecturers) to whom his remarks referred, disliked him. On the other side, G. Askaryan had many friends, including those who never saw him but were informed of his numerous witty statements.

During the third year of his education G. Askaryan proposed a new method of registration of fast charged particles. His idea was the following. Suppose, there is an overheated transparent liquid. A very small amount of energy is sufficient to make it boil. Let a fast charged particle penetrate through this overheated liquid. The particle expends its energy on ionization of atoms located near its trajectory. This energy loss is transformed into heat in amount which is sufficient to induce boiling along particle's trajectory. Then the trajectory becomes observable because many bubbles are created along it.

G. Askaryan discussed this proposal with some of his teachers and fellow students. No one objected. However, no one supported him, no one helped to realize the idea. G. Askaryan then was inexperienced in forms and methods of scientific investigation. He even did not publish his proposal. Several years later, in 1952, the same idea was set forth independently by an American physicist Donald Arthur Glazer. He put the idea into practice having assembled the device known now as bubble chamber. This instrument proved to be so useful in high energy physics that D. A. Glazer was awarded with the Nobel Prize in 1960. This event gave rise to Askaryan's deep concern. Of course, he was shaken that Nobel Prize was so near and, so to say, he let it slip. On the other hand, this event helped him to get faith in himself.

After graduating from University, G. Askaryan occupied the vacancy of "aspirant" (post-graduate student) in Institute of Chemical Physics, Russian Academy of Sciences. His scientific supervisor was the outstanding physicist Ya. B. Zel'dovich. After several short conversations with G. Askaryan, Zel'dovich gave him full independence in the choice of problems to be investigated. Askaryan began to work on problems connected with registration of high energy particles. However, after a year, Ya. B. Zel'dovich left Institute of Chemical Physics and moved to a secret place where the hydrogen weapon was created. As a consequence, The vacancy occupied by Askaryan was cancelled.

In 1953 Askaryan passed to P. N. Lebedev Physical Institute in the position "aspirant", as in previous institute. His supervisor here was M. S. Rabinovich, at that time an outstanding expert in the theory of accelerators. Later, his investigations in plasma physics also brought fame to him.

The head of the laboratory where Askaryan worked was V. J. Veksler, world known for his investigations of cosmic rays and discovery of the phase stability principle. Askaryan's supervisor, M. S. Rabinovich, was Veksler's deputy in laboratory, and after Veksler's passage to Joint Institute for Nuclear Research (Dubna), Rabinovich became the head. Here, as well as in Institute of Chemical Physics, Askaryan was given full freedom in his work. In this very laboratory Askaryan worked during all his subsequent life. In 45 years of fruitful work he obtained many important results.

Soon after his arrival in the laboratory, G. Askaryan began to work in close collaboration with V. J. Veksler. The latter was interested in all possible processes where strong electric fields might be obtained in order to use these fields for acceleration of charged particles. Askaryan studied different radiation effects such as transition radiation, Cherenkov radiation etc. Most detailed was his investigation of transition radiation at radio frequencies. Askaryan formulated conditions necessary for coherent radiation. He also found (together with M. L. Levin) a combination of auxiliary high-frequency fields which could secure stability of electron bunch during acceleration.

It is very difficult to enumerate Askaryan's achievements in different branches of physical sciences. Below, we mention several of his important contributions.

G. Askaryan discovered and investigated in details various effects accompanying passage of high energy particles through dense matter (liquids or solids). It was shown that hadron-electron-photon showers and even single fast particles may produce sound pulses. Ionization losses are quickly converted into heat, and the small region adjacent to trajectory undergoes quick thermal expansion thus generating sound waves. These results gave a new approach to the study of cosmic rays. Before, investigations of cosmic rays were based on direct interaction of cosmic ray particle with a detector. Askaryan's results made it possible to detect showers and single particles using sound receivers situated at some distance from the event. Later G. Askaryan showed that intense laser beam passing through matter also generates sound waves This effect may be used for processing and for destruction of matter. As a result of this series of investigations, a new branch of physics was created, radiation acoustics, and G. Askaryan was the founder. Several years ago, the registration of energetic particles and showers with sound detectors in sea water was planned as an important part of global monitoring.

G. Askaryan also showed that cosmic ray showers emit electromagnetic radiation, thus giving yet another way for their registration. Before Askaryan went after this problem, it was commonly assumed that electron-photon showers do not emit any electromagnetic radiation because electrons and positrons are created in pairs. As a consequence, it was accepted that shower had zero total charge and zero total current. G. Askaryan analyzed the problem and came to the conclusion that in an electron-photon shower there must be excess of negative charge (in other words, excess of electrons). The excess electrons are knocked out from atoms either by shower photons (photoeffect), or by shower electrons and positrons (ionization).

At the same time the number of positrons decreases because of the annihilation process. Therefore, an electric current is associated with shower, a current formed by excess electrons. The duration of the current is the same as the duration of the shower, and current varies along with development of shower. Variable current is the source of radiation. Therefore, every shower is the source of electromagnetic radiation. G. Askaryan estimated intensity of the radiation and showed that its field is well above the proper noise of the receiver. These investigations paved the way for distant registration of cosmic ray showers. Now many radio-astronomical stations are conducting observations on cosmic ray showers.

After discovery of lasers, G. Askaryan began to investigate interaction of laser beam with different substances. At that time physicists who worked with lasers, used to break through thin metal specimens (usually, razor blades) with laser beam. It was something like a game. G. Askaryan also rendered tribute to this game. He noticed that holes made by laser beam were of two kinds. When he used laser of moderate power, the edges of aperture were smooth, as if the aperture was melted through (indeed, it was melted). However, the hole made by powerful laser had rough uneven edges, as if the hole was broken through, not melted. At first G. Askaryan supposed that it was the light pressure which knocked out the part of razor blade in the light spot. However, simple estimates showed that the assumption was wrong. Light pressure was too weak to break through even a thin metal specimen. So, the different appearance of the holes made by lasers of different power remained obscure.

The problem was cleared up by G. A. Askaryan and E. M. Moroz. The explanation was the following. The beam from a powerful laser heats metallic surface so intensely that surface layer turns into a vapor before the heat penetrates into next layers. The vapor is ejected from the surface. Thereby, a force arises which acts on the part of surface within the spot. This force is numerically equal to the momentum of vapor ejected during a unit of time. Such is the reaction of vapor on the surface. And in the case of powerful laser this reaction is so strong that the metal within the spot is torn out. The reaction of the vapor gives pressure that is many orders greater than the light pressure. Vaporizing ablation is now used for compressing the nuclear fuel in the problem of laser induced controlled thermonuclear reactions.

One of Askaryan's brilliant achievements is discovery of self-focusing. It is well known that light beam in homogenous medium (and in vacuum) is always divergent —- its cross section grows with distance. It turned out that this rule is valid only for comparatively weak electromagnetic waves, when linear electrodynamics is applicable. G.Askaryan investigated propagation of a strong electromagnetic beam in a refractive medium. A strong field changes the electromagnetic properties of the medium. For instance in so called Kerr effect a strong electromagnetic field produces an increase of the dielectric constant (or, which is the same, an increase of refractive index). The addition to the dielectric constant is proportional to E^2, where E is electric field. This property is called Kerr non-linearity. Let us suppose that a strong electromagnetic beam propagates in a medium with Kerr non-linearity. Then the refractive index of the medium inside the beam is greater than the refractive index outside. If the field E is strong enough then the beam creates a dielectric waveguide in the medium, a waveguide for itself. This waveguide reduces or fully eliminates divergence of the beam. Askaryan called this effect self-focusing. Discovery of self-focusing opened a new chapter in non-linear electrodynamics.

G.Askaryan belonged to the category of physicists which one can now meet very seldomly: he was not only an outstanding theoretician, he was a skilled experimentalist. As a rule, he verified his ideas by carrying out a convincing experiment. In particular, he conducted a series of experiments on self-focusing of electromagnetic and acoustic waves and showed that for both types of wave field, self-focusing takes place.

The list of Askaryan's achievements given above is far from complete. However, here we are limited in space and cannot give more detailed information. In Russia a book was issued entitled "In memory of G. Askaryan" in which his selected papers were collected. The book contains about 70 of Askaryan's papers, less than one third of his scientific heritage. The selection was made in accordance with the author's opinion. He said once that just these papers brought him the most pleasure and most misfortune.

It is appropriate to add several words about G. Askaryan as a person. He was very sociable and many his colleagues were his friends. In turn, he was a constant and careful friend. On the other side, whenever G.Askaryan got to know about any violation of ethic in science or in everyday life, undeserved blame, or undeserved praise, he always started a battle in defence of justice. As a consequence, he had not only friends. Some people disliked him, and among them there were men of influence.

G. Askaryan was also very inventive in playing various tricks on his friends and colleagues. He was inexhaustible in playing tricks as well as in producing brilliant scientific ideas. I shall give below two examples of his numerous jokes.

The head of Askaryan's laboratory, Professor M. S. Rabinovich every day came to Institute with big and heavy portfolio. Every evening, when Rabinovich went home, he carried the portfolio away with him. Once, in a talk with colleagues, G. Askaryan expressed his opinion that M. S. Rabinovich never opens the portfolio. Listeners did not agree with his supposition. G. Askaryan laid a bet with opponents. One day, when M. S. Rabinovich went out of his office, Askaryan came in with several colleagues whom he called to witness. He opened Rabinovich's portfolio and laid there a brick wrapped in a newspaper. During two weeks M. S. Rabinovich transported the brick in his portfolio there and back. He did not notice anything. After two weeks Askaryan removed the brick, again, in the presence of several witnesses. So he won. M. S. Rabinovich was informed by Askaryan about all details of this adventure, and he laughed together with colleagues. At that time M. S. Rabinovich was overloaded with numerous duties. He carried the portfolio and hoped to find time, to open the bag and to begin the work on his papers. However, his numerous duties gave him no time for it.

During several years G. Askaryan shared his room in the Institute with Dr. Igor Danilkin. Every midday Danilkin went out for lunch in Institute canteen. After lunch he returned, sat down at his writing-table, opened central drawer where he kept his cigarettes, smoked out a cigarette, and thereafter resumed his work. One day, after I.Danilkin went out for lunch, G. Askaryan came to Danilkin's writing-table and removed handle of central drawer and door-knobs of pedestals. Then he turned the writing-table back to front and mounted handles on its rear side. Danilkin returned, sat down at his table and tried to open central drawer. He pulled the handle but the drawer did not open. Then he pulled stronger, again without success. At last I. Danilkin applied all his force, and then the writing table drifted toward him. Danilkin was confused, and some time passed before he came to understanding.

G. Askaryan was never married. He lived with his mother and sister. They both were in poor health and Askaryan spent much efforts to look after them. Usually in lunchtime he went home to feed (sometimes, to spoon-feed) his women. Then he returned to Institute and remained in the laboratory until late in the evening.

G. Askaryan received many invitations to participate in various conferences. However, he never left Moscow, never came abroad because he could not abandon mother and sister. Gurgen was 12 when his father Ashot Askaryan passed away. On his deathbed, the father asked him to take care of his mother and sister. Gurgen vowed to fulfill father's will. And he kept his vow. He was an excellent son and excellent brother.

Intensive scientific activity and arduous family duties gradually ruined Askaryan's health. He suffered heart disease. At the end of February, 1997, G. Askaryan said to his colleagues that he would work at home during one week. He asked to not disturb him during the next week. However, after the week passed, G. Askaryan did not come to the institute. His friends and colleagues came to him, ringed and knocked at the door, but Askaryan did not respond. On the 7-th of March several Askaryan's colleagues entered his flat. Askaryan was dead. He was sitting in the kitchen with a note-pad before him. He worked till last moment. According to the experts' decision, he died on the 2-nd of March, 1997, of ischemia. He was then 68.

FIGURE 1. Gurgen Ashotovich Askaryan

PHYSICS IMPLICATIONS OF THE HIGHEST ENERGY COSMIC RAYS

Ultra-High Energy Cosmic Rays and Superheavy Particles in the Universe.

V.A. Kuzmin

Yukawa Institute for Theoretical Physics, Kyoto University, Kyoto, Japan
and
Institute for Nuclear Research of Russian Academy of Sciences, Moscow, Russia
E-mails : kuzmin@ms2.inr.ac.ru, kuzmin@itp.phys.ethz.ch

Abstract

In this lecture we conjecture that the Ultra-High Energy Cosmic Rays (UHECR) with energies $E > E_{GZK}$, where $E_{GZK} \sim 5 \cdot 10^{19}$ eV is the Greisen–Zatsepin–Kuzmin cut-off energy of cosmic ray spectrum, may originate from decays of superheavy long-living X-particles populating the Universe thus providing a unique window into the very early epoch of the Universe, namely, that of preheating and reheating after inflation. These particles may constitute a considerable fraction of cold dark matter in the Universe. We argue that the unconventionally long lifetime of the superheavy particles, which should be in the range of $10^{10} - 10^{22}$ years, might require novel particle physics mechanisms of their decays, such as instantons. First of all I will describe a toy model illustrating the instanton scenario and then will describe the possible proper mechanism of creation of superheavy particles in the early Universe. It is my pleasure to emphasize that all I will talk about was done in collaboration with my friends V.A. Rubakov and I.I. Tkachev.

CP579, *Radio Detection of High Energy Particles*, edited by D. Saltzberg and P. Gorham
© 2001 American Institute of Physics 0-7354-0018-0/01/$18.00

1. Among the top-down mechanisms attempting to explain the observations [2] of ultra-high energy (UHE) cosmic rays beyond the Greisen–Zatsepin–Kuzmin cut-off [1] , decays of primordial heavy particles are most obvious possibility. It is clearly more conventional — at least at first sight — than scenarios invoking topological or non-topological defects [19], though the latter may be a viable alternative. Heavy particles with lifetime of the order of the age of the Universe or greater may constitute (a substantial fraction of) cold dark matter, so the two observed features of the Universe — ultra-high energy cosmic rays and dark matter — may be related to each other.

To get an idea of the range of properties of decaying particles (X-particles) that supposedly produce UHE cosmic rays, let us make the following simple observations.

First, assuming sizeable hadronic component (jets) among the decay products, the flux of protons or gammas of energy E on the Earth is estimated as

$$\frac{dF}{d\,lnE} = \frac{1}{4\pi}\frac{n_X}{\tau_X}R_{p,\gamma}N_j\frac{dN_{p,\gamma}(E)}{d\,lnE}, \tag{1}$$

where N_j is the number of jets in a typical decay, $R_{p,\gamma}$ is the effective distance to X-particles, n_X is the number density of X-particles at the scale $R_{p,\gamma}$, τ_X is the X-particle lifetime and $dN/d\,lnE$ is the fragmentation function. In the following estimates we take $N_j \sim 1-10$ (we will soon see that large jet multiplicity may be favored in some decay scenarios) and $dN_{p,\gamma}/d\,lnE \sim$ (a few) $\cdot (10-100)$ in the energy range of interest, $E >$ (a few) $\cdot 10^{10}$ GeV (the latter estimate comes from bold extrapolation of the fragmentation functions of ref.[20] to extremely high jet energies). For the effective distance we take $R \lesssim 100$ Mpc, with understanding that the actual value of R may be much smaller than 100 Mpc if X-particles are clumped. In fact, our conclusions will be fairly insensitive to the actual values of the above parameters.

The second relation is

$$m_X < n_X > = \Omega_X\rho_{crit}, \tag{2}$$

where $< n_X >$ is the average number density of X-particles and $\Omega_X \lesssim 1$. If X-particles are clumped, the density n_X entering Eq.(2) may be several orders of magnitude larger than $< n_X >$. Again, this uncertainty will not affect our main conclusions, and we set $n_X \sim\ < n_X >$ in what follows. In order to produce cosmic rays of energies $E \gtrsim$ (a few) $\cdot 10^{11}$ GeV, the mass of X-particles is to be very large, $m_X \gtrsim 10^{13}$ GeV.

Let us now estimate the range of X-particle densities required. From Eq.(2) we find a bound for the present X-particle density-to-entropy ratio

$$n_X/s \lesssim 10^{-21}. \tag{3}$$

On the other hand, to produce the observed flux of UHE cosmic rays, the density of X-particles should not be too small. Keeping in mind that their lifetime, τ_X, cannot be much smaller than the age of the Universe,

$$\tau \gtrsim 10^{10}\ yr, \tag{4}$$

we obtain from Eq.(3)

$$n_X/s \gtrsim 10^{-33}. \tag{5}$$

Even though the window for n_X is very wide, the estimates (3) and (5) raise the issue of the production of X-particles in the early Universe.

Alternatively, Eqs.(3) and (5) may be used to place an upper bound on the lifetime of X-particles,

$$\tau_X \lesssim 10^{22} \ yr. \tag{6}$$

Again, the window for τ_X is wide, but the estimates (5) and (6) indicate another problem, namely, that of the particle physics mechanism responsible for long but finite lifetime of very heavy particles.

In the rest of this paper we propose possible scenarios for i) generating the abundance of X-particles in the range (3), (5) by processes in the early Universe, and ii) explaining the lifetime of X-particles in the range (4), (6).

Before coming to our main points, let us stress that these two problems inherent in theories with very heavy and almost stable particles were realized long ago (see, e.g., ref.[21] and references therein) in different contexts. In ref.[21] it was proposed that the problem i) may be solved by large entropy generation in the Universe after the heavy particles freeze out of thermal equilibrium, while their long lifetime may be due to very large dimension of operators responsible for the decay. We leave for the reader to judge how exotic are alternative possibilities that we discuss below.

2. If the temperature in the early Universe at some epoch exceeded the mass of X-particles and then decreased smoothly without large entropy generation, the freeze-out density of X-particles would greatly exceed the bound (3). A way out of this problem is provided by inflation and subsequent reheating. To explain the small abundance of X-particles, the reheating temperature T_r must be much smaller than m_X, so that X-particles were never at thermal equilibrium after inflation. In that case X-particles-to-entropy ratio is exponentially small,

$$n_X/s = \text{const} \cdot \exp(-2m_X/T_r), \tag{7}$$

where the constant depends on a number of factors (the coupling constant responsible for pair production of X-particles, the effective number of degrees of freedom, the ratios m_X/T_r and M_{Pl}/m_X, etc.) and is of order 10^{-3} with several orders of magnitude uncertainty. As the dominant suppression comes from the exponential factor, the reheating temperature can be estimated from (3), (5) fairly precisely,

$$T_r = \left(\frac{1}{20} - \frac{1}{35} \right) m_X, \tag{8}$$

and should be in the range $10^{11} - 10^{15}$ GeV, depending on m_X. Note that this range is realistic in many scenarios of inflation.

Hence, inflationary scenario can easily explain small value of the present density of X-particles in the space. Conversely, the determination of m_X from measurements of the upper end of cosmic ray spectrum would allow for rather precise estimate of the reheating temperature. Ultra-high energy cosmic rays may indeed serve as a window to reheating epoch in the early Universe.

3. Explaining long lifetime of X-particles is much harder. Conventional perturbative mechanisms cannot be responsible for cosmologically large τ_X (unless very high dimension operators are involved [21]), so one turns naturally to non-perturbative phenomena. A well known example is instantons that produce exponentially small effects in weakly coupled theories. If instantons are responsible for X-particle decays, the lifetime is roughly estimated as

$$\tau_X \sim m_X^{-1} \cdot \exp(4\pi/\alpha_X), \tag{9}$$

where α_X is the coupling constant of the relevant (spontaneously broken) gauge symmetry. From Eqs.(4), (6) we find that the coupling constant (at the scale m_X) is

$$\alpha_X = \frac{1}{10} - \frac{1}{12}. \tag{10}$$

Hence, we are lead to introduce additional non-Abelian gauge interactions with fairly large coupling constant at high energy scale.

To illustrate this possibility let us consider a toy model with $SU(2)_X$ gauge interactions added to the Standard Model. The $SU(2)_X$ gauge symmetry is assumed to be broken at sufficiently high energy scale. Some conventional quarks and leptons carry non-trivial $SU(2)_X$ quantum numbers (say, $SU(2)_X$ may be right-handed subgroup of a left-right symmetric theory, or it may be a horizontal group with generations forming $SU(2)_X$ triplets). Let there be two[1] left-handed $SU(2)_X$ fermionic doublets X and Y and four right-handed singlets, all of which are singlets under $SU(2)_L \times SU(3)_c$ of the Standard Model. After $SU(2)_X$ breaks down, all X- and Y-particles acquire large masses in a manner similar to the Standard Model. We further assume that X and Y carry different global quantum numbers, so there is no mixing between them.

Under these assumptions the lightest of X-particles and the lightest of Y-particles (we call them X and Y at certain risk of confusing notations) are perturbatively stable. However, $SU(2)_X$ instantons induce effective interactions violating global quantum numbers of X and Y. Say, if X is heavier than Y, then $SU(2)_X$-instantons lead to the decay

$$X \to Y + quarks + leptons \tag{11}$$

with the rate estimate given by Eq.(9). It is this type of processes that may be responsible for the production of ultra-high energy cosmic rays.

Let us point out a few features of this scenario which seem generic.

i) Decays induced by instantons typically lead to multiparticle final states. The number of quarks (jets) produced in the process (11) should be rather large, of order 10, and their distribution in energy should be fairly flat. In principle, the spectrum of cosmic rays within this scenario should be distinguishable from the spectrum predicted by other mechanisms (like two-jet decays of heavy particles born in the interactions of topological defects). Also, there are necessarily hard leptons among the decay products in the process (11).

ii) If Y-particles are indeed perturbatively stable, they are also stable against instanton-induced interactions (because of energy conservation and instanton selection rules). Then the dark matter in the Universe may consist predominantly of Y-particles, while the admixture of X-particles is small. Since the abundance of Y-particles is given by the formula similar to Eq.(7), and since the density of X-particles is bounded from above, Eq.(5), the mass splitting between X- and Y-particles should not be large,

$$m_X < 2m_Y. \tag{12}$$

Alternatively, the Higgs sector and its interactions with fermions may be organized in such a way that Y-particles are in fact perturbatively unstable (while X-particles remain perturbatively stable). In that case the heavy candidates for dark matter are X-particles, and the approximate degeneracy (12) need not hold.

---

[1]The $SU(2)_X$ anomaly prevents the number of $SU(2)_X$ doublets from being odd.

iii) Because of the instanton selection rules, this scenario for the slow decay of heavy particles is rather restrictive. For example, X-particles cannot be colored (otherwise the $SU(2)_X$ instanton vertex would include at least three X-fields); if Y-particles are stable, they cannot be colored either. This fits nicely to the expectation that dark matter particles do not experience strong interactions. On the other hand, X-particles (and Y-particles, if stable) cannot be weak doublets for the same reason, so the heavy dark matter in this scenario does not have electroweak interactions, too.

To conclude, an explanation of ultra-high energy cosmic ray events beyond the GZK cut-off by decays of hypothetical heavy particles of cosmologically long lifetime is not unrealistic from both cosmological and particle physics points of view. Detailed study of the upper end of cosmic ray spectrum will provide insight into the decay mechanism involved, and allow for the determination of the mass of X-particles. As the latter has been argued to be related to the reheating temperature, the ultra-high energy cosmic rays may become a clue to the end-of-inflation epoch in the early Universe.

After this work was completed , we received a paper by Berezinsky, Kachelriess and Vilenkin [22] where the decays of heavy particles have been also considered as the origin of cosmic rays beyond the GZK cut-off. Their main point is that X-particles are expected to concentrate in the galactic halo, and in this way one easily avoids the constraints coming from the analysis of the cascade radiation [23]. Their proposal for long but finite lifetime of X-particles is that it is due to quantum gravity (wormhole) effects.

Now I would like to consider yet another problem concerning the population of the Universe with superheavy particles.

According to the Greisen-Zatsepin-Kuzmin [1] (GZK) bound, the Ultra High Energy (UHE) cosmic rays produced in any known candidate extra galactic source should have an exponential cut-off at energies $E \sim 5 \times 10^{10}$ GeV. On the other hand, the number of observed [2] cosmic rays events beyond the cut-off is growing and leads to a mounting paradox within standard frameworks of cosmological and particle physics models.

A wide variety of possible solutions were suggested. Resolution could be due to exotic particle which may be produced at cosmological distances were suitable conventional accelerators are found, be transmitted evading GZK bound, and yet which interact in the atmosphere like a hadron. A particle with correct properties was found in a class of supersymmetric theories [3]. Alternatively, high energy cosmic rays may have been produced locally. One possibility is connected to the events of destruction of (topological) defects [4], while another one to decays of primordial heavy long-living particles [5, 6]. The candidate particle must obviously obey constraints on the mass, density and lifetime.

In order to produce cosmic rays of energies $E \gtrsim 10^{11}$ GeV, the mass of X-particles has to be very large, $m_X \gtrsim 10^{13}$ GeV [5, 6]. The lifetime, τ_X, cannot be much smaller than the age of the Universe, $\tau \gtrsim 10^{10}$ yr. With this smallest value of the lifetime, the observed flux of UHE cosmic rays will be reproduced with rather low density of X-particles, $\Omega_X \sim 10^{-12}$, where $\Omega_X \equiv m_X n_X / \rho_{\text{crit}}$, n_X is the number density of X-particles and ρ_{crit} is the critical density. On the other hand, X-particles must not overclose the Universe, $\Omega_X \lesssim 1$. With $\Omega_X \sim 1$, the X-particles may play the role of cold dark matter and the observed flux of UHE cosmic rays can be matched if $\tau_X \sim 10^{22}$ yr. The allowed windows are quite wide [5], but on exotic side, which may rise problems.

The problem of the particle physics mechanism responsible for a long but finite lifetime of very heavy particles can be solved in several ways. For example, otherwise conserved quantum number carried by the X-particle may be broken very weakly due to instanton transitions [5], or quantum gravity (wormhole) effects [6]. If instantons are responsible for X-particle decays, the lifetime is roughly estimated as $\tau_X \sim m_X^{-1} \cdot \exp(4\pi/\alpha_X)$, where α_X is the coupling constant of the relevant (spontaneously broken) gauge symmetry. Lifetime will fit the window if the coupling constant (at the scale m_X) is $\alpha_X \approx 0.1$ [5].

X-particles can be produced in the right amount by usual collision and decay processes if the reheating temperature after inflation never exceeded m_X, but the temperature should be in the range $10^{11} \lesssim T_r \lesssim 10^{15}$ GeV, depending on m_X, [5, 6]. This is a rather high value of reheating temperature and will lead to the gravitino problem in generic supersymmetric models [7].

In the present paper we investigate another process of X-particle creation, namely the direct production from vacuum fluctuations during inflation.

Any viable modern cosmological model invokes the hypothesis of inflation [8]. During inflation the Universe expands exponentially, which solves the horizon and flatness problems of the standard Big-Bang cosmology. Inflation is generally assumed to be driven by the special scalar field ϕ known as the *inflaton*. Fluctuations generated at inflationary stage can have strength and the power spectrum suitable for generation of the large scale structure. This fixes the range of parameters in the inflaton potential. For example, the mass of the inflaton field has to be $m_\phi \sim 10^{13}$ GeV. During inflation, the inflaton field slowly rolls down towards the minimum of its potential. Inflation ends when the potential energy associated with the inflaton field became smaller than the kinetic energy. At that time all the energy of the Universe is contained entirely in the form of coherent oscillations of the inflaton field around the minimum of its potential. It is possible that a significant fraction of this energy is released to other Boson species after only a dozen or so inflaton field oscillations, in the regime of a broad parametric resonance [9]. This process was studied in details [10, 11]. Even particles with masses of order of magnitude larger than the inflaton mass can be produced quite abundantly. Applying these results to the case of our interest here, we find that stable very heavy particles, $m_X \gtrsim m_\phi$, generally will be produced in excess and will overclose the Universe.

However, if the parametric resonance is ineffective for some reason, and we estimate particle number density after inflation at the level of initial conditions used in Refs. [10], we find that Ω_X might prove to be of about the right magnitude. This level is saturated by the fundamental process of particle creation during inflation from *vacuum fluctuations* and it is the same process which generated primordial large scale density fluctuations. Parametric resonance for X particles is turned-off if X is either a fermion field or its coupling to inflaton is small, $g^2 \ll 10^4 (m_X/m_\phi)^4 (m_\phi/M_{\rm Pl})^2$ [10].

At some early epoch the metric of the Universe is conformally flat to a high accuracy, $ds^2 = a(\eta)^2 (d\eta^2 - d\mathbf{x}^2)$. We normalize the scale factor by the condition $a(0) = 1$ at the end of inflation. Number density of particles created in time varying cosmological background can be written as

$$n_X = \frac{1}{2\pi^2 a^3} \int |\beta_k|^2 k^2 dk \ , \tag{13}$$

where β_k are the Bogoliubov coefficients which relate "in" and "out" mode functions, and k is the comoving momentum. Massles conformally coupled particles (for scalars this means that $\xi = 1/6$ in the direct coupling to the curvature) are not created. For massive particles conformal invariance is broken. Therefore, for the power low (e.g., matter or radiation dominated) period of expansion of the Universe, one expects on dimensional grounds, $n_X \propto m_X^3/a^3$ at late times.

Indeed, it was found in Ref. [12]

$$n_X \approx 5.3 \times 10^{-4} m_X^3 \, (m_X t)^{-3/2} \, , \tag{14}$$

for the radiation dominated Universe, and $n_X \propto m_X^3 \, (m_X t)^{-3q}$ for $a(t) \propto t^q$. Note that all particle creation occur in the region $mt = qm/H \lesssim 1$. When $mt \ll 1$, the number density of created particles remains on the constant level $n_X = m_X^3/24\pi^2$ independent of q [12]. At $qm/H \gg 1$ particle creation is negligible. Here H is the Hubble constant, $H \equiv \dot{a}/a$.

For the radiation dominated Universe one finds, $\Omega_X \sim (m_X^2/M_{\mathrm{Pl}}^2)\sqrt{m_X t_e}$, where t_e is time of equal densities of radiation and matter in $\Omega = 1$ Universe. This gives $\Omega_X \sim m_9^{5/2}$, where $m_9 \equiv m_X/10^9$ GeV. Stable particles with $m_X \gtrsim 10^9$ GeV will overclose the Universe even if they were created from the vacuum during regular Friedmann radiation dominated stage of the evolution. It is possible to separate vacuum creation from creation in collisions in plasma since X-particles may be effectively sterile.

This restriction can be overcomed if evolution of the Universe, as it is believed, was more complicated than simple radiation dominated expansion from singularity. Hubble constant may have never exceeded m_X, which is the case of inflation, $H(0) \approx m_\phi$. Moreover, compared to the case considered above, density of X-particles created during inflation is additionally diluted by late entropy release in reheating after inflation.

Particle creation from vacuum fluctuations during inflation (or in de Sitter space) was extensively studied [13, 14]. Characteristic quantity which is usually cited, the variance of the field $\langle X^2 \rangle$, is defined by an expression similar to Eq. (13). In the typical case $\alpha_k \approx -\beta_k$ the difference is given by the factor $2\sin^2(\omega_k \eta)/\omega_k$ in the integrand, where $\omega_k^2 = k^2 + a^2 m_X^2$. If $m_X \sim H(0) \approx m_\phi$, one has on dimensional grounds $n_X = C m_\phi^3/2\pi^2 a^3$ where the coefficient C is expected to be somewhat smaller than unity. Both Fermions and Bosons are prodused by this mechanism, exact numerical value of C being dependent on spin-statistics. In general, C is the function of the ratio $H(0)/m_X$, the function of self-coupling of X and the coupling ξ, depends on details of the transition between inflationary and matter (or radiation) dominated phases, etc. For example, for the scalar Bose-field, $\langle X^2 \rangle = 3H(0)^4/8\pi^2 m_X^2$ if $m_X \ll H(0)$. For massless self-interacting field $\langle X^2 \rangle \approx 0.132 H(0)^2/\sqrt{\lambda}$. C is expected to decrease exponentially when $m_X > m_\phi$.

Let us estimate the today's number density of X-particles. We consider massive inflaton, $V(\phi) = m_\phi^2 \phi^2/2$. In this case inflation is followed by the matter domination stage. If there are light Bosons in a theory, $m_B \ll m_\phi$, even relatively weakly coupled to the inflaton, $g^2 \gtrsim 10^4 m_\phi^2/M_{\mathrm{Pl}}^2 \sim 10^{-8}$, this matter domination stage will not last long: inflaton will decay via parametric resonance and the radiation domination follows. This happens typically when the energy density in inflaton oscillations is redshifted by a factor $r \approx 10^{-6}$ compared to a value $m_\phi^2 M_{\mathrm{Pl}}^2$. Matter is still far from being in the thermal equilibrium, but it is still convenient to characterize this radiation dominated stage by an equivalent temperature, $T_* \sim r^{1/4}\sqrt{m_\phi M_{\mathrm{Pl}}}$. At this moment the ratio of energy density in X-particles to the total energy density retains its value reached at the end of inflation, $\rho_X/\rho_R \approx C m_\phi m_X/2\pi^2 M_{\mathrm{Pl}}^2$. Later on this ratio grows as $\propto T/T_*$ and reaches unity at $T = T_{\mathrm{eq}}$, where

$$T_{\mathrm{eq}} = \frac{C r^{1/4}}{2\pi^2} \left(\frac{m_\phi}{M_{\mathrm{Pl}}} \right)^{3/2} m_X \, . \tag{15}$$

Using relation $T_{\rm eq} = 5.6\Omega_X h^2$ eV we find that $10^{-12} \lesssim \Omega_X \lesssim 1$ if

$$10^{-23} \lesssim Cr^{1/4} m_X/m_\phi \lesssim 10^{-11}\,. \tag{16}$$

For $m_X \sim$ (a few) $\cdot$ m$_\phi$ this condition can be easily satisfied since the coefficient C is exponentially small. This condition may be satisfied even for $m_X \sim m_\phi$ since the coefficient $r^{1/4}$ (or equivalent reheating temperature) can be small too.

Our hypothesis has unique observational consequences. If UHE cosmic rays are indeed due to decay of superheavy particles which were produced from vacuum fluctuations during inflation, there has to be a new sharp cut-off in the cosmic ray spectrum at energy somewhat smaller m_X. Since the number density n_X depends exponentially upon m_X/m_ϕ, the position of this cut-off might be well predicted and has to be near $E_{\rm cut-off} < m_\phi \approx 10^{13}$ GeV, the very shape of the cosmic ray spectrum beyond the GZK cut-off being of quite generic form following from the QCD quark/gluon fragmentation. The Pierre Auger Project installation, the OWL project, the Telescope Array project might prove to be able to discover this fundamental phenomenon.

We conclude, observation of Ultra High Energy cosmic rays can probe the spectrum of elementary particles in its superheavy range and can be an additional opportunity (alongside with fluctuations in cosmic microwave background) to study the earliest epoch of the Universe evolution, starting from amplification of vacuum fluctuations during inflation through fine details of gravitational interaction and down to physics of reheating.

When our paper was at the very end of completion we became aware of the quite recent paper by Chung, Kolb and Riotto [18] where similar problems of superheavy dark matter creation were considered.

We are indebted to V. Berezinsky, G. Farrar, S. Khlebnikov, W. Ochs, G. Pivovarov, S. Sarkar, D. Semikoz, G. Sigl and L. Stodolsky for stimulating discussions. The work of V.K. was supported in part by Russian Foundation for Basic Research grant 95-02-04911a. The work of V.R. was supported in part by Russian Foundation for Basic Research grant 96-02-17449a and U.S. Civilian Research and Development Foundation for Independent States of FSU (CRDF) award RP1-187. V. A. Kuzmin and I. I. Tkachev thank Theory Division at CERN for hospitality where the major part of this work was done. I. I. T. was supported in part by the U.S. Department of Energy under Grant DE-FG02-91ER40681 (Task B) and by the National Science Foundation under Grant PHY-9501458.

References

[1] K. Greisen, Phys. Rev. Lett. **16**, 748 (1966); G. T. Zatsepin and V. A. Kuzmin, Pisma Zh. Eksp. Teor. Fiz. **4**, 114 (1966).

[2] N. Hayashida et.al., Phys. Rev. Lett. **73**, 3491 (1994); D. J.Bird et.al., Astroph. J. **424**, 491 (1994); **441**, 144 (1995); T. A.Egorov et.al., in: *Proc. Tokyo Workshop on Techniques for the Study of Extremely High Energy Cosmic Rays*, ed. M.Nagano (ICRR, U. of Tokyo, 1993).

[3] G. R. Farrar, Phys. Rev. Lett. **76**, 4111 (1996); D. J. Chung, G. R. Farrar, and E. W. Kolb, astro-ph/9707036.

[4] C. T. Hill, Nucl. Phys. **B224**, 469 (1983); C. T. Hill, D. N. Schramm and T. P. Walker, Phys. Rev. **D36**, 1007 (1987); G. Sigl, D. N.Schramm and P. Bhattacharjee, Astropart. Phys. **2**, 401 (1994); V. Berezinsky, X. Martin and A. Vilenkin, Phys. Rev. **D56**, 2024 (1997); V. Berezinsky and A. Vilenkin, astro-ph/9704257.

[5] V. A. Kuzmin and V. A. Rubakov, astro-ph/9709187.

[6] V. Berezinsky, M. Kachelriess and A. Vilenkin, Phys. Rev. Lett. **79**, 4302 (1997).

[7] J. Ellis, J. E. Kim, and D. V. Nanopoulos, Phys. Lett. **B145**, 181 (1984).

[8] For a review and list of references, see A. D. Linde, *Particle Physics and Inflationary Cosmology*, (Harwood Academic, New York, 1990); E. W. Kolb and M. S. Turner, *The Early Universe*, (Addison-Wesley, Reading, Ma., 1990).

[9] L. A. Kofman, A. D. Linde and A. A. Starobinsky, Phys. Rev. Lett. **73**, 3195 (1994).

[10] S.Yu. Khlebnikov and I.I. Tkachev, Phys. Rev. Lett. **77**, 219 (1996); Phys. Lett. **B390**, 80 (1997); Phys. Rev. Lett. **79**, 1607 (1997); Phys. Rev. **D56**, 653 (1997).

[11] L. A. Kofman, A. D. Linde and A. A. Starobinsky, Phys. Rev. **D56**, 3258 (1997).

[12] S. G. Mamaev, V. M. Mostepanenko, and A. A. Starobinskii, ZhETF **70**, 1577 (1976) [Sov. Phys. JETP **43**, 823 (1976)].

[13] N. A. Chernikov and E. A. Tagirov, Ann. Inst. Henri Poinare **9A**, 109 (1968); E. A. Tagirov, Ann. Phys. **76**, 561 (1973); T. S. Bunch and P. C. W. Davies, Proc. R. Soc. **A360**, 117 (1978).

[14] A. D. Linde, Phys. Lett. **116B**, 335 (1982); A. A. Starobinsky, Phys. Lett. **117B**, 175 (1982); A. Vilenkin and L. H. Ford, Phys. Rev. **D26**, 1231 (1982); B. Allen, Phys. Rev. **D32**, 3136 (1985).

[15] A. A. Starobinsky and J. Yokoyama, Phys. Rev. **D50**, 6357 (1994).

[16] D. Lyth and D. Roberts, hep-ph/9609441.

[17] M. Boratav, for Pierre Auger Collaboration, *The Pierre Auger Observatory Project: An Overview.*, Proc. of 25th International Cosmic Ray Conference, Durban, v. 5, p. 205 (1997).

[18] D. J. H. Chung, E. W. Kolb and A. Riotto, hep-ph/9802238.

[19] C.T. Hill, D.N. Schramm and T.P. Walker, Phys.Rev. **D36**, 1007 (1987);
G.Sigl, D.N.Schramm and P.Bhattacharjee, Astropart. Phys. **2**, 401 (1994);
V. Berezinsky, X. Martin and A.Vilenkin, Phys.Rev. **D56**, 2024 (1997);
For a review see G.Sigl, *Topological Defect Models of Ultra-High Energy Cosmic Rays*, astro-ph/9611190.

[20] Yu. L. Dokshizer, V.A. Khoze, A.H. Mueller and S.I. Troyan, *Basics of Perturbative QCD* (Editions Frontiers, 1991).

[21] J. Ellis, J.L.Lopez and D.V. Nanopoulos, Phys. Lett. **B247**, 257 (1990);
J. Ellis, G.B. Gelmini, J.L. Lopez, D.V. Nanopoulos and S. Sarkar, Nucl.Phys. **B373**, 399 (1992).

[22] V. Berezinsky, M. Kachelriess, and A. Vilenkin, *Ultra-High Energy Cosmic Rays without GZK Cutoff*, 1997, unpublished.

[23] R. Protheroe and T. Stanev, Phys.Rev.Lett. **77**, 3708 (1996);
G. Sigl, S. Lee and P.Coppi, *Highest Energy Cosmic Rays, Grand Unified Theories and the Diffuse Gamma-Ray Background*, astro-ph/9604093 and references therein.

Probing Physics and Astrophysics at Extreme Energies with Ultra High Energy Cosmic Radiation

Günter Sigl*

*Institut d'Astrophysique de Paris, 98bis Boulevard Arago, CNRS
75014 Paris, France

Abstract. The highest energy cosmic rays observed possess macroscopic energies and their origin is likely to be associated with the most energetic processes in the Universe. Their existence triggered a flurry of theoretical explanations ranging from conventional shock acceleration to particle physics beyond the Standard Model and processes taking place at the earliest moments of our Universe. Furthermore, many new experimental activities promise a strong increase of statistics at the highest energies and a combination with $\gamma-$ray and neutrino astrophysics will put strong constraints on these theoretical models. We give an overview of theoretical developments with a focus on neutrino fluxes and cross sections, an issue that may be of particular relevance for the radio detection method.

INTRODUCTION

Over the last few years, several giant air showers have been detected confirming the arrival of cosmic rays (CRs) with energies up to a few hundred EeV (1 EeV $\equiv$ 10^{18} eV) [1–4]. The existence of such extremely high energy cosmic rays (EHECRs) pose a serious challenge for conventional theories of CR origin based on acceleration of charged particles in powerful astrophysical objects. The question of the origin of these EHECRs is, therefore, currently a subject of much intense debate and discussions as well as experimental efforts; see Ref. [5] for recent brief reviews, and Ref. [6] for a detailed review.

The problems encountered in trying to explain EHECRs in terms of "bottom-up" acceleration mechanisms have been well-documented in a number of studies; see, e.g., Refs. [7–9]. It is hard to accelerate protons and heavy nuclei up to such energies even in the most powerful astrophysical objects such as radio galaxies and active galactic nuclei. Also, nucleons above $\simeq 70$ EeV lose energy drastically due to photo-pion production on the cosmic microwave background (CMB) — the Greisen-Zatsepin-Kuzmin (GZK) effect [10] — which limits the distance to possible sources to less than $\simeq 100$ Mpc [8]. Heavy nuclei are photodisintegrated in the CMB within

CP579, *Radio Detection of High Energy Particles*, edited by D. Saltzberg and P. Gorham
© 2001 American Institute of Physics 0-7354-0018-0/01/$18.00

a few Mpc [11]. There are no obvious astronomical sources within $\simeq 100$ Mpc of the Earth [12,8].

There are, of course, ways to avoid the distance restriction imposed by the GZK effect, provided the problem of energetics is somehow solved separately and provided one allows new physics beyond the Standard Model of particle physics; we shall discuss those suggestions in the last part of this paper.

In contrast, in the "top-down" scenarios, which will be discussed in the next section, the problem of energetics is trivially solved. Here, the EHECR particles are the decay products of some supermassive "X" particles of mass $m_X \gg 10^{20}$ eV, and have energies all the way up to $\sim m_X$. Thus, no acceleration mechanism is needed. The massive X particles could be metastable relics of the early Universe with lifetimes of the order the current age of the Universe or could be released from topological defects that were produced in the early Universe during symmetry-breaking phase transitions envisaged in Grand Unified Theories (GUTs). If the X particles themselves or their sources cluster similar to dark matter, the dominant observable EHECR contribution would come from the Galactic Halo and absorption would be negligible.

The main problem of non-astrophysical solutions of the EHECR problem in general is that they are highly model dependent. On the other hand, it is precisely because of this reason that these scenarios are also attractive — they bring in ideas of new physics beyond the Standard Model of particle physics (such as Grand Unification and new interactions beyond the reach of terrestrial accelerators) as well as ideas of early Universe cosmology (such as topological defects and/or massive particle production in inflation) into the realms of EHECRs where these ideas have the potential to be tested by future EHECR experiments. We discuss these scenarios here because they often lead to significant neutrino fluxes extending up to extreme energies $\gtrsim 10^{21}$ eV, where new techniques such as radio detection could play an important role.

The physics and astrophysics of EHECRs are intimately linked with the emerging field of neutrino astronomy (for reviews see Refs. [13,14]) as well as with the already established field of $\gamma-$ray astronomy (for reviews see, e.g., Ref. [15]). Indeed, all scenarios of EHECR origin, including the top-down models, are severely constrained by neutrino and $\gamma-$ray observations and limits. In turn, this linkage has important consequences for theoretical predictions of fluxes of extragalactic neutrinos above a TeV or so whose detection is a major goal of next-generation neutrino telescopes: If these neutrinos are produced as secondaries of protons accelerated in astrophysical sources and if these protons are not absorbed in the sources, but rather contribute to the EHECR flux observed, then the energy content in the neutrino flux can not be higher than the one in EHECRs, leading to the so called Waxman Bahcall bound [16]. If one of these assumptions does not apply, such as for acceleration sources that are opaque to nucleons or in the TD scenarios where X particle decays produce much fewer nucleons than $\gamma-$rays and neutrinos, the Waxman Bahcall bound does not apply, but the neutrino flux is still constrained by the observed diffuse $\gamma-$ray flux in the GeV range. This is because isospin symmetry in the production of charged and neutral pions leads to a comparable amount of energy in $\gamma-$rays and neutrinos produced.

RELICS FROM THE EARLY UNIVERSE

The apparent difficulties of bottom-up acceleration scenarios discussed earlier motivated the proposal of the "top-down" scenarios, where EHECRs, instead of being accelerated, are the decay products of certain sufficiently massive "X" particles produced by physical processes in the early Universe. Furthermore, particle accelerator experiments and the mathematical structure of the Standard Model of the weak, electromagnetic and strong interactions suggest that these forces should be unified at energies of about 2×10^{16} GeV, 4-5 orders of magnitude above the highest energies observed in CRs. The relevant GUTs predict the existence of X particles with mass m_X around the GUT scale of $\simeq 2 \times 10^{16}$ GeV. If their lifetime is comparable or larger than the age of the Universe, they would be dark matter candidates and their decays could contribute to EHECR fluxes today, with an anisotropy pattern that reflects the expected dark matter distribution. These models avoid the GZK cutoff because the EHECR flux is dominated by particle decays in the halo of our galaxy. However, in order to reproduce the observed EHECR flux, they require strong fine tuning between the density in units of the critical density, Ω_X, and lifetime, t_X, of the X particle, $\Omega_X \sim 10^{-11}(t_X/10^{10}\,\mathrm{yr})$. Details of these scenarios have been discussed in Ref. [17] However, in many GUTs supermassive particles are expected to have extremely small lifetimes and thus have to be produced continuously if their decays are to give rise to EHECRs. This can only occur by emission from topological defects which are relics of cosmological phase transitions that could have occurred in the early Universe at temperatures close to the GUT scale. Phase transitions in general are associated with a breakdown of a group of symmetries down to a subgroup which is indicated by an order parameter taking on a non-vanishing value. Topological defects occur between regions that are causally disconnected, such that the orientation of the order parameter cannot be communicated between these regions and thus will adopt different values. Examples are cosmic strings, magnetic monopoles, and domain walls. The defects are topologically stable, but in the case of GUTs time dependent motion can lead to the emission of GUT scale X particles.

One of the prime cosmological motivations to postulate inflation, a phase of exponential expansion in the early Universe [18], was to dilute excessive production of "dangerous relics" such as topological defects and superheavy stable particles. However, right after inflation, when the Universe reheats, phase transitions can occur and such relics can be produced in cosmologically interesting abundances, and with a mass scale roughly given by the inflationary scale. The mass scale is fixed by the CMB anisotropies to $\sim 10^{13}$ GeV [19], and it is not far above the highest energies observed in CRs, thus motivating a connection between these primordial relics and EHECRs which in turn may provide a probe of the early Universe.

Within GUTs the X particles typically decay into jets of particles whose spectra can be estimated within the Standard Model. Before reaching Earth, the injected spectra are reprocessed by interactions with the low energy photon backgrounds such as the CMB, and magnetic fields present in the Universe (see Ref. [6,20] for details). The "visible" spectrum of nucleons and $\gamma-$rays expected in top-down scenarios (see Fig. 1) shows that the observed flux is reproduced above 3×10^{19} eV; at lower energies where the Universe is transparent to nucleons, bottom-up mechanisms could explain the spectrum. The X particle sources are not necessarily expected to be associated with astrophysical objects, but their distribution has to be sufficiently continuous to be consistent with observed EHECR angular distribu-

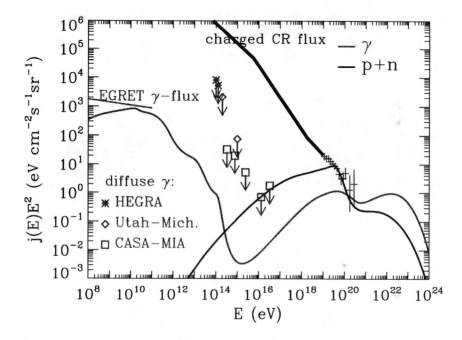

FIGURE 1. The "visible" spectra of nucleons and $\gamma-$rays for a top-down model involving the decay into two quarks of non-relativistic X particles of mass 10^{16} GeV, released from homogeneously distributed topological defects, assuming that quarks fragment into 10% nucleons and 90% pions with a spectrum given by a supersymmetric extension of the QCD fragmentation function. Also assumed was an intermediate estimate for the strength of the universal radio background and a large scale magnetic field $\lesssim 10^{-11}$ g (for details see Refs. [6,20]). One-sigma error bars show combined data from the Haverah Park [1], the Fly's Eye [2], and the AGASA [3] experiments above 10^{19} eV. Also shown are piecewise power law fits to the observed charged CR flux below 10^{19} eV, the measurement of the diffuse $\gamma-$ray flux between 30 MeV and 100 GeV by the EGRET instrument [21], as well as upper limits on the diffuse $\gamma-$ray flux from various experiments at higher energies (see Ref. [6] for more details).

tions.

The most characteristic features of top-down models are visible in Figs. 1 and 2: Electromagnetic cascades induced by interactions of the injected particles with the low energy photon backgrounds lead to a contribution to the diffuse γ−ray flux between 30 MeV and 100 GeV which is close to the flux measured by the EGRET detector flown on board the Compton γ−ray Observatory satellite [21]. The energy content in these γ−rays is comparable to the one in the ultra-high energy neutrino flux which should be detectable with next generation experiments

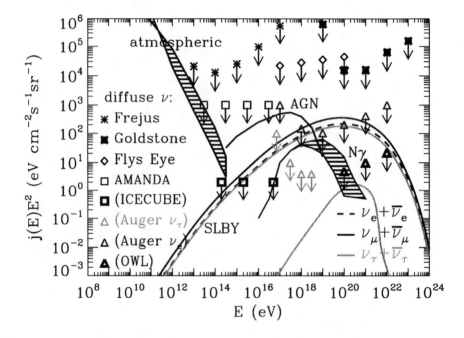

FIGURE 2. Neutrino fluxes for the top-down model of Fig. 1 (marked "SLBY"). Shown are experimental neutrino flux limits from the Frejus underground detector [22], the Fly's Eye experiment [23], the Goldstone radio telescope [24], and the Antarctic Muon and Neutrino Detector Array (AMANDA) neutrino telescope [25], as well as projected neutrino flux sensitivities of ICE-CUBE, the planned kilometer scale extension of AMANDA [26], the Pierre Auger Project [27] (for electron and tau neutrinos separately) and the proposed space based OWL [28] concept. For comparison also shown are the atmospheric neutrino background (hatched region marked "atmospheric"), and neutrino flux predictions for a model of AGN optically thick to nucleons ("AGN"), and for EHECR interactions with the CMB [29] ("$N\gamma$", dashed range indicating typical uncertainties for moderate source evolution). The top-down fluxes are shown for electron-, muon, and tau-neutrinos separately, assuming no (lower ν_τ-curve) and maximal $\nu_\mu - \nu_\tau$ mixing (upper ν_τ-curve, which would then equal the ν_μ-flux), respectively.

(see Fig. 2). The neutrino flux is hardly influenced by subsequent interactions, and thus directly represents the decay spectrum. In bottom-up scenarios neutrinos can only be produced as secondaries and for sources transparent to the primary nucleons, the neutrino flux is constrained by the Waxman-Bahcall bound [16] which does not apply to top-down scenarios where consequently the neutrino flux can be higher (see Fig. 2). This can also serve as a discriminator between the top-down and bottom-up concepts. Finally, top-down models predict a significant $\gamma-$ray component above $\sim 10^{20}$ eV, whereas nucleons would dominate at lower energies. This will be a strong discriminator as experiments will improve constraints on EHECR composition which currently favor nucleons [30].

Besides some uncertainties in the shape and chemical composition of the spectrum, possibly the most significant shortcoming of top-down scenarios is their lack of prediction of the absolute flux normalization. At least, the moderate rate of 10 decays per year in a spherical volume with radius equal to the Earth-Sun distance, the rate necessary to explain the EHECR flux, is not in a remote corner of parameter space for most scenarios. Dimensional and scaling arguments imply that topological defects release X particles with an average rate at cosmic time t of

$$\dot{n}_X(t) = \kappa\, m_X^p\, t^{-4+p}\,,\qquad(1)$$

where the dimensionless parameters κ and p depend on the specific top-down scenario [6]. For example, hybrid defects involving cosmic strings have $p = 1$ and normalization of predicted spectra both at EGRET energies and around 10^{20} eV (see Fig. 1), leads to $\kappa m_X \sim 10^{13} - 10^{14}$ GeV. For $\kappa \sim 1$, the resulting mass scale is again close to the inflation and GUT scales.

NEW PRIMARY PARTICLES AND NEW INTERACTIONS

A possible way around the problem of missing counterparts within acceleration scenarios is to propose primary particles whose range is not limited by interactions with the CMB. Within the Standard Model the only candidate is the neutrino, whereas in supersymmetric extensions of the Standard Model, new neutral hadronic bound states of light gluinos with quarks and gluons, so-called R-hadrons that are heavier than nucleons, and therefore have a higher GZK threshold, have been suggested [31].

In both the neutrino and new massive neutral hadron scenario the particle propagating over extragalactic distances would have to be produced as a secondary in interactions of a primary proton that is accelerated in a powerful AGN which can, in contrast to the case of EAS induced by nucleons, nuclei, or $\gamma-$rays, be located at high redshift. Consequently, these scenarios predict a correlation between primary arrival directions and high redshift sources. In fact, possible evidence for an angular correlation of the five highest energy events with compact radio quasars at redshifts between 0.3 and 2.2 was recently reported [32]. A new analysis with the somewhat larger data set now available does not support significant correlations [33]. This is currently disputed since another group claims to have found a correlation on the 99.9% confidence level [34]. Only a few more events could confirm or rule out the correlation hypothesis. Note that these scenarios would require the primary proton to be accelerated up to $\gtrsim 10^{21}$ eV, demanding a very powerful astrophysical accelerator. On the other hand, a few dozen such exceptional accelerators in the visible Universe may suffice.

New Neutrino Interactions

For the remainder we will focus on neutrinos as primary candidates since they have the advantage of being well established particles. However, within the Standard Model their interaction cross section with nucleons, whose charged current part can be parametrized by [35]

$$\sigma_{\nu N}^{SM}(E) \simeq 2.36 \times 10^{-32}(E/10^{19}\,\text{eV})^{0.363}\,\text{cm}^2 \quad (10^{16}\,\text{eV} \lesssim E \lesssim 10^{21}\,\text{eV}), \quad (2)$$

falls short by about five orders of magnitude to produce ordinary air showers. However, it has been suggested that the neutrino-nucleon cross section, $\sigma_{\nu N}$, can be enhanced by new physics beyond the electroweak scale in the center of mass (CM) frame, or above about a PeV in the nucleon rest frame. Neutrino induced air showers may therefore rather directly probe new physics beyond the electroweak scale.

Several possibilities have been discussed in the literature for which unitarity bounds need not be violated. We focus here on the possibility of a large increase in the number of degrees of freedom above the electroweak scale [36]. A specific implementation of this idea is given in theories with n additional large compact dimensions and a quantum gravity scale $M_{4+n} \sim \text{TeV}$ that has recently received much attention in the literature [37] because it provides an alternative solution (i.e., without supersymmetry) to the hierarchy problem in grand unifications of gauge interactions. The cross sections within such scenarios have not been calculated from first principles yet. Within the field theory approximation which should hold for squared CM energies $s \lesssim M_{4+n}^2$, the spin 2 character of the graviton predicts $\sigma_g \sim s^2/M_{4+n}^6$ [38] For $s \gg M_{4+n}^2$, several arguments based on unitarity within field theory have been put forward. Ref. [38] suggested

$$\sigma_g \simeq \frac{4\pi s}{M_{4+n}^4} \simeq 10^{-27} \left(\frac{M_{4+n}}{\text{TeV}}\right)^{-4} \left(\frac{E}{10^{20}\,\text{eV}}\right)\,\text{cm}^2, \quad (3)$$

where in the last expression we specified to a neutrino of energy E hitting a nucleon at rest. A more detailed calculation taking into account scattering on individual partons leads to similar orders of magnitude [39]. Note that a neutrino would typically start to interact in the atmosphere for $\sigma_{\nu N} \gtrsim 10^{-27}\,\text{cm}^2$, i.e. in the case of Eq. (3) for $E \gtrsim 10^{20}$ eV, assuming $M_{4+n} \simeq 1$ TeV. The neutrino therefore becomes a primary candidate for the observed EHECR events. However, since in a neutral current interaction the neutrino transfers only about 10% of its energy to the shower, the cross section probably has to be at least a few 10^{-26} cm^2 to be consistent with observed showers which start within the first 50 g cm^{-2} of the atmosphere [40,41]. A specific signature of this scenario would be the absence of any events above the energy where σ_g grows beyond $\simeq 10^{-27}$ cm^2 in neutrino telescopes based on ice or water as detector medium [14], and a hardening of the spectrum above this energy in atmospheric detectors such as the Pierre Auger Project [42] and the proposed space based AirWatch type detectors [28,43,44]. Furthermore, according to Eq. (3), the average atmospheric column depth of the first interaction point of neutrino induced EAS in this scenario is predicted to depend linearly on energy. This should be easy to distinguish from the logarithmic scaling expected for nucleons, nuclei, and $\gamma-$rays. To test such scalings one can, for example, take advantage of the fact that the atmosphere provides a detector medium whose column depth increases from ~ 1000 g/cm^2 towards the zenith to ~ 36000 g/cm^2 towards horizontal arrival directions. This probes cross sections in the range $\sim 10^{-29} - 10^{-27}$ cm^2. Due to the increased column depth water/ice detectors would probe cross sections in the range $\sim 10^{-31} - 10^{-29}$ cm^2 [45].

Within string theory, individual amplitudes are expected to be suppressed exponentially above the string scale M_s which for simplicity we assume here to be comparable to M_{4+n}. This can be interpreted as a result of the finite spatial extension of the string states. In this case, the neutrino nucleon cross section would be dominated by interactions with the partons carrying a momentum fraction $x \sim M_s^2/s$, leading to [40]

$$
\begin{aligned}
\sigma_{\nu N} &\simeq \frac{4\pi}{M_s^2} \ln(s/M_s^2)(s/M_s^2)^{0.363} \\
&\simeq 6 \times 10^{-29} \left(\frac{M_s}{\text{TeV}} \right)^{-4.726} \left(\frac{E}{10^{20}\,\text{eV}} \right)^{0.363} \\
&\quad \times \left[1 + 0.08 \ln \left(\frac{E}{10^{20}\,\text{eV}} \right) - 0.16 \ln \left(\frac{M_s}{\text{TeV}} \right) \right]^2 \text{cm}^2
\end{aligned}
\tag{4}
$$

This is probably too small to make neutrinos primary candidates for the highest energy showers observed, given the fact that complementary constraints from accelerator experiments result in $M_s \gtrsim 1$ TeV [46]. On the other hand, in the total cross section amplitude suppression may be compensated by an exponential growth of the level density [36]. It is currently unclear and it may be model dependent which effect dominates. Thus, an experimental detection of the signatures discussed in this section could lead to constraints on some string-inspired models of extra dimensions.

We note in passing that extra dimensions can have other astrophysical ramifications such as energy loss in stellar environments due to emission of real gravitons into the bulk. The strongest resulting lower limits on M_{4+n} come from the consideration of cooling of the cores of hot supernovae and read $M_6 \gtrsim 50\,\text{TeV}$, $M_7 \gtrsim 4\,\text{TeV}$, $M_8 \gtrsim 1\,\text{TeV}$, and $M_{11} \gtrsim 0.05\,\text{TeV}$ for $n = 2, 3, 4, 7$, respectively [47]. In addition, implications of extra dimensions for early Universe physics and inflation are increasingly studied in the literature, but much work is left to be done on the intersection of these research domains.

Independent of theoretical arguments, the EHECR data can be used to put constraints on cross sections satisfying $\sigma_{\nu N}(E \gtrsim 10^{19}\,\text{eV}) \lesssim 10^{-27}\,\text{cm}^2$. Particles with such cross sections would give rise to horizontal air showers. The Fly's Eye experiment established an upper limit on horizontal air showers [23]. The non-observation of the neutrino flux expected from pions produced by EHECRs interacting with the CMB the results in the limit [45]

$$\sigma_{\nu N}(10^{17}\,\text{eV}) \lesssim 1 \times 10^{-29}/\bar{y}^{1/2}\,\text{cm}^2$$
$$\sigma_{\nu N}(10^{18}\,\text{eV}) \lesssim 8 \times 10^{-30}/\bar{y}^{1/2}\,\text{cm}^2$$
$$\sigma_{\nu N}(10^{19}\,\text{eV}) \lesssim 5 \times 10^{-29}/\bar{y}^{1/2}\,\text{cm}^2\,, \tag{5}$$

where $\bar{y}$ is the average energy fraction of the neutrino deposited into the shower ($\bar{y} = 1$ for charged current reactions and $\bar{y} \simeq 0.1$ for neutral current reactions). Neutrino fluxes predicted in various scenarios are shown in Fig. 1. The projected sensitivity of future experiments such as the Pierre Auger Observatories and the AirWatch type satellite projects indicate that the cross section limits Eq. (5) could be improved by up to four orders of magnitude, corresponding to one order of magnitude in M_s or M_{4+n}. The radio detection technique could also provide an important contribution to test such cross sections beyond The Standard Model.

REFERENCES

1. See, for example, M. A. Lawrence, R. J. O. Reid, and A. A. Watson, *J. Phys. G Nucl. Part. Phys.* **17**, 733 (1991), and references therein; see also http://ast.leeds.ac.uk/haverah/hav-home.html.
2. D. J. Bird et al., *Phys. Rev. Lett.* **71**, 3401 (1993); *Astrophys. J.* **424**, 491 (1994); ibid. **441**, 144 (1995).
3. N. Hayashida et al., *Phys. Rev. Lett.* **73**, 3491 (1994); S. Yoshida et al., *Astropart. Phys.* **3**, 105 (1995); M. Takeda et al., **Phys. Rev. Lett. 81**, 1163 (1998); see also http://icrsun.icrr.u-tokyo.ac.jp/as/project/agasa.html.
4. for a review of the data see S. Yoshida and H. Dai, *J. Phys. G* **24**, 905 (1998).
5. J. W. Cronin, *Rev. Mod. Phys.* **71**, S165 (1999); A. V. Olinto, *Phys. Rept.* **333-334**, 329 (2000); X. Bertou, M. Boratav, and A. Letessier-Selvon, *Int. J. Mod. Phys.* **A15**, 2181 (2000).
6. P. Bhattacharjee and G. Sigl, *Phys. Rept.* **327**, 109 (2000).
7. A. M. Hillas, *Ann. Rev. Astron. Astrophys.* **22**, 425 (1984).
8. G. Sigl, D. N. Schramm, and P. Bhattacharjee, *Astropart. Phys.* **2**, 401 (1994).
9. C. A. Norman, D. B. Melrose, and A. Achterberg, *Astrophys. J.* **454**, 60 (1995).

10. K. Greisen, *Phys. Rev. Lett.* **16**, 748 (1966); G. T. Zatsepin and V. A. Kuzmin, *Pis'ma Zh. Eksp. Teor. Fiz.* **4**, 114 (1966) [*JETP. Lett.* **4**, 78 (1966)].

11. J. L. Puget, F. W. Stecker, and J. H. Bredekamp, *Astrophys. J.* **205**, 638 (1976); L. N. Epele and E. Roulet, *Phys. Rev. Lett.* **81**, 3295 (1998); *J. High Energy Phys.* **9810**, 009 (1998); F. W. Stecker, *Phys. Rev. Lett.* **81**, 3296 (1998); F. W. Stecker and M. H. Salamon, *Astrophys. J.* **512**, 521 (1999).

12. J. W. Elbert, and P. Sommers, *Astrophys. J.* **441**, 151 (1995).

13. T. K. Gaisser, F. Halzen, and T. Stanev, *Phys. Rept.* **258**, 173 (1995).

14. F. Halzen, e-print astro-ph/9810368; e-print astro-ph/9904216.

15. R. A. Ong, *Phys. Rept.* **305**, 95 (1998); M. Catanese and T. C. Weekes, e-print astro-ph/9906501, invited review, *Publ. Astron. Soc. of the Pacific*, Vol. 111, issue 764, 1193 (1999).

16. E. Waxman, J. Bahcall, *Phys. Rev. D.* **59**, 023002 (1999); J. Bahcall, E. Waxman, e-print hep-ph/9902383; K. Mannheim, R. J. Protheroe, J. P. Rachen, e-print astro-ph/9812398, to appear in *Phys. Rev. D.*; J. P. Rachen, R. J. Protheroe, K. Mannheim, e-print astro-ph/9908031.

17. V. Berezinsky, M. Kachelrieß , and A. Vilenkin, *Phys. Rev. Lett.* **79**, 4302 (1997); V. A. Kuzmin and V. A. Rubakov, *Phys. Atom. Nucl.* **61**, 1028 (1998).

18. see, e.g., E. W. Kolb, M. S. Turner, *The Early Universe* (Addison-Wesley, Redwood City, California, 1990).

19. for a brief review see V. Kuzmin, I. Tkachev, *Phys. Rept.* **320**, 199 (1999).

20. G. Sigl, S. Lee, P. Bhattacharjee, S. Yoshida, *Phys. Rev. D* **59**, 043504 (1999).

21. P. Sreekumar et al., *Astrophys. J.* **494**, 523 (1998).

22. W. Rhode et al., *Astropart. Phys.* **4**, 217 (1996).

23. R. M. Baltrusaitis et al., *Astrophys. J.* **281**, L9 (1984); *Phys. Rev. D* **31**, 2192 (1985).

24. P. W. Gorham, K. M. Liewer, C. J. Naudet, e-print astro-ph/9906504.

25. For general information see http://amanda.berkeley.edu/; see also F. Halzen, *New Astron. Rev.* **42**, 289 (1999).

26. For general information see http://www.ps.uci.edu/ icecube/workshop.html; see also F. Halzen, *Am. Astron. Soc. Meeting* **192**, # 62 28 (1998); AMANDA collaboration, e-print astro-ph/9906205, talk presented at the *8th Int. Workshop on Neutrino Telescopes*, Venice, Feb. 1999.

27. J. J. Blanco-Pillado, R. A. Vázquez, E. Zas, *Phys. Rev. Lett.* **78**, 3614 (1997); K. S. Capelle, J. W. Cronin, G. Parente, E. Zas, *Astropart. Phys.* **8**, 321 (1998); A. Letessier-Selvon, e-print astro-ph/0009444.

28. D. B. Cline, F. W. Stecker, OWL/AirWatch science white paper, e-print astro-ph/0003459; see also http://lheawww.gsfc.nasa.gov/docs/gamcosray/hecr/OWL/.

29. R. J. Protheroe, P. A. Johnson, *Astropart. Phys.* **4**, 253 (1996), and erratum *ibid.* **5**, 215 (1996).

30. F. Halzen, R. A. V'azques, T. Stanev, H. P. Vankov, *Astropart. Phys.* **3**, 151 (1995); M. Ave, J. A. Hinton, R. A. V'azques, A. A. Watson, E. Zas, *Phys. Rev. Lett.* **85** 2244 (2000).

31. G. R. Farrar, *Phys. Rev. Lett.* **76**, 4111 (1996); D. J. H. Chung, G. R. Farrar, and E. W. Kolb, *Phys. Rev. D* **57**, 4696 (1998).

32. G. R. Farrar and P. L. Biermann, *Phys. Rev. Lett.* **81**, 3579 (1998).

33. G. Sigl, D. F. Torres, L. A. Anchordoqui, and G. E. Romero, e-print astro-ph/0008363.

34. A. Virmani et al., e-print astro-ph/0010235

35. R. Gandhi, C. Quigg, M. H. Reno, and I. Sarcevic, *Astropart. Phys.* **5**, 81 (1996); *Phys. Rev. D* **58**, 093009 (1998).

36. G. Domokos and S. Kovesi-Domokos, *Phys. Rev. Lett.* **82**, 1366 (1999).

37. N. Arkani-Hamed, S. Dimopoulos, and G. Dvali, *Phys. Lett.* **B 429**, 263 (1998); I. Antoniadis, N. Arkani-Hamed, S. Dimopoulos, and G. Dvali, *Phys. Lett.* **B 436**, 257 (1998); N. Arkani-Hamed, S. Dimopoulos, and G. Dvali, *Phys. Rev. D* **59** 086004 (1999).

38. S. Nussinov and R. Shrock, *Phys. Rev. D* **59**, 105002 (1999).

39. P. Jain, D. W. McKay, S. Panda, and J. P. Ralston, *Phys. Lett.* **B484**, 267 (2000); J. P. Ralston, P. Jain, D. W. McKay, S. Panda, e-print hep-ph/0008153.

40. M. Kachelrieß and M. Plümacher, *Phys. Rev. D* **62**, 103006 (2000).

41. L. Anchordoqui et al., e-print hep-ph/0011097.

42. J. W. Cronin, *Nucl. Phys. B* (Proc. Suppl.) **28B**, 213 (1992); The Pierre Auger Observatory Design Report (2nd edition), March 1997; see also http://http://www.auger.org/ and http://www-lpnhep.in2p3.fr/auger/welcome.html.

43. See http://www.ifcai.pa.cnr.it/Ifcai/euso.html.

44. J. Linsley, in Proc. *25th International Cosmic Ray Conference*, eds.: M. S. Potgieter et al. (Durban, 1997)., Vol. 5, 381; ibid., 385; P. Attinà et al., ibid., 389; J. Forbes et al., ibid., 273; see also http://www.ifcai.pa.cnr.it/Ifcai/airwatch.html.

45. C. Tyler, A. Olinto, and G. Sigl, e-print hep-ph/0002257, to apprear in *Phys. Rev. D*.

46. see, e.g., S. Cullen, M. Perelstein, and M. E. Peskin, *Phys. Rev.D* **62**, 055012 (2000), and references therein.

47. S. Cullen and M. Perelstein, *Phys. Rev. Lett.* **83**, 268 (1999); V. Barger, T. Han, C. Kao, and R.-J. Zhang, *Phys. Lett. B* **461**, 34 (1999).

Signature Studies of Cosmic Magnetic Monopoles

Stuart D. Wick*,1, Thomas W. Kephart†, and Thomas J. Weiler†

* Department of Physics, University of Florida,
Gainesville, FL 32611
†Department of Physics and Astronomy, Vanderbilt University,
Nashville, TN 37235

Abstract. This talk explores the possibility that the Universe may be populated with relic magnetic monopoles. Observations of galactic and extragalactic magnetic fields, lead to the conclusion that monopoles of mass $\lesssim 10^{14}$ GeV are accelerated in these fields to relativistic velocities. The relativistic monopole signatures and features we derive are (i) the protracted shower development, (ii) the Cherenkov signals, (iii) the tomography of the Earth with monopoles, and (iv) a model for monopole airshowers above the GZK cutoff.

INTRODUCTION

Any symmetry breaking, after inflation, of a semisimple group to a subgroup leaving an unbroken $U(1)$ may produce an observable abundance of magnetic monopoles. The inferred strength and coherence size of existing extragalactic magnetic fields suggest that any free monopole with a mass near or less than 10^{14} GeV would have been accelerated in magnetic fields to relativisitic velocities. On striking matter, such as the Earth's atmosphere, these relativistic monopoles will generate a particle cascade. Here we investigate the associated shower signatures.

The free monopole flux is limited only by Parker's upper bound $F_P \sim 10^{-15}/\text{cm}^2/\text{s/sr}$ [1], which results from requiring that monopoles not short–circuit our Galactic magnetic fields faster than their dynamo can regenerate them. Since the Parker bound is several orders of magnitude above the observed highest–energy cosmic ray flux, existing cosmic ray detectors can meaningfully search for a monopole flux.

Because of their mass and integrity, a single monopole primary will continuously induce air–showers, in contrast to nucleon and photon primaries which transfer

---

1) Presenter at the First International Workshop on Radio Detection of High–Energy Particles

nearly all of their energy at shower initiation. Thus, the monopole shower is readily distinguished from non–monopole initiated showers. We also investigate the possibility that the hadronic cross–section of the monopole is sufficient to produce air–showers comparable to that from hadronic primaries, in which case existing data would already imply a meaningful limit on the monopole flux. One may even speculate that such monopoles may have been observed, as the primaries producing the enigmatic showers above the GZK cutoff at $\sim 5 \times 10^{19}$ eV [2,3].

CHARACTERISTICS OF A MONOPOLE FLUX

The flux of monopoles emerging from a phase transition is determined by the Kibble mechanism [4]. At the time of the phase transition, roughly one monopole or antimonopole is produced per correlated volume, ξ_c^3. The resulting monopole number density today is

$$n_M \sim 10^{-19} \, (T_c/10^{11}\text{GeV})^3 (l_H/\xi_c)^3 \, \text{cm}^{-3}, \tag{1}$$

where ξ_c is the phase transition correlation length, bounded from above by the horizon size l_H at the time when the system relaxes to the true broken–symmetry vacuum. Although minimal $SU(5)$ breaking gives monopoles of mass $\sim 10^{17}$ GeV, there are ample theoretical possibilities for producing monopoles with smaller mass while maintaining the possibility of strong interaction cross–sections that avoid proton decay [5–8]. Based on the Kibble mechanism for monopole production, bounds on the universe's curvature constrain the monopole mass to less than 10^{13} GeV, while a comparison of the Kibble flux to the Parker limit constrains the monopole mass to less than 10^{11} GeV. The general expression for the relativistic monopole flux may be written [3]

$$F_M = c \, n_M/4\pi \sim 2 \times 10^{-16} \left(\frac{M}{10^{11}\text{GeV}}\right)^3 \left(\frac{l_H}{\xi_c}\right)^3 \text{cm}^{-2} \, \text{sec}^{-1} \, \text{sr}^{-1}. \tag{2}$$

In higher dimensional cosmologies, the Kibble flux may be altered; then the straightforward Parker upper limit $F_P \leq 10^{-15}/\text{cm}^2/\text{sec}/\text{sr}$ becomes the only reliable bound on the monopole flux. In the spirit of generality, we take the monopole mass M to be a free parameter and the Kibble mechanism is a rough guide to F_M. We require that F_M obey the Parker limit and assume that proton decay is avoided in a way that does not restrict the parameter M.

Monopole Structure

Monopoles are topological defects with a non-trivial internal structure; the core of the monopole is a region of restored unified symmetry. Monopoles are classified [4] by their topological winding, but for the case of GUT monopoles this classification is too coarse. In an $SU(5)$ GUT the fundamental minimally-charged monopole is six-fold degenerate. For an appropriate Higgs potential there are four other types of stable bound states formed from the fundamental monopoles [9,10]. This work distinguishes between those monopoles with color–magnetic charge and those with only ordinary $U_{EM}(1)$ magnetic charge. Thus, we adopt the nomenclature "q–monopoles" for those monopoles with color–magnetic charge and "l–monopoles" for those with only the ordinary magnetic charge.

The possible confinement of q–monopoles has recently been considered [11] via the formation of Z_3 color–magnetic "strings." If such a mechanism were realized one result could be the formation of color–singlet "baryonic–monopoles" in which the fusion of three differently colored strings produces a baryon–like composite of q–monopoles. The internal structure of a baryonic–monopole would approximate that of an ordinary baryon in the QCD string model, but with q–monopoles in the place of the quarks. Thus, the baryonic–monopole structure is quite different from a single l–monopole and, as such, it is shown to have a very different cross–section and cosmic ray shower profile.

Monopole Acceleration

The kinetic energy imparted to a magnetic monopole on traversing a magnetic field along a particular path is [3]

$$E_K = g \int_{\text{path}} \vec{B} \cdot \vec{dl} \sim g\, B\, \xi\, \sqrt{n} \tag{3}$$

where

$$g = e/2\alpha = 3.3 \times 10^{-8} \text{ esu } (\text{or } 3.3 \times 10^{-8} \text{dynes/G}) \tag{4}$$

is the magnetic charge according to the Dirac quantization condition, B is the magnetic field strength, ξ specifies the field's coherence length, and $\sqrt{n}$ is a factor to approximate the random–walk through the n domains of coherent field traversed by the path. Galactic magnetic fields and magnetic fields in extragalactic sheets and galactic clusters range from about 0.1 to $100\mu G$, while their coherence lengths range from 10^{-4} to about 30Mpc [12,13]. These fields can accelerate a monopole from rest to the energy range 2×10^{20} to 5×10^{23} eV. For extragalactic sheets the number of random–walks can be roughly estimated to be of order $n \sim H_0^{-1}/50$ Mpc ~ 100, and so $E_{\max} \sim 5 \times 10^{24}$ eV. Hence, monopoles with mass below $\sim 10^{14}$ GeV may be relativistic. The rest of this talk is devoted to the novel phenomenology of relativistic monopoles. As a prelude to calculating monopole signatures in various detectors, we turn to a discussion of the interactions of monopoles with matter.

RELATIVISTIC MONOPOLE ENERGY LOSS

Both l–monopoles and baryonic–monopoles are conserved in each interaction because of their topological stability. However, as conjectured above, their different internal structures will lead to differing shower profiles and signatures. Because of space limitations, in this talk we only consider the electromagnetic interactions of l–monopoles and the hadronic interactions of baryonic–monopoles.

The shower profile of baryonic–monopoles is based upon a model [14] where the hadronic cross–section grows after impact and the net energy transfer is enough to stop the monopole very quickly. Since this mechanism is model dependent, we consider the baryonic–monopole signatures less reliable. Further discussion of the baryonic–monopole is postponed until the final section.

Our most reliable signatures are for l–monopoles (which are referred to as "monopoles" for the remainder of this talk) and are based upon well understood electromagnetic processes. At large distances and high velocities, a monopole mimics the electromagnetic interaction of a heavy ion of charge $Z \sim 1/2\alpha \simeq 68$. We view the monopole as a classical source of radiation, while treating the matter–radiation interaction quantum mechanically. In this way, the large electromagnetic coupling of the monopole is isolated in the classical field, and the matter–radiation interaction can be calculated perturbatively.

Electromagnetic Interactions

We consider here the energy loss of a monopole resulting from four electromagnetic processes: collisions (ionization of atoms), e^+e^- pair production, bremsstrahlung, and the photonuclear interaction. All of these processes involve the scattering of a virtual photon, emitted by an incident monopole, off of a target particle.

The monopole–matter electromagnetic interaction for monopole boosts $\gamma < 100$ is well reported in the literature [15,16]. Previous works include atomic excitations and ionization losses, including the density suppression effect. These processes are collectively referred to as "collisional" energy losses and are $\propto \ln \gamma$. The pair production ($MN \rightarrow MNe^+e^-$) and bremsstrahlung ($MN \rightarrow MN\gamma$) energy losses are $\propto \gamma$, where M, N, e and γ represent a monopole, nucleus, electron, and photon respectively. The photonuclear ($MN \rightarrow MNX$, where X are hadrons) energy loss [17] is roughly $\propto \gamma^{1.28}$. For large γ, the pair production and photonuclear interactions dominate while bremsstrahlung is suppressed by the large monopole mass as M^{-1}. (By comparison, the bremsstrahlung of a muon is of similar strength to other radiative energy loss processes.)

Here we only have space to collect the electromagnetic energy loss processes together and plot them, in fig. (1), for $M = 100$ TeV monopoles (see [14] for more details).

MONOPOLE ELECTROMAGNETIC SIGNATURES

Signature events for monopoles are discussed with a specific emphasis on 1) the general shower development, 2) the direct Cherenkov signal, 3) the coherent radio–

Cherenkov signal, and 4) the tomography of the Earth's interior. Monopoles will be highly penetrating primaries, interacting via the electromagnetic force and all the while maintaining their structural integrity. On average, there will be a quasi-steady cloud of secondary particles traveling along with the monopole. Thus, we will call this type of shower "monopole–induced."

Given a fast monopole passing through matter, the various electromagnetic processes can inject energetic photons, electrons, positrons, and hadrons into the absorbing medium. If the energy of these injected secondary particles is sufficient (roughly greater than $E_c \sim 100$ MeV), they may initiate a particle cascade. In terms of the inelasticity $\eta \equiv \Delta E/E$, the condition for electromagnetic shower development is $\eta \gtrsim E_c/E_0 \simeq 10^{-12} (E_0/10^{20}\text{eV})^{-1}$. Lower inelasticity events will contribute directly to ionization without intermediate particle production. The inelasticity per interaction and the subsequent shower development is best under-

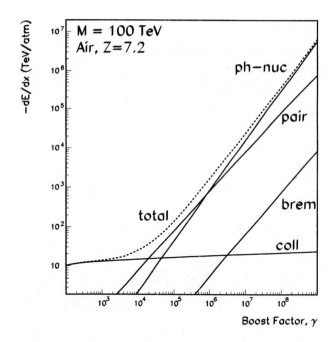

FIGURE 1. The electromagnetic energy loss from collisional, bremsstrahlung, electron–pair production, and the photonuclear interaction of a 100 TeV relativistic monopole in air. Collisional, pair production, and the photonuclear interaction are roughly independent of the monopole mass whereas bremsstrahlung is $\propto M^{-1}$. The units of energy loss are given in TeV per atmosphere.

stood for pair production. Detailed calculations [14] show that for $\gamma \gtrsim 10^4$ all of the monopole energy lost via pair production goes into the electromagnetic shower.

The contribution of the photonuclear process to the *electromagnetic* shower is indirect. The photonuclear interaction injects high energy hadrons into the monopole–induced shower. A subshower initiated by a high energy hadron will produce π^0's as secondaries, which each decay to 2 γ's. If these γ's have $E > E_c$, they may initiate an electromagnetic shower. So, only a fraction the energy lost via the photonuclear interaction contributes to the electromagnetic shower in the end.

Given the arguments above, it is reasonable to assume that pair production alone provides a lower bound to the electromagnetic shower size and that the pair production plus photonuclear interaction provides an upper bound. We plot the

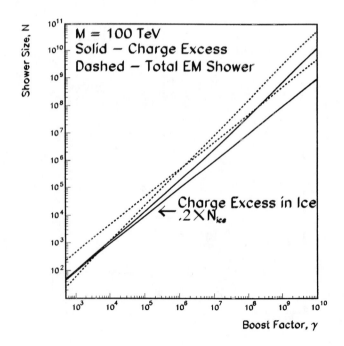

FIGURE 2. The monopole-induced quasi-steady shower size in ice for a monopole of mass 100 TeV. The shower size is the total number of electron, positrons, and photons. The dashed line $\propto \gamma$ is for pair production alone and the dashed line $\propto \gamma^{1.28}$ is for the photonuclear interaction alone. The solid lines show the electric charge excess (roughly 20% of the shower size) for pair production alone ($\propto \gamma$) and pair production plus photonuclear ($\propto \gamma^{1.28}$).

pair production and photonuclear processes separately (dashed lines) in fig. (2).

The electromagnetic shower sweeps a net charge excess from the medium into the shower of roughly 20% the shower size. For the charge excess we are again justified in using pair production alone as a lower bound and pair production plus photonuclear as an upper bound. This is reflected in fig. (2) by plotting pair production alone (the solid curve $\propto \gamma$) and by plotting pair production plus the photonuclear interaction (the solid curve $\propto \gamma^{1.28}$).

The lateral profile is approximately uniform out to a lateral cutoff given by the Molière radius

$$R_{\mathrm{M}} = 7.4 \, \frac{\mathrm{g}}{\mathrm{cm}^2} \, \left(\frac{\xi_e}{35\mathrm{g/cm}^2}\right) \left(\frac{100\mathrm{MeV}}{E_c}\right), \tag{5}$$

where ξ_e is the electron radiation length. As defined, the Molière radius is independent of the incident primary energy, being determined only by the spread of low energy particles resulting from multiple Coulomb scattering. Within a distance R_{M} of the monopole path will be $\sim 90\%$ of the shower particles [18].

Monopole Cherenkov Signatures

When a charge travels through a medium with index of refraction n, at a velocity $\beta > 1/n$, Cherenkov radiation is emitted. The total power emitted in Cherenkov radiation per unit frequency ν and per unit length l by a charge Ze is given by the Frank-Tamm formula

$$\frac{d^2W}{d\nu \, dl} = \pi\alpha Z^2\nu \left[1 - \frac{1}{\beta^2 n^2}\right]. \tag{6}$$

The maximal emission of the Cherenkov light occurs at an angle $\theta_{\mathrm{max}} = \arccos(1/n\beta)$ where θ is measured from the radiating particle's direction. The interaction of a magnetic charge with bulk matter requires the replacement of factors of ϵ with the Maxwell dual factors μ. But μ and ϵ are related by the index of refraction. The replacement in the electric charged–particle interaction formulae (for $Z = 1$) adequate for magnetic monopoles is $\alpha \to n^2/4\alpha$, and leads to an enhancement factor of 4700 for monopoles interacting in vacuum and 8300 for monopole interactions in water. However, in matter a relativisitic monopole is accompanied by an extensive cloud of charged particles it continually produces, so the difference in monopole electromagnetic interactions caused by the index of refraction factor is totally obscured.

The monopole-induced shower also contributes to the Cherenkov signal. In particular, the electric charge excess (of roughly 20% the shower size as shown in fig. (2)) will emit coherent Cherenkov for radio wavelengths, $\lambda >> R_{\mathrm{M}}$. For coherent radio–Cherenkov the Z^2 factor could be large, with $Z^2 \stackrel{<}{\sim} 10^{18}$, while the shower size is expected to remain roughly constant as the monopole traverses a large–scale ($\sim$ km^3) detector. Thus, a monopole signature event is clearly distinct from that of a neutrino event in the RICE array or similar large–scale detectors. The non–detection of a monopole event after one year of observation in a $\sim$ km^3 detector can, conservatively, set a flux limit of

$$F_M \overset{<}{\sim} 10^{-18} \text{ cm}^{-2} \text{ sec}^{-1} \text{ sr}^{-1} \tag{7}$$

which is significantly below the Parker limit.

Earth Tomography with Relativistic Monopoles

Direct knowledge about the composition and density of the Earth's interior is lacking. Analysis of the seismic data is currently the best source of information about the Earth's internal properties [19]. However, another potential probe would be the study of highly penetrating particles which could pass through the Earth's interior and interact differently depending upon the composition and density of material traversed. Thus, it may be possible to directly measure the density profile of the Earth's interior [20]. Over a range of masses, $M \sim 10^{4\pm1}$ TeV, and initial kinetic energies, monopoles can pass through the Earth's interior and emerge with relativistic velocities and, therefore, function as such probe. See [14] for more details.

BARYONIC–MONOPOLE AIR SHOWERS

The natural acceleration of monopoles to energies above the GZK cutoff at $E_{GZK} \sim 5 \times 10^{19}$ eV, and the allowed abundance of a monopole flux at the observed super–GZK event rate motivates us to ask whether monopoles may contribute to the super–GZK events. As a proof of principle, we have studied a simple model of a baryonic–monopole interaction in air which produces a shower similar to that arising from a hadronic primary. To mimic a hadron–initiated shower the baryonic–monopole must transfer nearly all of its energy to the shower over roughly a hadronic interaction length, $\lambda_0 \sim 80$ g cm^{-2}. The large inertia of a massive monopole makes this impossible if the cross–section is typically strong, ~ 100 mb [21]. The cross–section we seek needs to be much larger.

We model our arguments on those of [11] where three q–monopoles are confined by Z_3 strings of color–magnetic flux to form a color–singlet baryonic–monopole. We further assume that 1) the cross–section for the interaction of the baryonic–monopole with a target nucleus is geometric; in its unstretched state (before hitting the atmosphere) the monopole's cross–section is roughly hadronic, $\sigma_0 \sim \Lambda^{-2}$ (where $\Lambda \equiv \Lambda_{QCD}$); 2) each interaction between the baryonic–monopole and an air nucleus transfers an $O(1)$ fraction of the exchanged energy into stretching the chromomagnetic strings; 3) the chromomagnetic strings can only be broken with the formation of a monopole–antimonopole pair, a process which is highly suppressed and therefore ignored; other possible relaxation processes of the stretched string are assumed to be negligible; 4) the energy transfer per interaction is soft, $\Delta E/E \equiv \eta \sim \Lambda/M$.

The color–magnetic strings have a string tension $\mu \simeq \Lambda^2$. Therefore, when $O(1)$ of the energy transfer ($\gamma\Lambda$) stretches the color–magnetic strings (assumption 2), the length $l \sim \Lambda^{-1}$ increases by $\delta l = dE/\mu$, so that the fractional increase in length is $\delta l/l = \gamma$. Consequently, the geometrical cross–section grows $\propto \gamma\Lambda^{-2}$ after each interaction. The energy loss for baryonic-monopoles can then be approximated as

$$\frac{dE}{dx}(x) \simeq -\frac{\gamma\Lambda}{\lambda(x)} \simeq -\gamma\Lambda\, n_N\, \sigma(x), \tag{8}$$

where the strong cross–section $\sigma(x)$ is explicitly a function of column depth x and n_N is the number density of target nucleons. From assumption (4) we infer that the total number of monopole-nucleus interactions required to transfer most of the incoming kinetic energy is roughly η^{-1}. From the above discussion, the geometrical cross–section after n interactions is

$$\sigma_n \sim \frac{1+\sum_{i=1}^{n}\gamma_i}{\Lambda^2} = \frac{1+n\gamma}{\Lambda^2}, \tag{9}$$

where we have approximated $\gamma_n \sim (1-\eta)^n\,\gamma \sim \gamma$. The mean-free-path $\lambda \equiv 1/\sigma n_N$ after the n-th interaction is therefore

$$\lambda_n \sim \frac{\Lambda^2}{n_N\gamma n}, \qquad n \geq 1, \tag{10}$$

and the total distance traveled between the first interaction and the $(\eta^{-1})^{\text{th}}$ interaction is

$$\Delta X \sim \sum_{n=1}^{\eta^{-1}} \lambda_n \sim \frac{\Lambda^2}{n_N\,\gamma}\ln\left(\frac{M}{\Lambda}\right) << \lambda_0 \tag{11}$$

for $\eta^{-1} \gg 1$. Thus the stretchable chromomagnetic strings of the baryonic–monopole provide an example of a very massive monopole which nevertheless transfers $O(1)$ of its kinetic energy to an air shower over a very short distance. In conclusion, the baryonic–monopole's air–shower signature roughly mimics that of a hadronic primary.

ACKNOWLEDGMENTS.

This work was supported in part by the U.S. Department of Energy grants no. DE-FG05-86ER40272 (SDW), DE-FG05-85ER40226 (TWK & TJW), and the Vanderbilt University Research Council.

REFERENCES

1. E. N. Parker, Astrophys. J. **160**, 383 (1970).
2. N.A. Porter, Nuovo Cim. **16**, 958 (1960); E. Goto, Prog. Theo. Phys. **30**, 700 (1963).
3. T.W. Kephart and T.J. Weiler, Astropart. Phys. **4** 271 (1996); Nucl. Phys. (Proc. Suppl.) **51B**, 218 (1996).
4. T. W. Kibble, Phys. Rept. **67**, 183 (1980), and references therein.
5. S. F. King and Q. Shafi, Phys. Lett. **B422**, 135 (1998)[hep-ph/9711288].
6. D.K. Hong, J. Kim, J.E. Kim, and K.S. Soh, Phys. Rev. **D27**, 1651 (1983).
7. N. G. Deshpande, B. Dutta and E. Keith, Nucl. Phys. Proc. Suppl. **52A**, 172 (1997)[hep-ph/9607307]; Phys. Lett. **B388**, 605 (1996)[hep-ph/9605386]; Phys. Lett. **B384**, 116 (1996)[hep-ph/9604236].
8. P. H. Frampton and B. Lee, Phys. Rev. Lett. **64**, 619 (1990); P. H. Frampton and T. W. Kephart, Phys. Rev. **D42**, 3892 (1990).
9. C.L. Gardner and J.A. Harvey, Phys. Rev. Lett. **52**, 879 (1984).
10. T. Vachaspati, Phys. Rev. Lett. **76**, 188 (1996)[hep-ph/9509271]; H. Liu and T. Vachaspati, Phys. Rev. **D56**, 1300 (1997)[hep-th/9604138].
11. A. S. Goldhaber, Phys. Rept. **315**, 83 (1999)[hep-th/9905208].
12. For a review see P. P. Kronberg, Rept. Prog. Phys. **57**, 325 (1994).
13. D. Ryu, H. Kang, and P.L. Biermann, Astron. & Astrophys. **335**, 19 (1998)[astro-ph/9803275].
14. S. D. Wick, T.W. Kephart, T.J. Weiler, and P.L. Biermann, submitted to Astropart. Phys. [astro-ph/0001233].
15. G. Giacomelli, in: *Theory and Detection of Magnetic Monopoles in Gauge Theories*, ed. N. Craigie p.407 (Singapore: World Scientific Publishing Co., 1986).
16. S. P. Ahlen, Rev. Mod. Phys. **52**, 121 (1980).
17. S. Iyer Dutta, M. H. Reno, I. Sarcevic, and D. Seckel, [hep-ph/0012350].
18. Particle Data Group, Phys. Rev. D **50**, 1173 (1994). Prentice-Hall, 1952). 485 (1940).
19. T. Lay and T. C. Wallace, *Modern Global Seismology* (New York: Academic Press, 1995); Properties of the Solid Earth, in Rev.Geophys. **33**, (1995).
20. This idea has been exploited in neutrino physics as neutrinos are sufficiently weakly interacting to pass through the earth largely unimpeded for neutrino energies $\lesssim 10^{15}$ eV.
21. R.N. Mohapatra and S. Nussinov,Phys. Rev. **D57**, 1940 (1998); in I.F.M. Albuquerque, G. Farrar, and E.W. Kolb, Phys. Rev. **D59**, 015021 (1999) it is noted that a baryon mass above 10 GeV produces a noticeably different shower profile, and a baryon mass above 50 GeV is so different as to be ruled out.

"Signature" neutrinos from photon sources at high redshift

Marieke Postma

Department of Physics and Astronomy, UCLA, Los Angeles, CA 90095-1547
E-mail: postma@physics.ucla.edu

Abstract. The temperature of the cosmic microwave background increases with red-shift; at sufficiently high redshift it becomes possible for ultrahigh-energy photons and electrons to produce muons and pions through interactions with background photons. At the same time, energy losses due to interactions with radio background and inter-galactic magnetic fields are negligible. The energetic muons and pions decay, yielding a flux of "signature" neutrinos with energies $E_\nu \sim 10^{17}$eV. Detection of these neutrinos can help understand the origin of ultrahigh-energy cosmic rays.

The origin of ultrahigh-energy cosmic rays [1] with energies beyond the Greisen-Zatsepin-Kuzmin (GZK) cutoff [2] remains an outstanding puzzle [3]. Most of the proposed explanations can be categorized in two main classes. In the *bottom-up/acceleration* scenarios, charged particles are accelerated to ultrahigh energies in giant astrophysical "accelerators", such as active galactic nuclei and radio galaxies [4,5]. In the *top-down/decay* scenario on the other hand, massive objects such as topological defects (TD) [6–9] and superheavy relic particles [10–12] decay, emit-ting mainly ultrahigh-energy (UHE) photons. TD and relic particles can exist at much higher redshift than the astrophysical candidate sources, which are formed only after the onset of galaxy formation. To understand the origin of the ultrahigh-energy cosmic rays (UHECR), it is crucial to distinguish between these two different scenarios. In this talk we will show that the observation of a diffuse flux of "sig-nature" neutrinos with energies $E_\nu \sim 10^{17}$eV [13] can provide evidence for UHE photon sources at high redshift — for members of the *top-down/decay* class.

At small redshift UHE photons lose their energy mainly through scattering off the radio background (RB), and in the subsequent electromagnetic cascade [14,3]. In addition, the cascade electrons lose energy through synchrotron radiation in the intergalactic magnetic field (IGMF). This picture changes completely at high redshift, where both radio background and magnetic fields are small. At redshift z the density of CMB photons is higher by a factor $(1 + z)^3$, while the density of radio background is either constant or, more likely, lower. At some redshift z_R the scattering of high-energy photons off CMBR dominates over their scattering off RB.

CP579, *Radio Detection of High Energy Particles,* edited by D. Saltzberg and P. Gorham
© 2001 American Institute of Physics 0-7354-0018-0/01/$18.00

Based on the analyses of refs. [15,16] we take $z_R \sim 5$. Also, before galaxy formation the IGMF is weak, and for $z > z_M \sim 5$ synchrotron losses are not significant.

But the absence of RB and IGMF at early times is not the only difference: at redshift z the temperature of the cosmic microwave background radiation increases by a factor $(1 + z)$. Because of this, at high redshift the production of muons and pions in interactions of UHE photons and electrons with the CMBR becomes possible. Muons and charged pions decay into neutrinos, which can reach us today. The threshold for muon production is $\sqrt{s} > 2m_\mu = 0.21\text{GeV}$, or

$$E_{\gamma,e} > E_{\text{th}}(z) = \frac{10^{20}\text{eV}}{1 + z} \tag{1}$$

But will muons, and thus neutrinos, indeed be produced? To answer this question one has to look closely at the propagation of UHE photons at $z \gtrsim 5$. Photons scatter off CMB photons and, if their energy is above the threshold for muon pair production, given in eq. (1), they can either produce a muon pair, an electron pair or a double electron pair. Among these processes electron pair production (PP) has the highest cross section for photon energies $E_\gamma \lesssim 5 \times 10^{20}\text{eV}/(1 + z)$. Since the energies of the two interacting photons are vastly different, either the electron or the positron from PP has energy close to that of the initial photon. At higher photon energies, double pair production (DPP) becomes more important [17]. Four electrons, each carrying about $1/4$ of the initial photon energy, are produced in this reaction. Thus, after an initial $\gamma\gamma_{CMB}$ interaction one ends up with one or more UHE electrons.

These electrons in turn scatter off CMBR. For electron energies above the muon threshold, inverse Compton scattering is negligible compared to higher order processes [3], such as triplet production (TPP) $e\gamma_{CMB} \rightarrow e\,e^+e^-$ and muon electron-pair production (MPP) $e\gamma_{CMB} \rightarrow e\,\mu^+\mu^-$. For $\sqrt{s} > 2m_{\pi^\pm} = 0.28\text{GeV}$ charged pion production may also occur through $e\gamma_{CMB} \rightarrow e\,\pi^+\pi^-$. The TPP cross section is larger than that for MPP [18,19]. However, for center of mass energies $s \gg m_e^2$ the inelasticity for TPP is very small: one of the electrons produced in TPP carries 99.9% of the incoming electron's energy. It can interact once again with the CMBR. As a result, the leading electron can scatter many times before losing a considerable amount of energy; and with every scattering there is another chance to produce a muon pair. So instead of comparing cross sections, one should compare the *energy attenuation* length for TPP with the *interaction* length for MPP. As the latter is much larger, all electrons with energies above muon threshold will produce muons (or pions).

Topological defects and relic particles emit UHE photons. One can parameterize the time-dependence of their photon production rate as $\dot{n}_\gamma = \dot{n}_{\gamma,0}(t/t_0)^{-m}$, where t_0 is the present time. For decaying relic particles $m = 0$, for ordinary strings, monopolonium and necklaces $m = 3$, and for superconducting strings $m \geq 4$ [6–8]. Integrating $\dot{n}_\gamma = \dot{n}_{\gamma,0}$ over time, taking redshift into account, yields a neutrino flux:

$$n_\nu = \xi \int_{z_{\text{min}}}^{z_{\text{max}}} dt\; \dot{n}_\gamma(z)\;(1 + z)^{-4}$$

54

$$= \xi \frac{3}{-2a} \dot{n}_{\gamma,0} \, t_0 \left[(1 + z_{\min})^a - (1 + z_{\max})^a \right]. \tag{2}$$

Here $a = (3m - 11)/2$, $z_{\min} = \max(z_R, z_M) \approx 5$ the minimum redshift at which both RB and IMGF are negligible, $z_{\max} \sim 3 \times 10^3$ the redshift at which the universe becomes transparent to UHE neutrinos [20], and ξ is the number of neutrinos produced per UHE photon. We take $\xi \approx 4$, as one UHE photon produces one UHE electron, which generates a pair of muons whose decay gives four neutrinos. This is probably an underestimate because DPP produces more than one UHE electron. Also, the electron produced alongside the muon pair in MPP may have enough energy for a second round of muon pair-production.

For $m < 11/3$, $a < 0$, and, according to eq. (2), most of neutrinos come from red shift $z \sim z_{\min} \approx 5$. All these neutrinos are produced by photons with energies $E_\gamma > E_{\min} = 10^{20} \mathrm{eV}/(1 + z_{\min}) \sim 2 \times 10^{19} \mathrm{eV}$. If decaying TD or relic particles are the origin of the UHECR today, one can use the observed UHECR flux to fix the overall normalization constant $\dot{n}_{\gamma,0}(E > E_{\min})$. Using the photon fluxes calculated in [8,21,22], we get for the neutrino flux:

$$\phi_\nu \sim \begin{cases} 10^{-21} \mathrm{cm}^{-2} \mathrm{s}^{-1} \mathrm{sr}^{-1}, & \text{relic particles } (m = 0), \\ 10^{-18} \mathrm{cm}^{-2} \mathrm{s}^{-1} \mathrm{sr}^{-1}, & \text{monopolonium } (m = 3), \\ 10^{-16} \mathrm{cm}^{-2} \mathrm{s}^{-1} \mathrm{sr}^{-1}, & \text{necklaces } (m = 3). \end{cases} \tag{3}$$

The difference in the fluxes produced by monopolonium and by necklaces lies in their different clustering properties: whereas monopolonium (and relic particles) clusters in galaxies, necklaces are uniformly distributed throughout the universe.

The energy of these neutrinos can be estimated by

$$E_\nu \sim \frac{10^{20} \mathrm{eV}}{1 + z} \times \frac{1}{4} \times \frac{1}{3} \times \frac{1}{1 + z}. \tag{4}$$

As the photon spectrum of TD is a sharply falling function of energy, most neutrinos will come from photons with energies just above the threshold. In MPP, each muon produced gets about $1/4$ of the incoming electron energy [19]. A muon decays into an electron and two neutrinos, each getting approximately $1/3$ of the muon energy. Furthermore, the neutrino energy is redshifted by a factor $1/(1 + z)$. As already noted before, most neutrinos come from $z \sim z_{\min} \approx 5$; their energy is $E_\nu \sim 10^{17} \mathrm{eV}$.

Neutrinos from slowly decaying relic particles or some other $m = 0$ source are too sparse to be detectable in the foreseeable future. However, the flux from $m = 3$ sources, especially from necklaces, may be detected soon. This flux exceeds the background flux from the atmosphere and from pion photo-production on CMBR at this energy [23–25], as well as the fluxes predicted by a number of models [26]. TD can produce a large flux of primary neutrinos. However, the primary flux peaks at $E_\nu \sim 10^{20} \mathrm{eV}$, while the secondary flux peaks at $E_\nu \sim 10^{17} \mathrm{eV}$ and creates a distinctive "bump" in the spectrum. Models of active galactic nuclei (AGN) have predicted a similar flux of neutrinos at these energies [5]. The predictions of these models have been a subject of debate [27]. However, everyone agrees that AGN cannot produce neutrinos with energies of 10^{20}eV. So, an observation of 10^{17}eV neutrinos accompanied by a comparable flux of 10^{20}eV neutrinos would be a signature of a TD rather than an AGN.

Rapidly evolving $m \geq 4$ sources, $e.g.$ superconducting strings, have been ruled out as the origin of UHECR by the EGRET bounds on the flux of low-energy γ-rays [3]. Although their density today is constrained, in the early universe these sources might have existed in large enough numbers to produce a detectable flux of neutrinos. Neutrinos with $E_\nu \sim 10^{17}$eV are probably the only observable signature of such rapidly evolving sources that were active at high red shift but are "burned out" by now.

To conclude, we have shown that sources of ultrahigh-energy photons that operate at red shift $z \gtrsim 5$ produce neutrinos with energy $E_\nu \sim 10^{17}$eV. The flux depends on the evolution index m of the source. A distinctive characteristic of this type of neutrino background is a cutoff below 10^{17}eV due to the universal radio background at $z < z_{\min}$. Detection of these neutrinos can help understand the origin of ultrahigh-energy cosmic rays.

REFERENCES

1. M. Takeda $et\ al.$, Phys. Rev. Lett. **81**, 1163 (1998); M.A. Lawrence, R.J. Reid and A.A. Watson, J. Phys. G **G17**, 733 (1991); D. J. Bird $et\ al.$, Phys. Rev. Lett. **71**, 3401 (1993); Astrophys. J. **424**, 491 (1994).

2. K. Greisen, $Phys.\ Rev.\ Lett.$ **16**, 748 (1966); G. T. Zatsepin and V. A. Kuzmin, $Pisma\ Zh.\ Eksp.\ Teor.\ Fiz.$ **4**, 114 (1966).

3. For review, see, $e.g.$, P. L. Biermann, J. Phys. G **G23**, 1 (1997); P. Bhattacharjee and G. Sigl, Phys. Rept. **327**, 109 (2000).

4. E. Waxman, Phys. Rev. Lett. **75**, 386 (1995); G. R. Farrar and P. L. Biermann, Phys. Rev. Lett. **81**, 3579 (1998); A. Dar, A. De Rujula and N. Antoniou, astro-ph/9901004; G. R. Farrar and T. Piran, Phys. Rev. Lett. **84**, 3527 (2000); E. Ahn, G. Medina-Tanco, P. L. Biermann, and T. Stanev, astro-ph/9911123.

5. K. Mannheim, R. J. Protheroe and J. P. Rachen, astro-ph/9812398; astro-ph/9908031.

6. A. Vilenkin, Phys. Rept. **121**, 263 (1985); A. Vilenkin and E. P. S. Shellard, $Cosmic\ strings\ and\ other\ topological\ defects$, Cambridge University Press, Cambridge, England, 1994; M. B. Hindmarsh and T. W. Kibble, Rept. Prog. Phys. **58**, 477 (1995).

7. C. T. Hill and D. N. Schramm, Phys. Rev. **D31**, 564 (1985); C. T. Hill, D. N. Schramm, and T. P. Walker, Phys. Rev. **D36**, 1007 (1987); P. Bhattacharjee, C. T. Hill, and D. N. Schramm, Phys. Rev. Lett. **69**, 567 (1992).

8. V. Berezinsky, P. Blasi and A. Vilenkin, Phys. Rev. **D58**, 103515 (1998).

9. V. S. Berezinsky and A. Vilenkin, hep-ph/9908257.

10. V. Berezinsky, M. Kachelriess, and A. Vilenkin, Phys. Rev. Lett. **79**, 4302 (1997); V. A. Kuzmin and V. A. Rubakov, Phys. Atom. Nucl. **61**, 1028 (1998) [Yad. Fiz. **61**,

1122 (1998)]; V. Kuzmin and I. Tkachev, JETP Lett. **68**, 271 (1998); M. Birkel and S. Sarkar, Astropart. Phys. **9**, 297 (1998); D.J. Chung, E.W. Kolb, and A. Riotto, Phys. Rev. Lett. **81**, 4048 (1998); Phys. Rev. **D59**, 023501 (1999); K. Benakli, J. Ellis, and D. V. Nanopoulos, Phys. Rev. **D59**, 047301 (1999); Phys. Rev. **D59**, 123006 (1999).

11. G. Gelmini and A. Kusenko, Phys. Rev. Lett. **84**, 1378 (2000). J. L. Crooks, J. O. Dunn, and P. H. Frampton, astro-ph/0002089.

12. For review, see, *e.g.*, V. A. Kuzmin and I. I. Tkachev, Phys. Rept. **320**, 199 (1999).

13. A. Kusenko and M. Postma, hep-ph/0007246.

14. V. Berezinsky, Sov. J. of Nucl. Phys. **11**, 222 (1970).

15. J. J. Condon, Astrophys. J. **284**, 44 (1984).

16. P. Madau et al., Mon. Not. R. Astron. Soc. **283**, 35 (1996); C.C. Steidel, Proc. Nat. Acad. Sci. **96**, 4232, 1999; T. Miyaji, G. Hasinger, and M. Schmidt, astro-ph/9809398; G. Pugliese, H. Falcke, Y. Wang, and P.L. Biermann, Astron. and Astrophys., **358**, 409 (2000).

17. R. W. Brown, W. F. Hunt, K. O. Mikaelian and I. J. Muzinich, Phys. Rev. **D8**, 3083 (1973).

18. A. Borsellino, Nuovo Cimento **4**, 112 (1947); K. J. Mork Phys. Rev. **160**, 1065 (1967); A. Mastichiadis, A. P. Marscher and K. Brecher.

19. Astrophys. Journal **300**, 178 (1986); V. Anguelov, S. Petrov, L. Gurdev and J. Kourtev, J. Phys. **G25**, 1733 (1999).

20. P. Gondolo, G. Gelmini and S. Sarkar, Nucl. Phys. **B392**, 111 (1993).

21. R. J. Protheroe and P. A. Johnson, Astropart. Phys. **4**, 253 (1996) [astro-ph/9506119].

22. R. J. Protheroe and T. Stanev, Phys. Rev. Lett. **77**, 3708 (1996) [astro-ph/9605036].

23. C. T. Hill and D. N. Schramm, Phys. Lett. **B131**, 247 (1983).

24. F. W. Stecker, C. Done, M. H. Salamon, and P. Sommers, Phys. Rev. Lett. **66**, 2697 (1991); erratum: *ibid.*, **69**, 2738 (1992).

25. T. Stanev, R. Engel, A. Mucke, R. J. Protheroe and J. P. Rachen, Phys. Rev. **D62**, 093005 (2000) [astro-ph/0003484].

26. D. B. Cline and F. W. Stecker, astro-ph/0003459.

27. E. Waxman and J. Bahcall, Phys. Rev. **D59**, 023002 (1999).

Extreme-Energy Cosmic Rays: Puzzles, Models, and Maybe Neutrinos

Dept. of Physics & Astronomy, Vanderbilt University, Nashville, TN 37235, USA

Abstract. The observation of twenty cosmic-ray air-showers at and above 10^{20} eV poses fascinating problems for particle astrophysics: how the primary particles are accelerated to these energies, how the primaries get here through the 2.7K microwave background filling the Universe, and how the highest-energy events exhibit clustering on few-degree angular scales on the sky when charged particles are expected be bent by cosmic magnetic fields. An overview of the puzzles is presented, followed by a brief discussion of many of the models proposed to solve these puzzles. Emphasis is placed on (i) the signatures by which cosmic ray experiments in the near future will discriminate among the many proposed models, and (ii) the role neutrino primaries may play in resolving the observational issues. It is an exciting prospect that highest-energy cosmic rays may have already presented us with new physics not accessible in terrestrial accelerator searches.

I   INTRODUCTION

An unsolved astrophysical mystery, now forty years old, is the origin and nature of the extreme energy cosmic ray primaries (EECRs) responsible for the observed events at highest energies, $\sim 10^{20}$ eV [1]. About twenty events at $\sim 10^{20}$ eV have been observed by five different experiments [2]. The origin of these events is a mystery, for there are no visible source candidates within 50 Mpc except possibly M87, a radio-loud AGN at ~ 20 Mpc, and Cen-A (NGC5128), a radio galaxy at 3.4 Mpc, and neither of these is in the direction of any of the observed events [3]. Since the observed events display a large-scale isotropy, many sources rather than one source seem to be required. The nature of the primary particle is also mysterious, because interactions with the 2.73K cosmic microwave background (CMB) renders the Universe opaque to nucleons at $E_{\rm GZK} \sim 5 \times 10^{19}$ eV, and double pair production on the cosmic radio background (CRB) renders the Universe opaque to photons at even lower energies. The theoretical prediction of the end of transparency for nucleons at $E_{\rm GZK} \sim 0.5 \times 10^{20}$ eV is the famous "GZK cutoff" [4]. Figure I shows a recent compilation of the AGASA data set, clearly extending beyond $E_{\rm GZK}$.

The main theory challenges in attempting to understand the super-GZK data are: (i) What cosmic source could have accelerated the primary particles to such extremely high energies? and (ii) If the sources are distant ($\gtrsim 100$ Mpc), then how could their primaries have propagated through the cosmic background radiation

CP579, *Radio Detection of High Energy Particles,* edited by D. Saltzberg and P. Gorham

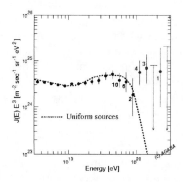

FIGURE 1. Extreme-energy cosmic ray spectrum as observed by AGASA. Error bars correspond to 68 % C.L. and the numbers count the events per energy bin. The dashed line revealing the GZK cutoff is the spectrum expected from uniformly distributed astrophysical sources (from the AKENO website [12]).

without substantial energy loss? The acceleration mechanism either requires a Zevatron accelerator ($1\text{ZeV} \equiv 10^{21}$ eV) [5], distant because such a source could not be missed if it were nearby; or speculative decaying super-massive particles (SMPs) or topological defects (TDs) with mass-scale $\gtrsim 10^{22}$ eV, clustered nearby; or possibly magnetic monopoles accelerated by the cosmic magnetic fields. From distant Zevatrons, only neutrinos among the known particles can propagate unimpeded to earth. Exploiting this fact are the "Z-burst model," and the strongly-interacting neutrino model. Adding to the drama and mystery at present is the observed *large-scale isotropy* and *small-scale anisotropy*. Surely these characteristics hint at a solution to the mystery of origin; protons, nuclei, and magnetic charges bend in cosmic magnetic fields, whereas photons and neutrinos do not. With more data above E_{GZK}, several distinct telltale signatures including the isotropy/anisotropy will allow one to discriminate among the many models proposed for the origin and nature of the EECRs.

II EVENT CLUSTERING

One revealing signature is already evident in the existing data sample. This is the pairing of events on the celestial sky. The AGASA experiment has already presented data strongly suggesting that directional pairing is occurring at higher than chance coincidence [6]. Of the 47 published AGASA events above 4×10^{19} eV, 9 are contained in three doublets and one triplet with separation angle less than the angular resolution of 2.5°. The chance probability of this clustering occurring in an isotropic distribution is less than 1%. The chance probability for the triplet alone is only 5%. Of the seven events above 10^{20} eV, three are counted among the doublet events. Most recently, AGASA has reported two more events above E_{GZK}

[7]. Each aligns in direction with a previous event, reducing further the probability for random clustering to 0.07% at 3σ. Comparisons of event directions in a combined data sample of four experiments further supports non-chance coincidences, especially in the direction of the SuperGalactic Plane [8].

Whether the pairing is random or dynamical, we shall also know in the near future. If the pairing turns out to be dynamical, I would argue that neutrinos are a favorite candidate for the primary particles. This is because photons have such a short (~ 10 Mpc) absorption length, and protons are bent by cosmic magnetic fields during their extragalactic journey.

If neutrinos are the primaries, they should point back to their sources, thereby enabling point-source astronomy for the most energetic sources of flux at and above 10^{20} eV. It was reported [9] that the first five events at 10^{20} eV did in fact point toward extragalactic compact radio-loud quasars, just the class of objects which could accelerate EECRs to ZeV energies via shock mechanisms. With the inclusion of subsequent data, this association is controversial [10]. The jury awaits further evidence. A random distribution is obtained by tossing n events randomly into $N \simeq (\Omega/\pi\theta)^2 = 1045\,(\Omega/1.0\text{ sr})(\theta/1.0°)^{-2}$ angular bins, where Ω is the solid angle on the celestial sphere covered by the experiment and θ is the bin half-angle. Each resulting event distribution is specified by the partition of the n total events into a number m_0 of empty bins, a number m_1 of single hits, etc., among the N angular bins. The probability to obtain a given event topology is [11]

$$P(\{m_i\}, n, N) = \frac{1}{N^n} \frac{N!\; n!}{\prod_{j=0} m_j!\, (j!)^{m_j}} \, . \tag{1}$$

The variables in the probability are not all independent, as $\sum_{j=1} j \times m_j = n$ and $\sum_{j=0} m_j = N$. It is useful to use these constraints to rewrite this probability as

$$P(\{m_i\}, n, N) = \frac{N!}{N^N} \frac{n!}{n^n} \prod_{j=0} \frac{(\overline{m_j})^{m_j}}{m_j!} \, , \tag{2}$$

where we have defined

$$\overline{m_j} \equiv N \left(\frac{n}{N}\right)^j \frac{1}{j!} \, . \tag{3}$$

When $N \gg n \gg 1$, a limit valid for the AGASA, HiRes, Auger and Telescope Array experiments, one finds

$$P(\{m_i\}, n, N) \approx \prod_{j=2} \frac{(\overline{m_j})^{m_j}}{m_j!} e^{-\overline{m_j}\, r^j (j-2)!} \, , \tag{4}$$

where $r \equiv (N - m_0)/n \approx 1$. The non-Poisson nature of eq. (4) is reflected in the factorials and powers of r in the exponents.

Typical values of effective area A (km^2 sr), celestial solid angle Ω, and angular resolution $\theta_{\min}$ for the existing and proposed EECR experiments [12] are shown in the following table, where the incident flux $F(\geq E_{\text{GZK}}) = 10^{-19}\text{cm}^{-2}\text{s}^{-1}\text{sr}^{-1}$ has been used to estimate the number of events (n/yr) above $E_{\text{GZK}} = 5 \times 10^{19}$ eV.

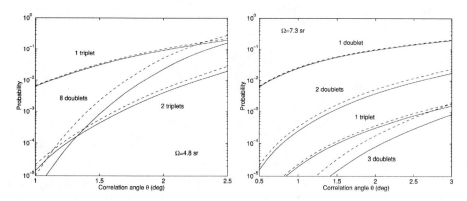

FIGURE 2. Inclusive probabilities for various clusters in a 100 event sample at Auger (left) and 20 events at HiRes (right). The solid line is the exact result, the dashed line is the Poisson approximation.

	AGASA	HiRes	Auger/TA	EUSO/OWL
$km^2\,sr$	150	800	6×10^3	3×10^5
n/yr	5	30	200	10^4
Ω (sr)	4.8	7.3	4.8	4π
θ_{min}	3.0°	0.5°	1.0°	1.0°

In Figure 2 are shown inclusive probabilities for the Auger experiment with 100 events, as determined by formulae eq. (4). Note that "inclusive probability" means the stated number of j-plets *plus any other clusters*. The 8-doublet probability is extremely sensitive to the angular binning; observation of a flatter dependence on angular bin-size could signal a non-chance origin for the clustering. The observation of two triplets with angular binning of less than 2° would constitute 3-sigma evidence for cosmic dynamics. The same conclusion would hold if a quadruplet within 2.5° is observed among the first 100 Auger events.

Without some modification, our analytic formula may not be directly relevant to HiRes data. Clear, moonless nights are required for detection of atmospheric fluorescence, and summer nights ($\sim$ 18 hrs right ascension) are effectively 40% shorter than winter nights ($\sim$ 6 hrs right ascension) for monocular HiRes [13]. Accordingly, the HiRes efficiency versus Galactic longitude varies significantly, roughly as $N(RA) = \overline{N}(1 + \epsilon \sin(2\pi RA/24\text{hrs}))$, with $\epsilon \approx 0.25$. One must ask whether this sinusoidally-varying efficiency invalidates the analytic approach with its assumption of a constant efficiency. In the Monte Carlo approach there is an easy method to generate "background" data sets with the experimental efficiency properly included – one randomly permutes the RA coordinates of the real data to remove any dynamical correlation among the events. Fortunately, there is also a simple method to estimate the efficiency correction to an analytic Poisson distribution, if one assumes that although N and n vary with right ascension, $\xi \equiv n/N$ is constant.

Then it is easy to show that the Poisson distribution

$$P(j, m) = \left(\frac{N\xi^j}{j!} \right)^m \frac{1}{m!} \, e^{-N\xi^j/j!} \tag{5}$$

is corrected upwards by the factor

$$\mathcal{I} = \frac{1}{2\pi} \int_0^{2\pi} d\phi (1 + \epsilon \sin(\phi))^m \, e^{-\overline{N}\epsilon \sin(\phi)\xi^j/j!} . \tag{6}$$

For $\frac{\epsilon\xi^{j-2}}{j!} \frac{n^2}{N} \ll 1$, the exponential is near unity and the correction has a closed form, $\mathcal{I}_m \approx (1 - \epsilon^2)^{m/2} P_m(\frac{1}{\sqrt{1-\epsilon^2}})$, with P_m a Legendre polynomial. As a series, the correction is

$$\mathcal{I}_m = 2^{-m} \sum_{k=0}^{[m/2]} \frac{(-)^k (2m - 2k)!}{k!(m - k)!(m - 2k)!} (1 - \epsilon^2)^k . \tag{7}$$

In particular, for $n^2 \ll \overline{N}$, we have $\mathcal{I}_1 = 1$ independent of ϵ, $\mathcal{I}_2 = 1 + \epsilon^2/2, = 1.03$ for HiRes, $\mathcal{I}_3 = 1 + 3\epsilon^2/2, = 1.09$ for HiRes, and $\mathcal{I}_4 = 1 + 3\epsilon^2 + 3\epsilon^4/8, = 1.189$ for HiRes. The lesson is that HiRes efficiency corrections appear negligible for small numbers of clusters, and we may proceed with our analytic analysis for the HiRes experiment, for which about 20 events at 10^{20} eV are expected when the first full year's data is analyzed. We display in Figure 2 the inclusive probabilities for one or more, two or more, and three or more doublets; and one or more triplets, over a range of angular binning.

Note that for all except the 3 doublet configuration, the Poisson approximation using the mean values in Eq.(3) provides an estimate good to within 50% of the non-approximate form; for the (much suppressed) 3 doublet configuration, it over-estimates the probability by about a factor of 3 in much of the angular region. For angular binning tighter than 2°, an observation of two doublets among the first 20 events has a chance probability of less than 0.5%. Thus the observation of this topology could be construed as evidence (at the 3σ level) for clustering beyond statistical. The observation of a triplet within $\leq 3°$ has a random probability of less than 10^{-3}, and hence observation of such a triplet would most likely signify clustered or repeating sources, or magnetic focusing effects. With the accumulation of 40 events (not shown in the figure), the appearance of two doublets has a probability of less than 0.5% for a correlation angle of 1° or less. This illustrates how the good angular resolution of HiRes may be used to detect non-statistical clustering with only a few observed clusters.

Projected event rates for the EUSO/OWL/AW experiments present a pleasant problem for our analytic formula. In the case where $n > N \gg 1$, relevant for the EUSO/OWL/AW experiments after a year or more of running, higher j-plets are common and the distribution of clusters can be rather broad in j. From eq. (3) we have $(\overline{m_j}/\overline{m_{j-1}}) = (n/jN) \sim (\pi n\theta^2/j\Omega)$. Already at $j = 1\,(2)$, Stirling's approximation to $j!$ is good to 8% (4%), and so for $j \geq 1$ we may approximate

$\overline{m_j} \approx (N^3/2\pi en)^{\frac{1}{2}} (en/jN)^{j+\frac{1}{2}}$. Extremizing this expression with respect to j, one learns that The most populated j-plet occurs near $j \sim n/N$. Combining this result with the broad distribution expected for large n/N, one expects clusters with j up to several $\times \frac{n}{N}$ to be common in the EUSO/OWL/AW experiments. Probabilities and meaning for frequent, large clusters are somewhat difficult to assess numerically.

The random distributions displayed in fig. 2 are expected from some models, such as randomly-situated decaying super-massive particles (SMPs), or charged-particle or magnetic monopole primaries with directions randomized by incoherent cosmic magnetic fields. A complementary approach to the random probabilities shown here is to consider specific source models generating non-random angular distributions. Steps along this line of inquiry have recently been taken [14]. Future progress in the field will involve comparisons of the random and non-random model predictions with the data.

III MODELS FOR SUPER-GZK EVENTS

The conjectured origins of the super-GZK particles fall into four basic categories. These are (i) nearby accelerators, (ii) exotic primaries, (iii) exotic physical law, and (iv) neutrino primaries. Other reviews [1] have emphasized the first three categories (especially (i) and (ii)), so here we will be brief with those, and put more emphasis on the neutrino-primary option.

Nearby accelerators: Several types of sources have been proposed to exist within our Galactic halo. These include highly-ionized relativistic dust grains, Galactic supershocks, young neutron stars, magnetars (highly magnetized pulsars), decaying SMPs with GUT masses of order 10^{15} GeV or with inflation-motivated masses of order $10^{13\pm1}$ GeV, topological defects [15] such as strings, Q-balls and vortons, and annihilating monopole-antimonopole bound states (monopolonium) [16]. For rare sources emitting charged particles, such as magnetars, it is necessary to postulate that the primaries are iron nuclei to ease the acceleration requirement and to isotropize the flux in our galactic magnetic field. It is intriguing that the observed magnitude of CMB fluctuations fixes the reheat temperature following inflation to $10^{13\pm1}$ GeV, which allows gravitational and thermal production of TDS and/or SMPs of just this mass (if they exist) [17]; this mass scale is just right for producing 10^{20} secondary particles via decay.

For some TD models, dimensional arguments and scaling laws seem to give an emission rate to short-lived SMPs consistent with the observed EECR rate without tuning exotic parameters. However, for most decaying sources such as SMPs, TDs, and monopolonium, it is necessary to tune the lifetime to be longer but not too much longer than the age of the Universe in order to maintain an appreciable secondary particle emission rate today. Discrete gauged symmetries [18] or hidden sectors [19] are introduced to stabilize the heavy particle. Then rather esoteric physics is needed to break the new symmetry super-softly to maintain the long lifetime. High-dimension operators, wormholes [20], and instantons [21] have been invoked for this purpose.

Models with sources mainly in the Galactic plane are disfavored by the lack of any observed planar anisotropy in the data. Models with sources clustered in the Galactic halo predict a dipole enhancement in the direction of the Galactic center [22], which will be tested by the Auger Observatory in a few years.

Possible sources outside our halo, but still relatively nearby, include the radio-loud quasar M87 at 18 Mpc [23], the similar Cen-A at 3.4 Mpc [24], rare nearby GRBs, and now-dormant rare AGNs (also called massive dark objects – MDOs). These sources are few at best, and strong magnetic fields must be postulated to isotropize and/or confine their emissions. In the case of GRBs, the identification of the red-shift of their host galaxies at typically $\overset{>}{\sim} 1$ renders a local occurrence of a GRB highly improbable [25].

Finally, primordial black holes (PBHs) have also been suggested, but any sensible initial mass spectrum is unable to provide a sufficient number of PBHs in the final stage of decay today.

Exotic primaries: The GZK cutoff can be raised by simply postulating a primary hadron slightly heavier than the proton. The reason is kinematical – the cutoff energy varies as the square of the mass of the first excited resonant state. For this reason, Farrar has proposed that light supersymmetric baryons, made from a light gluino plus the usual quarks and gluons, may be the primary EECRs [26]. Such a scenario renders the source-energetics issue even more challenging. In any event, terrestrial experiments seem to have recently closed the window on a possible light gluino.

Another interesting possibility for the primary EECR is the *magnetic monopole* [27,28]. Any breaking of a semisimple gauge symmetry which occurs after inflation and which leaves unbroken a $U(1)$ symmetry group may produce an observable abundance of magnetic monopoles. The monopole mass is expected to be $\sim \alpha^{-1}$ times the temperature T_c of the symmetry breaking. At the time of the phase transition, roughly one monopole or antimonopole is produced per correlated volume. The resulting monopole number density today is

$$n_M \sim 10^{-19} \, (T_c/10^{11}\text{GeV})^3 (l_H/\xi_c)^3 \, \text{cm}^{-3}, \tag{8}$$

where ξ_c is the phase transition correlation length, bounded from above by the horizon size l_H at the time of the transition. In a second order or weakly first order phase transition, the correlation length is comparable to the horizon size. In a strongly first order transition, the correlation length is considerably smaller than the horizon size.

The kinetic energy imparted to a magnetic monopole on traversing a magnetic field B is $E_K \sim gB\xi$, where $g = e/2\alpha$ is the magnetic charge according to the Dirac quantization condition, and ξ specifies the field's coherence length. Given the magnitude and coherence length data for the cosmic magnetic fields, monopole kinetic energies in the range 10^{20} to 10^{23} eV are expected; the acceleration problem is naturally solved. Monopoles with $M \stackrel{<}{\sim} 10^{14}$ GeV should be relativistic, and carry the appropriate energy to qualify as candidates for the EECR primaries [28,29]. Within field theory there exist many possibilities for an intermediate unification scale and intermediate-mass monopoles.

The propagation problem is also naturally solved. The scattering cross-section for the monopole on the 3K and diffuse photon backgrounds is just classical Thomson, valid even for strong coupling: $\sigma_T = 8\pi\alpha_M^2/3M^2 \sim 2 \times 10^{-43}\,(M/10^{10}\mathrm{GeV})^{-2}\,\mathrm{cm}^2$. The resulting mean free path for inverse Compton scattering is many orders of magnitude larger than the Hubble size of the Universe.

The relativistic monopole flux is simply

$$F_M = c\,n_M/4\pi \sim 2 \times 10^{-19}\,(M/10^{10}\mathrm{GeV})^3(l_H/\xi_c)^3 \tag{9}$$

per cm^2-s-sr, which compares favorably with the integrated flux above 10^{20} eV, $F_{\mathrm{data}}(> 10^{20}\mathrm{eV}) \sim 2 \times 10^{-20}\mathrm{cm}^{-2}\mathrm{s}^{-1}\mathrm{sr}^{-1}$, and is comfortably below Parker's upper bound $F_{\mathrm{Parker}} = 10^{-15}\mathrm{cm}^{-2}\mathrm{s}^{-1}\mathrm{sr}^{-1}$ for a cosmic monopole flux.

Signatures for EECR monopoles in our atmosphere and in ice are discussed at length in [29]. Of particular interest as a model for the super-GZK primaries is the "baryonic monopole" [30,29]. It is a bound state of three colored monopoles, confined by chromomagntic strings. Chromomagnetic strings can stretch when excited, but cannot break into quark-antiquark pairs. On first interaction in the atmosphere, the baryonic monopole of mass M and energy $E = \gamma M$ stretches to create a huge geometrical cross-section of order $\gamma/\Lambda_{\mathrm{QCD}}^2 \sim (\gamma/10^6) \times 10^7$ mb. Consequently, nearly all of the initial monopole energy is transferred to the atmospheric shower in a very short distance. A recent simulation of the baryonic-monopole showed good agreement with the lateral muon and hadron content of the highest-energy Yakutsk event, but less than good agreement with the longitudinal profile of the Fly's Eye event [31].

Any confirmed directional pairing of events would appear difficult to achieve with the monopole model. Also, in the context of a model of the Galactic magnetic field, it has been shown that some memory of the local spiral arm direction, and an energy spectrum flatter than the observed one, are expected in the data if monopoles are the primaries [32]. Although the directional criterion may not survive inclusion of extragalactic fields, the flatness of the spectrum probably does.

Exotic physical laws: The most remarkable proposals posit a breakdown of Lorentz invariance [33] or a breakdown of general relativity above some high scale. In the case of broken GR, it could be spacetime fluctuations (expected in a theory of quantum gravity) wiggling the on-shell dispersion relation [34] or appearing as $1/M_P$ operators which alter the physics. String theory provides motivation for GR-breaking at a possibly lower scale, the string scale M_S. Certainly the energy window of EECRs ($10^{20}\mathrm{eV}/M_P \sim 10^{-8}$) is beyond that of terrestrial accelerators, and so ripe for speculation on new high-energy physics.

Signatures of models with no photo-pion production above E_{GZK} include the absence of a proton pile-up below E_{GZK}, the absence of a cosmogenic neutrino flux, and possibly undeflected pointing of the primary back to its source.

Neutrino primaries: Turning to the possibility that the primaries may be neutrinos, one encounters an immediate obstacle: the SM neutrino cross-section is down from that of an electromagnetic or hadronic interaction by six orders of magnitude. This implies a low air-shower rate, and an accumulation of events at low altitudes ("penetrating" events) where the target density is highest. On the other hand, the neutrino-primary hypothesis is supported by the observed clustering discussed earlier. Two solutions to the small cross-section problem for primary neutrinos have been proposed.

1) Neutrino annihilation to Z-bursts: Here it is proposed that the primary particles which propagate across cosmic distances above the GZK cutoff energy are neutrinos, which then annihilate with the cosmic neutrino background (CNB) within the GZK zone ($D < D_{\text{GZK}} \sim 50$ Mpc) to create a "local" flux of nucleons and photons above E_{GZK}, as shown in Fig. 3. It was noted many years ago that a cosmic ray neutrino arriving at earth from a cosmically distant source has an annihilation probability on the relic–neutrino background of roughly $3.0\,h_{65}^{-1}\,\%$ (neglecting cosmic expansion) [35]. The probability for a neutrino with resonant energy to annihilate to a Z-burst within distance D_{GZK} is then 2.5×10^{-4} for a homogeneous CNB [36]. The annihilation rate depends upon the CNB density, reliably predicted in the mean by Big Bang cosmology, and on the Standard Model (SM) of particle physics. The local annihilation rate is larger if our matter-rich portion of the Universe clusters neutrinos [36,37], or if there is an intrinsic CP-violating $\nu - \bar{\nu}$ asymmetry [38].

Each resonant neutrino annihilation produces a Z boson with a 70% branching ratio into hadrons known to include on average about one baryon–antibaryon pair, seventeen charged pions, and ten neutral pions [39]. The ten π^0's decay to produce twenty high–energy photons. For m_ν in the range ~ 0.1 to 2 eV, the energy in this resonant "Z-burst" is fortuitously situated sufficiently above E_{GZK} at

$$E_\nu^R = M_Z^2/2m_\nu = 4\,(\text{eV}/m_\nu) \times 10^{21}\text{eV} \tag{10}$$

so as to produce photons and nucleons with energies exceeding E_{GZK}.[1] The mean energies of the ~ 2 baryons and ~ 20 photons produced in the Z decay are easily estimated. Distributing the Z-burst energy among the mean multiplicity of 30 secondaries in Z-decay [39], one has

$$\langle E_p \rangle \sim \frac{E_R}{30} \sim 1.3 \left(\frac{\text{eV}}{m_j}\right) \times 10^{20}\text{eV}\,. \tag{11}$$

---

[1] The resonant-energy width is narrow, reflecting the narrow width of the Z-boson: at FWHM $\Delta E_R/E_R \sim \Gamma_Z/M_Z = 3\%$.

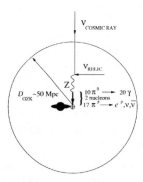

FIGURE 3. Schematic diagram showing the production of a Z-burst resulting from the resonant annihilation of a cosmic–ray neutrino on a relic (anti)neutrino. If the Z–burst occurs within the GZK zone (~ 50 to 100 Mpc) and is directed towards the earth, then photons and nucleons with energy above the GZK cutoff may arrive at earth and initiate super–GZK air–showers.

The photon energy is further reduced by an additional factor of 2 to account for their origin in two-body π^0 decay:

$$\langle E_\gamma \rangle \sim \frac{E_R}{60} \sim 0.7 \left(\frac{\text{eV}}{m_j}\right) \times 10^{20}\text{eV}\,. \tag{12}$$

Even allowing for energy fluctuations about mean values, it is clear that in the Z-burst model the relevant neutrino mass cannot exceed ~ 2 eV. On the other hand, the neutrino mass cannot be too light or the predicted primary energies will exceed the observed event energies. In this way, one obtains the approximate 0.1 eV lower limit on the neutrino mass, when allowance is made for an order of magnitude energy-loss for those secondaries traversing 50 to 100 Mpc. The challenging issue of how experiments might actually determine the absolute neutrino mass is discussed in [40].

If the Z–burst points in the direction of earth and occurs within the GZK distance, then one or more of the photons and nucleons in the burst may initiate a super–GZK air–shower at earth [36,37]. For a sufficient cosmic neutrino flux, the hypothesis successfully explains the observed air–showers above E_{GZK}. Comparisons of the model predictions to super-GZK data are available in [41].

The existence of neutrino mass in the desired range seems nearly guaranteed from the tritium decay upper bound [42] and the lower bounds inferred from the terrestrial neutrino oscillation experiments. The simplest explanation for the atmospheric neutrino results is neutrino oscillations driven by a mass–squared difference of $\delta m^2_{\text{atm}} \sim 3 \times 10^{-3}\text{eV}^2$ [43], which implies a neutrino mass of *at least* 0.05 eV. Also, the recent LSND measurement appears to indicate a mass–squared difference $\delta m^2_{\text{LSND}} \gtrsim 0.2\text{eV}^2$ [44], from which one deduces a neutrino mass of at least 0.5 eV.

From these lower bounds on neutrino mass, one gets upper bounds on the Z–burst energy of 10^{23} and 10^{22} eV, respectively, just right for extending the air-shower spectrum an order of magnitude or two beyond the GZK cutoff!

A considerable cosmic neutrino flux above E_{GZK} is required for the Z–burst hypothesis to successfully explain the super–GZK events. The requirement is that the product of the resonant energy times the neutrino flux at the reso-nant energy per flavor, times the annihilation probability within the GZK zone (which may be as large as 0.025% to 1% due to neutrino clustering), times the photon and nucleon multiplicity per burst (~ 20), is comparable to the ob-served flux at 10^{20} eV. The resulting requirement on the neutrino flux is roughly $E_R F_{\nu_j}(E_R) \sim 10^{-18.5\pm1}/\mathrm{cm}^2/\mathrm{s/sr}$. Such a neutrino flux at $E \sim 10^{22}$ eV is directly measurable in a teraton (10^{12} ton) detector like EUSO/OWL/AirWatch, and pos-sibly in a search for radio pulses produced by high energy neutrinos penetrating a small column–density of matter in the limb of the moon [45].

While certainly large, this required neutrino flux violates no existing limits. It has been pointed out [46] that this flux cannot extrapolate as E^{-2} to 10^{17} eV, for then it would violate the Fly's Eye bound arising from nonobservation of pene-trating horizontal (i.e., neutrino induced) air showers at that energy. It has also been pointed out [47] that local neutrino clustering is required to avoid generating a density of 30 MeV to 100 GeV photons in excess of the EGRET experimen-tal bound [48], from distant Z-bursts undergoing electromagnetic cascading. The extreme-energy neutrino flux implied by the Z-burst model probably requires un-usual source dynamics [49]. Among the reasons to hope that Nature obliges is that resonant neutrino annihilation provides the best hope at present to actually measure the relic neutrino density.

2) Strong ν Cross-section at $\stackrel{>}{\sim} E_{\mathrm{GZK}}$: It is interesting and suggestive that the observed EECR flux beyond E_{GZK} is well matched by the flux predicted for cos-mogenic neutrinos. This is not a complete coincidence. With the GZK cutoff, any continued nucleon flux beyond E_{GZK} is degraded in energy, photo-producing pions which in turn decay to produce cosmogenic neutrinos. The number of produced neutrinos compensates for their lesser energy, with the result that the neutrino flux matches well to the observed super-GZK flux. One may entertain the notion that the cosmogenic neutrinos <u>are</u> the super-GZK primaries, and that these neutrinos acquire a strong cross-section at $\sim 10^{20}$ eV.

Limits on the strength of the neutrino cross-section at 10^{20} eV can be inferred from existing data. Heuristically, one argues as follows. The GZK process ensures that there is a flux of cosmogenic neutrinos at 10^{20} eV, with an easily calculated flux. If the neutrino cross-section were weak, an experiment looking for penetrating air-showers initiated by the cosmogenic neutrinos would see nothing. If the cross-section were strong enough, the neutrino could not penetrate the atmosphere at all. So the fact that the Fly's Eye experiment saw no penetrating showers tells us that the neutrino cross-section is either strong or weak; the mid-range is excluded. The vertical depth of our atmosphere is $x_{\mathrm{v}} = 1033\,\mathrm{g/cm}^2$, and the horizontal depth

x_h is about 36 times greater. In terms of the mean free path λ of a particle with cross-section σ, one has $x_v/\lambda = \sigma/1.6$mb, and $x_h/\lambda = \sigma/44\mu$b. Thus, an estimate of the excluded cross-section is $\sim 40\mu$b to 1mb. A more careful calculation has been performed, with the result that $\sim 20\mu$b to 1mb is excluded [50]. Hypothetical high-energy neutrino cross-sections in excess of a mb remain viable.

The idea that neutrinos, indeed, all particles, may have a strong interaction at a high but observable energy scale is not new [51]. However, some recent ideas concerning new interactions relate well to the 10^{20} eV scale. One idea is that leptons are bound states of dual QCD gluons, which reveal themselves just above the electroweak (EW) scale at parton-parton $\sqrt{s} \sim$ TeV [52]. Another idea is that grand unification occurs precociously at $\sqrt{s} \sim$ TeV, because of extra dimensions or other reasons, and a neutrino above this threshold becomes strongly-interacting via leptoquark resonances [53]. A third idea is that the exchange of a towers of Kaluza-Klein (KK) modes from extra compactified dimensions lead to a strong neutrino cross-section above $\sqrt{s} \sim$ TeV [54]. In all three cases, it is the combination of a low $\sim$ TeV scale for radically new physics and a quickly rising spectrum of new states (possibly increasing exponentially, $\rho \sim e^{\sqrt{s/s_0}}$) that provides a rapid turn-on of a strong cross-section for the neutrino. Through unitarity, the new threshold at $\sim$ TeV has consequences for cross-sections at lower energies [55], but they are not dramatic.

The KK exchange model may fail [56] in that the KK modes couple to neutral currents, and the scattered neutrino carries away 90% of the incident energy per interaction, thereby elongating the shower profile. But if the neutrino cross-section can be made large enough, $\stackrel{>}{\sim} 20$ mb, then multiple scattering within a nucleus will effect a sufficiently large energy transfer and save the model [57]. Independent of the neutral current issue, the dual QCD and TeV-scale unification models seem to provide viable explanations of the super-GZK data. However, a recent calculation for the rate of rise of the low-scale unification cross-section in a string context is not encouraging [58].

Signatures for these models include directional pointing back to the EECR source, longitudinal shower profiles differing somewhat from those of a proton or a gamma, and a strong correlation between observed energies and zenith angle. The latter signature should show an inverse proportionality between the neutrino-air interaction length and the rising (with energy) neutrino cross-section.

IV MODEL SIGNATURES

There are several telltale discriminators to be sought in higher statistics data. These will eventually eliminate most (perhaps all!) of the models so far proposed for the super-GZK events. We list some discriminating signatures and discuss them.

Small-scale anisotropies and pointing: The discriminatory power of small-scale clustering was discussed already in §II. Here we add some detail to the discussion. In traversing a distance D, a charged particle interacting with magnetic domains having coherence length λ will bend through an energy-dependent angle[2]

$$\delta\theta \sim 0.5° \times \frac{Z\,B_{nG}}{E_{20}}\sqrt{D_{\mathrm{Mpc}}\,\lambda_{\mathrm{Mpc}}}\,. \tag{13}$$

Here B_{nG} is the magnetic field in units of nanogauss, E_{20} and Z are the particle energy in units of 10^{20} eV and charge, and the lengths D and λ are given in units of Mpc. It is thought likely that coherent extragalactic fields are nanogauss in magnitude [59], in which case super-GZK primaries from $\overset{<}{\sim} 50$ Mpc will typically bend only a few degrees (but note that protons at 10^{19} eV will bend through $\sim 30°$). Thus, local models either postulate many invisible sources isotropically-distributed with respect to the Galaxy to provide the roughly isotropic flux observed above E_{GZK}, or postulate a large extragalactic magnetic field to isotropize over our Northern Hemisphere the highest-energy particles from a small number of sources [60]. Among the latter category, some models postulate helium or iron nuclei as the primaries, to increase the bending by the charge factors 2 and 26, respectively. For those models invoking randomly distributed, decaying super-massive particles (SMPs) [61] or topological defects (TDs) [62] as sources, and models invoking a large magnetic field with considerable incoherent component, one may expect a nearly chance distribution of observed events on the sky. However, there may be some clustering even in these models, due to possible small-scale density fluctuations in the local SMP or TD distributions [63], or due to possible caustics in the projection of large-scale extragalactic magnetic fields on our sky (assuming the incoherent magnetic fields are sufficiently small) [64]. In the SMP and TD models, a high photon fraction in the primary composition further enhances clustering possibilities.

From the point of view of opening a new window to astronomy, those models in which the primaries do point back to their distant, active sources are the most interesting. These models are few in number. They are the Z-burst model, the strongly-interacting neutrino model, and the quantum-gravity/LI-violating models.

Large-scale anisotropies: On large scales, one seeks associations of the CR directions with the Galactic halo (to be revealed by a dipole anisotropy favoring the direction of the Galactic center) or the local galactic magnetic field, with matter distributions in nearby galactic or super-galactic clusters such Virgo, or with possible large coherent galactic or extragalactic magnetic fields [65]. For large-scale studies, the Southern hemisphere Auger experiment will prove invaluable for several reasons. It offers coverage of potential sources and matter distributions, and

---

[2] On average, half of the interactions of a super-GZK nucleon with the CMB change the isospin. At energies for which $c\tau$ of the neutron is small compared to the interaction mfp of ~ 6 Mpc, the neutron decays back to a proton with negligible energy loss and the bending-angle formula is unchanged. However, at the energy 6×10^{20}, $c\tau$ for the neutron is comparable to the interaction mfp, so at higher energies the nucleon bending-angle is reduced by $\overset{<}{\sim} 2$.

galactic and extragalactic magnetic fields, not available from the North. Moreover, it offers a view of our Galactic center which will provide a North-South dipole discriminator for or against a halo-centered population of sources such as magnetars or halo-bound SMPs. Southern Auger will also discriminate the M87-source model [23] wherein EECRs are channeled by a hypothesized galactic magnetic wind into the Northern hemisphere, and the Cen-A source model [24] which also yields a dipole anisotropy. Of course, an orbiting experiment with 4π vision like EUSO/OWL will be an even better instrument for multipole analysis.

Energy-direction-time correlations: Because bending of charged particle trajectories by intervening magnetic fields increases as particle energy decreases, one may learn about the strength and geometry of extragalactic fields from relative time delays and angular correlations of particles from a common source. One may also learn about the source. Quantitatively, the increase in path length due to bending leads to a relative increase in travel time of $\delta t/t \sim (\delta\theta)^2$, for small bending angle. Adding the contributions from the coherent magnetic domains then yields

$$\delta t \sim 300\, D_{\text{Mpc}} \left(\frac{Z\, B_{nG}\lambda}{E_{20}} \right)^2 \text{ yrs} \tag{14}$$

for the time delay. The time separation at earth is obtained by taking differences in eq. (14); to first order in δE it is already large:

$$t_1 - t_2 \sim 600\, D_{\text{Mpc}} \left(\frac{Z\, B_{nG}\lambda}{E_{20}} \right)^2 \left(\frac{\delta E}{E_{20}} \right) \text{ yrs}. \tag{15}$$

The correlation in energy and time becomes even more significant when it is remembered that the higher energy primary has an even higher mean energy in transit, before losses on the 2.7K background.

Surprisingly, one AGASA event-pair has the higher-energy 1.06×10^{20} eV primary arriving about 3 years *after* a 0.44×10^{20} eV primary. Assuming these primaries originate form a common source, a possible explanation is that their source has a duration of at least $3/(1 + z)$ years (the red-shift factor is due to time-dilation). Such a long-lived source does not occur in one-time burst models (e.g. GRBs). It may occur in decay models (e.g. SMPs, TDs, monopolonium) if sub-clustering exists on small-angular scales within halo clusters [66]. A recent paper [67] notes that such sub-clusters may actually be observable as micro-lenses for stars and background galaxies. Counter-intuitive reverse pairing with the *earlier* arrival time for the lower energy charged-primary can also occur in certain magnetic field configurations, as shown in [68].

Composition of the primaries: Another signature to be sought is a statistical identification of the nature of the primaries as a function of their energies. Three methods have been identified to distinguish photon-initiated showers from hadron-initiated showers. One method relies on the longitudinal profile of the event, particularly the depth at shower maximum x_{max}. The longitudinal profile of the Fly's Eye event at 3×10^{20} eV (well-measured by its nitrogen fluorescence trail) is ill-fit by a photon primary, well-fit by an iron primary, and somewhat fit by a proton primary [69]. The second method relies on measurement of muon number, with a high muon number purportedly favoring a hadron primary over a photon, and vice versa. A recent study of the muon content of showers above 10^{19} eV seems to favor nucleons over photons [70]. However, caution is warranted with this method, in that some simulations show little difference in the muon-content of showers from nucleon primaries versus photon primaries [71]; and photons have themselves a significant partonic component at high energy. Nevertheless, it is safe to say that photon primaries appear disfavored at present, but more data is needed before conclusions are drawn. The third method of gamma identification will will rely on a predicted characteristic N-S vs. E-W gamma asymmetry. This quadrupole asymmetry results from the polarization-dependent interaction of the gammas with the earth's magnetic field.

It is interesting to mention that the highest-energy Yakutsk event has an unusually high muon number; the only model so far which successfully explains this invokes a magnetic baryonic-monopole as the primary particle [31]. Basically, a relativistic baryonic-monopole of mass M and incident energy E showers like a giant nucleus of atomic number $A_{eff} \sim M/\Lambda_{QCD} \sim 10^6 \, (M/\text{PeV})$ and the same $\gamma = E/M$. Intermediate mass monopoles therefore generate many, many charged pions which decay to muons.

All models wherein the primaries arise from QCD jets produce many more pions than baryons. The neutral pions in turn produce gammas. In these models, the ratio of gammas to baryons is typically of order ten at the source. Even allowing for the shorter attenuation length of photons relative to nucleons, the measurement of the primary composition for super-GZK events becomes an excellent discriminator for models with jet-producing sources. These models include Z-bursts, and decaying SMPs and TDs.

Possible E_{max} energy cutoff: The predicted of a cutoff at E_{GZK} is wrong. Does Nature provide an alternative cutoff within our reach? Or do the data continue beyond our reach? The shock-jock experts claim that it is difficult for conventional shock-acceleration mechanisms to produce ZeV proton energies; for this class of model, an E_{max} below 10^{22} eV is certainly expected [72]. Decaying SMP models also have a natural cutoff, at half of the SMP mass. This could be as high as $E_{max} \sim 10^{24}$ eV for a long-lived GUT-mass particle, but would be lower for other postulated SMPs. In the Z-burst model there is a natural cutoff related to the tiny mass of neutrinos: $E_{max} = 4 \, (\text{eV}/m_\nu)$ ZeV. Implications from the atmospheric SK data are that this cutoff is at most 7×10^{22} eV.

CR flux above vs. below E_{GZK}: A "smoothness" variable such as $R_j \equiv F_j(E > E_{GZK})/F_j(E < E_{GZK})$ for each primary species j=nucleon, photon, iron nuclei, neutrino, etc. may be revealing [73]. For primaries with a GZK cutoff, $F(E < E_{\mathrm{GZK}})$ samples sources from the whole volume of the Universe, and may even include cascade products from $F(E > E_{\mathrm{GZK}})$, whereas $F(E > E_{\mathrm{GZK}})$ samples just the GZK volume; for primaries without a GZK cutoff, $F(E < E_{\mathrm{GZK}})$ and $F(E > E_{\mathrm{GZK}})$ sample sources from the whole volume of the Universe. Lumps, bumps, and gaps in the spectrum near E_{GZK} are a consequence of some models. For example, hadron and neutrino pile-ups just below E_{GZK} are expected from the photo-pion production process which occurs above E_{GZK} [74]. Other sources of neutrino pile-ups have also been suggested [36,75]. Present data show continuity. Smoothness studies of various models require simulation, and are just beginning.

Spectral index above E_{GZK}: One means of achieving more events above E_{GZK} is to postulate a flattening of the primary proton spectrum at highest energies. With more data, the extreme-energy spectrum will be measured. A flattening of the $E^{-2.7}$ power law inferred from just below E_{GZK} would indicate and constrain new sources.

Measurable neutrino flux above E_{GZK}: If there is a new source of primary neutrinos above E_{GZK}, or if the neutrino cross-section becomes strong at super-GZK energies, then there is the possibility that the primary neutrino flux can be measured in an EUSO/OWL/AW-sized detector.

Diffuse ∼GeV gamma-ray flux: An upper limit on the diffuse gamma-ray flux between 30 MeV and 100 GeV has been published by the EGRET experiment of the now defunct Gamma Ray Observatory [48]. This limit has serious discriminatory power for models where SMPs or TDs or extremely boosted massive particles decay to quark-antiquark jets which then hadronization to produce the EECRs [76]. This is because QCD jets via π^0 production and decay produce very high energy gammas which initiate an electromagnetic cascade on the cosmic radio, microwave, infrared, and magnetic field backgrounds. For models with jet-production distributed over cosmic distances, such as some TD models and the Z-burst model with a homogeneous distribution of relic neutrino targets, the cascade has the distance to reach completion; the end result is gamma power (energy/time) in the EGRET range roughly an order of magnitude below the total power of the original sources, and comparable to the power in EECR neutrinos. Such models are disfavored. More local models, such as SMPs bound to our halo, and the Z-burst model with a local over-density of relic neutrinos, are not impacted at present, but may be tested with GLAST, the next generation gamma-ray observatory.

V CONCLUSIONS

The ultimate explanation for the puzzles in EECRs will provide a surprise at a minimum, and possibly radically new physics at a maximum. The extreme energies of events already observed cannot be approached by terrestrial accelerators. Thus, there is ample motivation to build the next generation of CR detectors, Auger and Telescope Array, and to plan even farther beyond for a teraton detector like EUSO, OWL and AirWatch. At 10^{20} eV, AGASA provides about an event per year, and HiRes about an event per month. Auger and TA will see two such events per week, while EUSO/OWL/AW has the potential to collect such an event every two hours. As the data sample grows, statistical studies will reveal signatures that discriminate among the many galactic and extragalactic sources so far proposed to resolve the EECR puzzles.

It is quite possible that neutrino primaries are responsible for the EECRs. If so, it appears that the weakly-interacting neutrino either grows a very strong cross-section at 10^{20} eV, or it annihilates on the relic neutrinos left-over from the hot phase of the Big Bang. Another possibility is that free magnetic charges exist and are the EECR primaries; for magnetic monopoles, the Universe is transparent and cosmic magnetic fields provide a natural acceleration mechanism.

Clearly, we live in exciting EECR times, and we possess the technology to prove this is so. The resolution of our puzzles is forthcoming, as on-going and future experiments will provide us with the statistics to discriminate among the many interesting models.

ACKNOWLEDGMENTS

We acknowledge fruitful collaboration with P. Biermann, H. Goldberg, T. Kephart, H. Paes, and Dipthe Wick on some of the material presented here; and discussions with G. Farrar, G. Gelmini, A. Kusenko, D. McKay, A. Olinto, J. Ralston, G. Sigl, F. Stecker, and E. Zas. This work was supported in part by the U.S. Department of Energy grant no. DE-FG05-85ER40226.

REFERENCES

1. Recent reviews include: P. Biermann, J. Phys. G23, 1 (1997); P. Bhattacharjee and G. Sigl, Phys. Rept. 327, 109 (2000), astro-ph/9811011; A.V. Olinto, "David Schramm Memorial Volume" of Phys. Rept. 333, 329 (2000); X. Bertou, M. Boratov, and A. Letessier-Selvon, Int. J. Mod. Phys. A15, 2182 (2000); A. Letessier-Selvon, Lectures at "XXVIII International Meeting on Fundamental Physics", Cadiz, Spain (2000), astro-ph/0006111; M. Nagano and A.A. Watson, Rev. Mod. Phys. 72, 689 (2000); A. Olinto, astro-ph/0011106; M. Kachelriess, astro-ph/0011231; G. Sigl, Science 291, 73 (2001); F.W. Stecker, astro-ph/0101072.
2. M. Takeda et al. (AGASA Collab.), Phys. Rev. Lett. 81, 1163 (1998), astro-ph/9807193; D.Bird et al. (Fly's Eye Collab.), Astrophys. J. 424, 491 (1994), and ibid 441, 144 (1995); M. Lawrence, R.J.O.Reid and A. Watson (Haverah Park Collab.), J. Phys. G 17,773 (1991), and M. Ave et al., Phys. Rev. Lett. 85, 2244 (2000); N. Efimov et al. (Yakutsk Collab.), Proc. "Astrophysical Aspect of the Most Energetic

Cosmic Rays," p. 20, eds. M. Nagano and F. Takahara, World Sci., Singapore, 1991; D. Kieda *et al.* (HiRes Collab.), Proc. of the 26th ICRC, Salt Lake City, Utah, 1999.

3. J. Elbert and P. Sommers, Astrophys. J. 441, 151 (1995).

4. Named after the pioneering work of Greisen, Kuzmin, and Zatsepin in the 1960's; recent detailed explorations of the GZK cutoff include S. Lee, Phys. Rev. D58, 043004 (1998); A. Achterberg *et al.*, astro-ph/9907060; T. Stanev *et al.*, astro-ph/0003484.

5. A.M. Hillas, Ann. Rev. Astron. Astrophys. 22, 425 (1984); C.A. Norman, D.B. Melrose, and A. Achterberg, AStrophys. J. 454, 60 (1995); R.D. Blandford, Physica Scripta T85, 191 (2000).

6. N. Hayashida *et al.* (AGASA Collab.), Phys. Rev. Lett. 77, 1000 (1996); M. Takeda *et al.*, Astrophys. J. 522, 225 (1999), astro-ph/9902239 and astro-ph/0008102.

7. M. Teshima, reported at "Radio Detection of High Energy Neutrinos 2000", UCLA, Nov. 2000, eds. P. Gorham and D. Saltzberg.

8. Uchihori *et al.*, Astropart. Phys. 13, 151 (2000), astro-ph/9908193.

9. P.Biermann and G.Farrar, Phys. Rev. Lett. 81, 3579 (1998); ibid 83, 2478(E) (1999).

10. G.Sigl, D.F.Torres, L.A.Anchordoqui and G.E.Romero, astro-ph/0008363; A. Virmani *et al.*, astro-ph/0010235.

11. H. Goldberg and T.J. Weiler, astro-ph/0009378.

12. http://ast.leeds.ac.uk/haverah/hav-home.shtml for Haverah Park;
http://www-akeno.icrr.u-tokyo.ac.jp/AGASA/ for AKENO;
http://hires.physics.utah.edu for HiRes;
http://www.auger.org, http://www-lpnhep.in2p3.fr/auger/welcome.html for Auger;
http://www-ta.icrr.u-tokyo.ac.jp/ for Telescope Array;
http://ifcai.pa.cnr.it/ifcai/euso.html for EUSO;
http://owl.gsfc.nasa.gov for OWL; http://ifcai.pa.cnr.it/~AirWatch/ for AirWatch.

13. W. Springer and J. Belz, private communication.

14. E. Waxman, K. Fisher and T. Piran, Astrophys. J. 483,1 (1997); M. Lemoine, G. Sigl, A. Olinto and D.N. Schramm, Astrophys. J. 486, L115 (1997); G. Sigl, M. Lemoine and A. Olinto, Phys. Rev. D56, 4470 (1997); G. Sigl and M. Lemoine, Astropart. Phys. 9, 65 (1998); G. Medina-Tanco, Astrophys. J. L71, 495 (1998), and astro-ph/9707054; V. Berezinsky, hep-ph/0001163; S. Dubovsky, P. Tinyakov and I. Tkachev, Phys. Rev. Lett. 85, 1154 (2000); Z.Fodor and S. Katz, hep-ph/0007158.

15. Topological Defect physics is reviewed in A. Vilenkin and E.P.S. Shellard, *Cosmic Strings and Other Topological Defects*, Cambridge Pr., 1994; M.B. Hindmarsh and T.W.B. Kibble, Rep. Prog. Phys. 58, 477 (1995).

16. C.T. Hill, Nucl. Phys. B224, 469 (1983); J.J. Blanco-Pillado and K.D. Olum, Phys. Rev. D60, 083001 (1999).

17. A review is given in V. Kuzmin and I. Tkachev, Phys. Rept. 320, 199 (1999).

18. K. Hamaguchi, Y. Nomura, and T. Yanagida, Phys. Rev. D58, 103503 (1998); K. Hamaguchi, K. Izawa, Y. Nomura, and T. Yanagida, Phys. Rev. D60, 125009 (1999).

19. J. Ellis, J.L. Lopez, and D.V. Nanopoulos, Phys. Lett. B247, 257 (1990); K. Benakli, J. Ellis, and D.V. Nanopoulos, Phys. Rev. D59, 047301 (1999).

20. V. Berezinsky, M. Kachelriess, and A. Vilenkin, Phys. Rev. Lett. 79, 4302 (1997).

21. V.A. Kuzmin and V.A. Rubakov, Phys. Atom. Nucl. 61, 1028 (1998).

22. S.L. Dubovsky and P.G. Tinyakov, JETP Lett. 68, 107 (1998); V. Berezinsky, P.

Blasi, and A. Vilenkin, Phys. Rev. D 58, 103515 (1998).

23. E-J Ahn, G. Medina-Tanco, P.L. Biermann, T. Stanev, astro-ph/9911123.

24. M. Hillas, Nature 395, 15 (1998); G. Farrar and T. Piran, astro-ph/0010370; L.A. Anchordoqui, H. Goldberg, and T.J. Weiler, in progress.

25. F.W. Stecker, Astropart. Phys. 14, 207 (2000).

26. G.R.Farrar, Phys. Rev. Lett. 76, 4111 (1996); D.J.H.Chung, G.R.Farrar and E.W.Kolb, Phys. Rev. D57, 4606 (1998); I.F.M. Albuquerque, G.R. Farrar and E.W. Kolb, Phys. Rev. D59, 015021 (1999).

27. N.A. Porter, Nuovo Cim. 16, 958 (1960); E. Goto, Prog. Theo. Phys. 30, 700 (1963).

28. T.W. Kephart and T.J. Weiler, Astropart. Phys. 4, 271 (1996).

29. S.D. Wick, T.W. Kephart, T.J. Weiler and P.L. Biermann, Astropart. Phys. (to appear), astro-ph/0001233.

30. A. S. Goldhaber, in Phys. Rept. 315, 83 (1999) and hep-th/9905208; an unrelated idea for a strong monopole cross-section is given in E. Huguet and P. Peter, Astropart. Phys. 12, 277 (2000).

31. L. Anchordoqui et al., hep-ph/0009319.

32. C.O. Escobar and R.A. Vazquez, Astropart. Phys. 10, 197 (1999).

33. Mestres-Gonzales, Proc. 25th ICRC, 1997, Durban, So. Africa, (World Sci., Singapore), and Workshop on Observing Giant CR Air-Showers from $> 10^{20}$ eV Particles from Space, ed. J.F. Krizmanic et al., 1998 (AIP Conf. Proc. No. 433, Woodbury, NY); S. Coleman and S.L. Glashow, hep-ph/9808446, and Phys. Rev. D59, 116008.

34. Y.J. Ng, D.S. lee, M.C. Oh, and H. van Dam, hep-ph/0010152.

35. T.J. Weiler, Phys. Rev. Lett. 49, 234 (1982); Astrophys. J. 285, 495 (1984); E. Roulet, Phys. Rev. D47, 5247 (1993);

36. T.J. Weiler, Astropart. Phys. 11, 303 (1999), and ibid. 12, 379E (2000) [for corrected receipt date].

37. D. Fargion, B. Mele and A. Salis, Astrophys. J. 517, 725 (1999).

38. G. Gelmini and A. Kusenko, Phys. Rev. Lett. 82, 5202 (1999).

39. Review of Particle Physics, Euro. Phys. J. C15, 1 (2000).

40. H. Paes and T.J. Weiler, Phys. Rev. D, to appear, hep-ph/0101091.

41. S. Yoshida, G. Sigl, and S. Lee, Phys. Rev. Lett. 81, 5505 (1998); G. Gelmini, hep-ph/0005263.

42. V. Barger, T.J. Weiler, and K. Whisnant, Phys. Lett. B442 255 (1998).

43. Review by H. Sobel (SuperK Collab.) at Neutrino 2000, Sudbury, Canada, 6/2000.

44. Review by G. Mills (LSND Collab.) at Neutrino 2000, Sudbury, Canada, 6/2000.

45. P. Gorham, K. Liewer, and C. Naudet, astro-ph/9906504 and Proc. 26th Int. CR Conf., Salt Lake City, Utah, Aug. 1999.

46. J.J. Blanco-Pillado, R.A. Vazquez, and E.Zas, Phys. Rev. D61, 123003 (2000).

47. S. Yoshida, G. Sigl, and S. Lee, in [41].

48. P. Sreekumar et al., Astrophys. J. 494, 523 (1998).

49. G. Gelmini and A. Kusenko, Phys. Rev. Lett. 84, 1378 (2000); J.L. Crooks, J.O. Dunn, and P.H. Frampton, astro-ph/0002089.

50. C. Tyler, A.V. Olinto, and G. Sigl, hep-ph/0002257.

51. V.S. Berezinsky and G.T. Zatsepin, Phys. Lett. 28B, 423 (1969); G. Domokos and S. Nussinov, Phys. Lett. B 187, 372 (1987);

52. J. Bordes *et al.*, hep-ph/9705463; Astropart. Phys. 8, 135 (1998).

53. G. Domokos, S. Kovesi-Domokos, and P.T. Mikulski, hep-ph/0006328.

54. S. Nussinov and R. Shrock, Phys. Rev. D59, 105002 (1999); G. Domokos and S. Kovesi-Domokos, Phys. Rev. Lett. 82, 1366 (1999); P. Jain, D.W. McKay, S. Panda and J.P. Ralston, Phys. Lett. B484, 267 (2000).

55. H. Goldberg and T. J. Weiler, Phys. Rev. 59, 113005 (1999).

56. M. Kachelriess and M. Plumacher, Phys. Rev. D62, 103006 (2000); L. Anchordoqui *et al.*, hep-ph/0011097.

57. A. Jain, P. Jain, D.W. McKay, and J.P. Ralston, hep-ph/0011310.

58. F. Cornet, J.I. Illana, and M. Masip, hep-ph/0102065.

59. P.P. Kronberg, Rep. Prog. Phys. 57, 325 (1994); J.P. Vallee, Fund. Cosmic Phys. 19, 1 (1997); P. Blasi, S. Burles, and A.V. Olinto, Astrophys. J. 514, L79 (1999).

60. There is new evidence for large μG magnetic fields filling the space within galaxy clusters: T.E. Clarke, P.P Kronberg, and H. Boehringer, astro-ph/0011281.

61. V. Berezinsky, M. Kachelriess and A. Vilenkin, Phys. Rev. Lett 79, 4302 (1997); V.A.Kuzmin and V.A.Rubakov, Phys. Atom. Nucl. 61, 1028 (1998); M. Birkel and S. Sarkar, Astropart. Phys. 9, 297 (1998); P.Blasi, Phys. Rev. D60, 023514 (1999); S. Sarkar, hep-ph/0005256.

62. P. Bhattacharjee, C. Hill, and D. Schramm, Phys. Rev. Lett. 69, 567 (1992); G. Sigl, S. Lee, P. Bhattacharjee, and S. Yoshida, Phys. Rev. D59, 043504 (1999).

63. P. Blasi and R. Sheth, Phys. Lett. B486, 233 (2000).

64. G. Sigl, M. Lemoine, and P. Biermann, Astropart. Phys. 10, 141 (1999), and astro-ph/9903124; D.Harari, S. Mollerach and E. Roulet, J. High Energy Phys. 08, 022 (1999); S. Mollerach and E. Roulet, astro-ph/9910205.

65. T. Stanev *et al.*, Phys. Rev. Lett. 75, 3056 (1995); G. Medina-Tanco, Astrophys. J. 510, 91 (1999); J. Bahcall and E. Waxman, hep-ph/9912326; O.E. Kalashev, V.A. Kuzmin, and D.V. Semikoz, astro-ph/0006349; M. Blanton, P. Blasi, and A. Olinto, astro-ph/0009466.

66. P. Blasi and R.K.Sheth, astro-ph/0006316.

67. A. Kusenko and V.A. Kuzmin, astro-ph/0012040.

68. D.Harari, S. Mollerach and E. Roulet, J. High Energy Phys. 02, 035 (2000).

69. F. Halzen, R. Vazquez, T. Stanev, and H. Vankov, Astropart. Phys. 3, 151 (1995); M. Nagano and A. Watson, fig. 16 by Heck, in [1]; Ave *et al.*, in [2]; T. Gaisser (fig. courtesy of T. Stanev), Proc. Int'l Workshop on Observing UHE Cosmic-Rays from Space and Earth, Metepec, Puebla, Mexico, Aug. 2000, eds. A. Zepeda *et al.*

70. M. Ave *et al.*, Phys. Rev. Lett. 85, 2244 (2000).

71. F.A. Aharonian, B.L. Kanevsky, and V.A. Sahakian, J. Phys. G17, 1909 (1991).

72. See, e.g., the review of P. Biermann in ref. [1], and of R. Blandford in ref. [5].

73. Discussed in Farrar and Piran [24].

74. J.L. Puget, F.W. Stecker, and J. Bredkamp, Astrophys. J. 205, 638 (1976); C.T. Hill and D.N. Schramm, Phys. Rev. D31, 564 (1985); V.S. Berezinsky and S.I. Grigoreva, Astron. Astrophys. 199, 1 (1988); also, T. Stanev *et al.* in [4].

75. A. Kusenko and M. Postma, hep-ph/0007246; M. Postma, hep-ph/0102106.

76. S. Lee; R.J Protheroe and P.A. Johnson, Nucl. Phys. B Proc. Suppl. 48, 485 (1996); R.J Protheroe and T. Stanev, Phys. Rev. Lett. 77, 3708 (1996).

RADIO DETECTION OF
EXTENSIVE AIR SHOWERS

Extensive Air Shower Radio Detection: Recent Results and Outlook

Jonathan L. Rosner and Denis A. Suprun

Enrico Fermi Institute and Department of Physics
University of Chicago, Chicago, IL 60637 USA

Abstract. A prototype system for detecting radio pulses associated with extensive cosmic ray air showers is described. Sensitivity is compared with that in previous experiments, and lessons are noted for future studies.

I INTRODUCTION

The observation of the radio-frequency (RF) pulse associated with extensive air showers of cosmic rays has had a long and checkered history. In the present report we describe an attempt to observe such a pulse in conjunction with the Chicago Air Shower Array (CASA) and Michigan Muon Array (MIA) at Dugway, Utah. Only upper limits on a signal have been obtained at present, though we are still processing data and establishing calibrations.

In Section 2 we review the motivation and history of RF pulse detection. Section 3 is devoted to the work at CASA/MIA, while Section 4 deals with some future possibilities, including ones associated with the planned Pierre Auger observatory. We conclude in Section 5. This report is an abbreviated version of a longer one now in preparation [1].

II MOTIVATION AND HISTORY

A Auxiliary information on shower

Present methods for the detection of an extensive air shower of cosmic rays leave gaps in our information. The height above ground at which showers develop cannot be provided by ground arrays, though stereo detection by air fluorescence detectors is useful. Composition of the primary particles, another unknown, is correlated with shower height, with heavy primaries leading to showers which begin higher in the atmosphere.

CP579, *Radio Detection of High Energy Particles*, edited by D. Saltzberg and P. Gorham
© 2001 American Institute of Physics 0-7354-0018-0/01/$18.00

Radio detection can help fill such gaps. The electric field associated with a charge $|e|$ undergoing an apparent angular acceleration $\ddot{\theta}$ is $|\mathcal{E}| = 1.5 \times 10^{-26} \ddot{\theta}$ V/m, where time is measured in seconds [2,3]. For typical showers the charges of radiating particles are expected to be able to act coherently to give a pulse with maximum frequency component $\nu_{max}(\text{MHz}) \simeq 10^6/R^2(\text{m})$, where R is the distance of closest approach of the shower axis to the antenna. Showers originating higher in the atmosphere are expected to have higher-frequency components. Thus RF detection may be able to add information on shower height and primary composition, and to provide a low-cost auxiliary system in projects such as the Pierre Auger array [4].

B   Pulse generation mechanisms

Several possibilities have been discussed for generation of a pulse by air showers. Cosmic rays could induce the atmosphere to act as a giant spark chamber, triggering discharges of the ambient field gradient [5]. Compton scattering and knock-on electrons can give rise to a negative charge excess of some 10 to 25% at shower maximum [6]. Separation of positive and negative charges can occur in the Earth's magnetic field as a result of a $q\mathbf{v} \times \mathbf{B}$ force [7]. This last mechanism is thought to be the dominant one accounting for atmospheric pulses with frequencies in the 30–100 MHz range [2], and will be taken as the model for the signal for which the search was undertaken. The charge-excess mechanism is probably the major source of an RF signal in a dense material such as polar ice [8], but is expected to be less important in the atmosphere.

C   Early measurements

The first claim for detection of the charge-separation mechanism utilized narrow-band techniques at 44 and 70 MHz [9,10]. A Soviet group reported signals at 30 MHz [11], while a University of Michigan group at the BASJE Cosmic Ray Station on Mt. Chacaltaya, Bolivia [12] studied pulses in the 40–90 MHz range.

The collaboration of H. R. Allan et al. [2] at Haverah Park in England studied the dependence of signals on primary energy E_p, perpendicular distance R of closest approach of the shower core, zenith angle θ, and angle α between the shower axis and the magnetic field vector. Their results indicated that the electric field strength per unit of frequency, $\mathcal{E}_\nu$, could be expressed as

$$\mathcal{E}_\nu = s \frac{E_p}{10^{17}\ \text{eV}} \sin \alpha \cos \theta \exp\left(-\frac{R}{R_0(\nu,\theta)}\right) \quad \mu\text{V}\ m^{-1}\ \text{MHz}^{-1} \quad , \tag{1}$$

where R_0 is an increasing function of θ, equal (for example) to (110 ± 10) m for $\nu = 55$ MHz and $\theta < 35°$. The constant s was originally claimed to be 20. The Haverah Park observations were recalibrated to yield $s = 1.6$ (0.6 μV m^{-1} MHz^{-1} for a 10^{17} eV shower at $R = 100$ m) while observations in the U.S.S.R. gave $s = 9.2$

(3.4 μV m^{-1} MHz^{-1} at $R = 100$ m) [13]. To estimate the corresponding signal strength in [9,10], we note that the signal power for showers of average primary energy $E = 5 \times 10^{16}$ eV was measured to be about 4 times that of galactic noise, for which [2] $\mathcal{E}_\nu^{\mathrm{Gal}} \simeq 1$–2 μV m^{-1} MHz^{-1}. Thus, for such showers, one expects $\mathcal{E}_\nu \simeq 2$–4 μV m^{-1} MHz^{-1}. Similarly, the estimate [10] of an average pulse power $V_{\mathrm{peak}}^2/2R = 10^{-12}$ W gives $V_{\mathrm{peak}} = 10$ μV for $R = 50$ Ω. Using the relation between pulse voltage and $\mathcal{E}_\nu$ [2]

$$V = 30 G^{1/2}(\delta\nu/\nu)\mathcal{E}_\nu \quad , \qquad (2)$$

where G is the antenna gain, and $\delta\nu$ is the bandwidth centered at frequency ν, we find for an assumed $G = 5$ (7 dB) (it is not quoted in Ref. [9]) and $\delta\nu/\nu = 2.75/44$ [9], a value of $\mathcal{E}_\nu \simeq 2.4$ μV m^{-1} MHz^{-1} at a primary energy of 5×10^{16} eV, or about 5 μV m^{-1} MHz^{-1} at 10^{17} eV if $\mathcal{E}_\nu$ scales linearly with primary energy [2]. For $G = 5$ the data of Refs. [9,10] thus would favor the higher field-strength claims of the U.S.S.R. group cited in Ref. [13].

More recent claims include pulses with components at or below several MHz [15–18], and at VHF frequencies [17,18]. The Gauhati University group has reviewed evidence for pulses at a wide range of frequencies [18].

D Pulse characteristics

The Haverah Park observations are consistent with a model in which the pulse's onset is generated by the start of the shower at an elevation of about 10 km above sea level, while its end is associated with the greater total path length (shower + signal propagation distance) associated with the shower's absorption about 5 km above sea level. If a vertical shower is observed at a distance of 100 m from its core, the pulse should rise and fall back to zero within about 10 ns, with a subsequent longer-lasting negative component. High frequencies should be less visible far from the shower axis. Heavy primaries should lead to showers originating higher in the atmosphere, with consequent higher-frequency RF components as a result of the geometric aspect ratio with which they are viewed by the antenna, and possibly a greater $\mathcal{E}_\nu$ for a given primary energy [2]. The polarization of the pulse should be dictated by the mechanism of pulse generation: e.g., perpendicular to the line of sight with component along $\mathbf{v} \times \mathbf{B}$ for the charge-separation mechanism.

E RF backgrounds

Discharges of atmospheric electricity will be detected at random intervals at a rate depending on local weather conditions and ionospheric reflections. Man-made RF sources include television and radio stations, police and other communications services, broad-band sources (such as ignition noise), and sources within the experiment itself. The propagation of distant noise sources to the receiver is a strong function of frequency and of solar activity.

Galactic noise can be the dominant signal in exceptionally radio-quiet environments for frequencies in the low VHF (30–100 MHz) range [2]. For higher frequencies in such environments, thermal receiver noise becomes the dominant effect.

III   THE INSTALLATION AT CASA

The CASA/MIA detector is located about 100 km southwest of Salt Lake City, Utah, at the Dugway Proving Ground [14]. The Chicago Air Shower Array (CASA) is a rectangular grid of 33×33 stations on on the desert's surface. The inter-station spacing is 15 m. A station has four 61 cm $\times$ 61 cm $\times$ 1.27 cm sheets of plastic scintillator each viewed by its own photomultiplier tube (PMT). When a signal appears on 3 of 4 PMTs in a station, a "trigger request pulse" of 5 mA with 5 μs duration is sent to a central trailer, where a decision is made on whether to interrogate all stations for a possible event. Details of this trigger are described in Ref. [14]. When this experiment was begun the CASA array had been reconfigured to remove the 4 westernmost "ribs" of the array. For runs performed in 1998, the easternmost rib had also been removed.

The University of Michigan designed and built a muon detection array (MIA) to operate in conjunction with CASA. It consists of sixteen "patches," each having 64 muon counters, buried 3 m below ground at various locations in the CASA array. Each counter has lateral dimensions 1.9 m $\times$ 1.3 m. Four of the patches (numbered 1 through 4), each about 45 m from the center of the array, lie on the corners of a skewed rectangle; four (numbered 5 through 8), each about 110 m from the center of the array, lie on a rectangle with slightly different skewed orientation, and eight (numbered 9 through 16) lie on the sides and corners of a rectangle with sides $x \simeq \pm180$ m and $y \simeq \pm185$ m, where x and y denote East and North coordinates.

A   Expected integral rates

The CASA trigger threshold is a few $\times 10^{14}$ eV and corresponds to a trigger rate of 10–20 Hz. The expected rates above (10^{15}, 10^{16}, 10^{17}, 10^{18}) eV are about (1 Hz, 1 per 2 min, 1 per 4 hr, and 2 per mo), respectively, over the 1/4 km^2 area of the array. A primary energy of at least 10^{17} eV seems to be needed if radio signals are to exceed the galactic noise level of 1–2 μV m^{-1} MHz^{-1}. Such pulses could be generated by a 10^{17} eV vertical shower with axis 100 m from the antenna under the most optimistic estimates. Since the whole array should see such showers only every few hours, and most have axes farther from the antenna than 100 m, the possibility of accidental noise pulses during such long time intervals reduces the expected sensitivity considerably.

B The "radio shack" at CASA

A survey of the CASA/MIA site determined that within the array, broad-band noise associated with computers, switching power supplies, and other electronics was so intense that no RF searches could be undertaken. The same was true at any position within the perimeter of the array. Consequently, an antenna was mounted on top of a mobile searchlight tower at a height of 10 m about 24 m east of the eastern edge of the array, corresponding to $x = 263.8$ m, $y = 0$ m. The antenna, a 9-element portable log-periodic antenna manufactured by Dorne and Margolin, was acquired from FairRadio Co. in Lima, Ohio, for about $60. Its nominal bandwidth is 26–76 MHz but it was measured to have usable properties up to 170 MHz.

The signal was fed through 60 ft. of RG-58U cable, filtered by a high-pass filter admitting frequencies above 23 MHz, preamplified using a Minicircuits ZFL-500LN preamplifier with 26 dB of gain, low-pass-filtered to admit frequencies below 250 MHz, and fed to the oscilloscope at a sensitivity of 5 mV/division. This constituted the "wide-band" configuration used for most data acquisition runs. The filters were Minicircuits BNC coaxial models. A "narrow-band" configuration with response between 23 and 37 MHz and two preamplifiers had substantially poorer signal-to-noise ratio in distinguishing transient signals from background.

A trigger based on the coincidence of seven of the eight outer muon "patches" was set to select large showers. Each muon patch was set to produce a trigger pulse when at least 5 of its 64 counters registered a minimum-ionizing pulse within 5.2 μs of one another. The pulses were then combined to produce a summed pulse, fed to a discriminator, whose output was amplified and sent over a cable (with measured delay time 2.15 μs) to the RF trailer. No evidence for pickup of the trigger pulse from the antenna was found. The trigger corresponded to a minimum shower energy somewhat below 10^{16} eV, based on the integral rate [19] at 10^{18} eV of $0.17/\mathrm{km}^2/\mathrm{day}/\mathrm{sr}$.

A Tektronix TDS-540B digitizing oscilloscope registered filtered and preamplified RF data on a rolling basis. These data were then captured upon receipt of a large-event trigger and stored on hard disk using a National Instruments GPIB interface. Data were taken using separate computers (at different times), allowing for analysis both at the University of Washington and at Chicago. The Washington system used a Macintosh Quadra 950 running Labview, while the Chicago system used either a Dell XPS200s Pentium desktop computer or a Dell Latitude LM laptop computer running a C program adapted from those provided by National Instruments. Each trigger caused 50 μs of RF data, centered around the trigger and acquired at 1 GSa/s, to be saved.

The total trigger rate ranged between about 20 and 50 events per hour, depending on intermittent noise sources in the trigger system. Concurrently, the CASA on-line data acquisition system was instructed to write files of events in which at least 7 out of the 8 outermost muon patches produced a patch pulse. These files typically overlapped with the records taken at the RF trailer to a good but not perfect extent as a result of occasional noise on the trigger line.

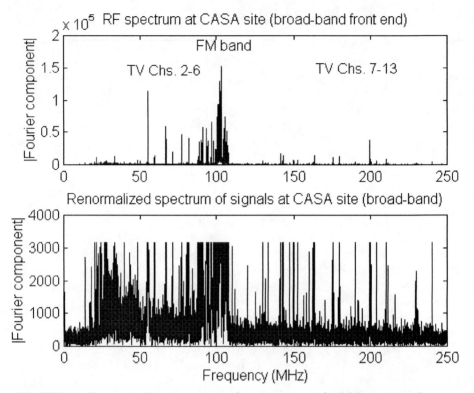

FIGURE 1. Top panel: Fourier spectrum (in arbitrary units) of RF signals at Dugway site. Prominent features include video and audio carriers for TV Channels 2, 4, 5, 7, and 11, and the FM broadcast band between 88 and 108 MHz. Bottom panel: Fourier spectrum (same vertical scale) after renormalization of large Fourier components to an arbitrary maximum magnitude. The continuum between 23 and 88 MHz was not detectable in Chicago; TV and FM signals were found to be almost 40 dB stronger there, so gain was reduced correspondingly.

C   Raw data and RF backgrounds

To remove strong Fourier components associated with narrow-band RF signals which were approximately constant over the duration of each data record, a MATLAB routine performed the fast Fourier transform of the signal and renormalized the large Fourier components to a given maximum intensity. Fig. 1 shows the fast Fourier transform of a typical RF signal before and after this procedure was applied. In each case the data were acquired using the "wide-band" filter configuration mentioned above, whose response cuts off sharply below 23 MHz.

The effect of digital filtering on detectability of a transient is illustrated in Fig. 2. The top panel shows the RF record whose Fourier transform was given in Fig. 1, on which has been superposed a simulated transient of peak amplitude 14.5 digitization

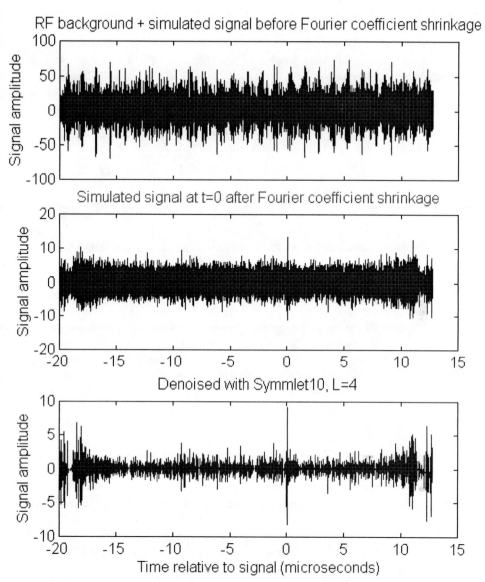

FIGURE 2. Effect of Fourier coefficient shrinkage on detectability of a transient. Top panel: raw RF record (in arbitrary units) with simulated signal superposed. Middle panel: record (same scale) after Fourier coefficient shrinkage. Here a maximum Fourier coefficient magnitude of 10^3 (in the units of Fig. 1) has been imposed. Bottom panel: the same record after denoising with a wavelet routine.

units. (The data acquisition scale ranges from -128 to $+127$ digitization units; one scale division on the oscilloscope corresponds to 25 units.) The transient is invisible beneath the large amplitude associated with television and FM radio signals. The middle panel shows the result after application of the Fourier coefficient shrinkage algorithm. The bottom panel shows the same record after denoising with a wavelet routine [21]. The records in Figs. 1 and 2 were obtained for 32,768 data points obtained at a 1 ns sampling interval, with the trigger at the 20,000th point.

D   Signal simulation

To quantify signal processing efficiency, we generated the expected signal, fed it through the same preamplifier and filter configurations used for data acquisition, and superposed it on records otherwise free of transients. We successively reduced the amplitude of the superposed test signal until it was no longer detectable, thereby obtaining an estimate of sensitivity.

A Hewlett-Packard Arbitrary Waveform Generator was used to generate signals whose typical characteristics are illustrated in Fig. 3. These signals were taken to have the form $f(t) = \theta(t)At^2(e^{-Bt} - Ce^{-Dt})$ with the coefficient C chosen so that $f(t)$ has no DC component, and D corresponding to a long duration of the negative-amplitude component. For all pulses we chose $D = B/20$, so that $C = (8000)^{-1}$ cancels the DC component. The Fourier components of the test pulse fall off smoothly with frequency. The initial t^2 behavior was chosen so that both the test pulse and its first derivative vanish at $t = 0$, as might be expected for a pulse from a developing shower. We chose $B = 0.8, 0.4, 0.2, 0.1$ corresponding to a time difference between pulse onset and maximum of $\delta = 2.5, 5, 10, 20$ ns and simulated both narrow-band (23–37 MHz) and broad-band (23–250 MHz) configurations.

The shape of the pulse of Fig. 3 is affected by preamplification and filtration as shown in Fig. 4 for the broad-band example. (The narrow-band configuration leads to a longer ringing time.) The noise is associated with the system used to generate the test pulse, and the fact that the Fourier transform is taken over a much longer time than the duration of the pulse. The sharp feature at 125 MHz is a local artifact.

Systematic studies of signal-to-noise ratios have been performed so far only for the simulated pulses with $\delta = 5$ ns applied to the broad-band front end. A typical pulse of this type gave a front end output of 21 mV peak-to-peak, acquired at an oscilloscope sensitivity of 5 mV per division. Since each division corresponds to 25 digitization units, the peak-to-peak range is about 104 digitization units, or slightly less than half the dynamic range (255 units, or 8 bits). Positive and negative peaks are thus about 52 digitization units each.

The stored test signal is then multiplied by a scale factor and added algebraically to a collection of RF records in which, in general, randomly occurring transients will be present. One then inspects these records to see if the transient can be distinguished from random noise.

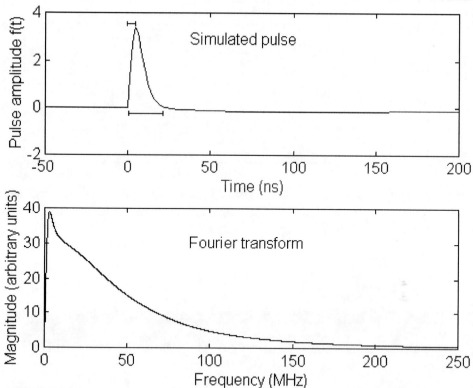

FIGURE 3. Analytic depiction of typical pulse presented to filter-preamplifier configuration. Top panel: time dependence of pulse $f(t) = \theta(t)t^2[e^{-0.4t} - e^{-0.02t}/8000]$ (t in ns); bottom panel: Fourier spectrum of pulse for $-20\ \mu s \leq t \leq 12.768\ \mu s$ (calculated analytically). In the top panel, the short bar above the pulse denotes δ, the time difference between onset and maximum, while the longer bar below the pulse denotes the duration of the positive component.

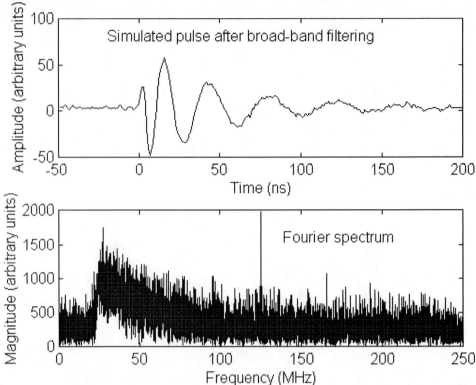

FIGURE 4. Test pulse of Fig. 3 after broad-band filtration (23–250 MHz) and preamplification. Top panel: time dependence of pulse; bottom panel: Fourier spectrum of recorded pulse for $-20 \ \mu s \leq t \leq 12.768 \ \mu s$.

For the broad-band data we estimated that pulses with input voltages corresponding to about 1/5 the original test pulse can be distinguished from average noise (not from noise spikes!). Since the original test pulse had a peak value of 1.3 mV, this corresponds to sensitivity to an antenna output of about $V_{pk} \simeq 260 \ \mu V$. The ability to detect such a pulse with an effective bandwidth of about 30 MHz corresponds to a threshold sensitivity at the level of order 3 $\mu V/m/MHz$ [1].

Preliminary studies of simulated pulses applied to the narrow-band front end suggest a considerably poorer achievable signal-to-noise ratio, despite the expectation that the signal should have a large portion of its energy between 23 and 37 MHz. It appears difficult to detect a pulse from the antenna below about 0.7 mV, which for a bandwidth of 14 MHz corresponds to a threshold sensitivity of 7 $\mu V/m/MHz$ [1]. Studies of possible improvements of the analysis algorithm for the narrow-band data are continuing.

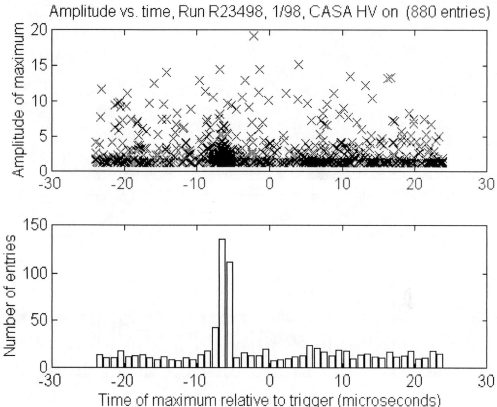

FIGURE 5. Top panel: time-vs.-amplitude plot for maxima of 880 events recorded in January 1998 (RF run 23498 only) with CASA HV supplied to all stations. All events recorded with East-West antenna polarization. Bottom panel: time distribution of transients.

E   Transients detected under various conditions

Several means were used to characterize transients. One method with good time resolution involved the shrinkage of large Fourier coefficients. One can then search for peaks of each data record, plotting their amplitude against time relative to the trigger. One such plot is shown in Fig. 5 for a data run in which CASA HV was delivered to all boxes. A strong accumulation of transients, mostly with amplitude just above the arbitrarily chosen threshold (mean + 3 σ), is visible at times -5 to -7 μs relative to the trigger. In a comparable plot for a run in which CASA HV was completely disabled (Fig. 6), only a small accumulation at times -6 to -7 μs is present. This excess appears due to transients with predominantly high-frequency components (over 100 MHz). Since signal pulses are expected to have more power below 100 MHz (see Fig. 4, bottom) we believe that this accumulation is not due

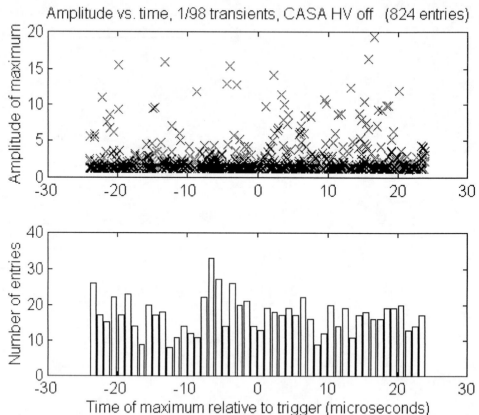

FIGURE 6. Top panel: time-vs.-amplitude plot for maxima of 824 events recorded in January 1998 with CASA HV disabled. All events recorded with East-West antenna polarization. Bottom panel: time distribution of transients.

to shower radiation, but most likely arises from the muon patches, one of which is within 75 m of the antenna.

A typical transient occurring in a run with CASA HV on is shown in Fig. 7. The transients are highly suppressed (though not in all runs) when CASA boxes within 100 m of the antenna are disabled.

The time distribution of event maxima above an arbitrary threshold for 880 events taken with CASA HV on (one run from January 1998) is shown in the bottom panel of Fig. 5. The mean arrival time is about 6 μs before the trigger, with a distribution which is slightly broader for pulses arriving earlier than the mean. This broadening may correspond to some jitter in forming the trigger pulse from the sum of muon patch pulses.

As mentioned earlier, the time for the trigger pulse to propagate from the central station to the RF trailer was measured to be 2.15 μs. One expects a similar

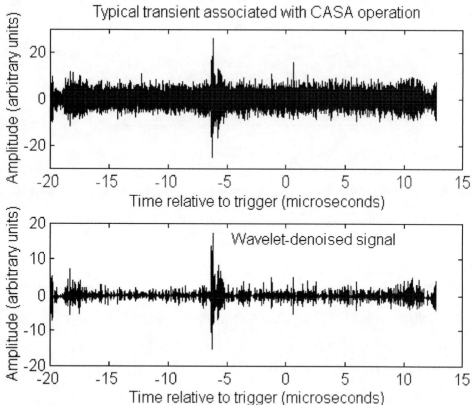

FIGURE 7. Signal of a typical transient associated with CASA operation. Top panel: before denoising; bottom panel: after denoising.

travel time for pulses to arrive from muon patches to the central station. Moreover, the muon patch signals are subjected to delays so that they all arrive at the central station at the same time for a vertically incident shower. Thus, the peak in Fig. 5 is consistent with being associated with the initial detection of a shower by CASA boxes. This circumstance was checked by recording CASA trigger request signals simultaneously with other data; they coincide with transients such as those illustrated in Fig. 7 within better than $1/2$ μs.

The RF signals from the shower are expected to arrive around the same time as, or at most several hundred nanoseconds before, the transients associated with CASA operation. They would propagate directly from the shower to the antenna, whereas transients from CASA stations are associated with a slightly longer total path length from the shower via the CASA station to the antenna. There will also be some small delay at a CASA station in forming the trigger request pulse. Thus, we expect a genuine signal also to show up around 6–7 μs before the trigger. However, for data recorded with CASA boxes disabled, no significant peak with

TABLE 1. Broad-band data recorded under lowest-noise conditions.

Antenna Polarization	CASA HV on	CASA HV off	Partial CASA HV	Total events
East-West	4503	857	1834	7194
North-South	677	641	366	1821
Total events	5180	1498	2200	9015

the expected frequency spectrum is visible in this time window. The upper limit on the rate of events giving rise to such a peak can be used to set a limit on RF pulses associated with air showers.

In Table 1 we summarize the triggers taken under optimum conditions, corresponding to broad-band data acquisition under conditions of minimum ambient noise. These triggers represent about 50 hours of data. Data were taken with both East-West and North-South antenna polarizations. Since noise from CASA boxes was found to be a significant source of RF transients, data were taken with some or all CASA boxes disabled by turning off high voltage (HV) supply to the photomultipliers.

F   Sensitivity estimate

The main difficulty associated with pulse detection is that signal pulses are not easily distinguishable from large spurious pulses originating from atmospheric discharges. Both air shower pulses and these background noise pulses can considerably exceed the average noise level. We estimate that all signal pulses should arrive between 7 and 6 μs before the trigger, while the time distribution of noise pulses is assumed to be uniform. Hence, a sufficiently large relative accumulation of pulse maxima in that time bin was adopted as a key criterion in search of signal pulses. We also employed several criteria to increase the fraction of signal pulses: The pulse should be larger than some specified magnitude threshold, bandwidth limited and approximately uniform within its limited bandwidth. To date, no significant accumulation has been detected.

At present we are only able, after accounting for the rate of accidental noise pulses, using Monte-Carlo simulation, and taking an antenna gain of $G = 2.5$ and a cable attenuation factor of 1.4 dB, to set an upper limit of $s = 54$ in Eq. (1). This is to be compared with the original Haverah Park result $s = 20$ [2], the recalibrated result $s = 1.6$ [13], and the Soviet group's result $s = 9.2$ [13]. The noise level at Dugway is too high and the acquired sample is too limited in statistics and dynamic range to allow us to place upper limits strict enough to check the claims of the two groups. We hope that further improvement of the data processing technique will reduce the noise contribution. The magnetic dip angle γ is much smaller at the

Auger site in Argentina (34° versus 68° at Utah), leading one to expect bigger electric fields for vertical showers and facilitating the detection of shower radiation. We also expect that the Argentina site will be quieter than Dugway and some clarity as regards the calibrating factor will be established.

IV OUTLOOK

A Further possibilities for processing present data

We are still hoping to improve our sensitivity to the point that we can see a true RF signal from a shower even when CASA HV is not disabled. Such a signal should precede CASA-related transients by at least the delay of formation of a phototube signal. An important calibration will be achieved if we can determine whether we are sensitive to galactic noise, which may be responsible for the continuum between 23 and 88 MHz visible in Fig. 1.

We are still exploring improved methods for removing constant RF signals from our records. In this respect we are limited by the 8-bit dynamic range of the TDS-540B oscilloscope. Data taken with various configurations of boxes near the antenna disabled may help us to better characterize the CASA-related transients.

The triggered data may be useful in a rather different context. It has been proposed that radar methods be used to detect ion trails associated with extensive air showers [22]. In our case we may be able to investigate sudden enhancements of the signals of distant television signals (on Channels 3, 6, 8, and 12) correlated with receipt of a large-event trigger.

B Considerations for Auger site

A number of questions have been suggested by the present investigation if RF pulse detection is to be considered as an adjunct to the Auger array.

- How far from the shower axis can antennas detect pulses from showers with energies above 10^{18} or 10^{19} eV? The answer determines whether a sparse array (e.g., one with the same density as Auger stations) would be sensitive to the RF pulse.

- What dynamic range for a data-acquisition system is needed so that a transient signal survives digital filtering? Apparently 8-bit range is not enough.

- What RF interference exists at each site? Surveys would be desirable. They could pinpoint not only narrow-band sources due to broadcasting stations but potentially dangerous broadband sources from switching power supplies, computers, etc. It would be best to undertake such studies after prototype systems are in place.

- What power budget would an RF detection system need? Each Auger solar-powered station is limited to a total budget of 10 watts. Presumably an RF system would have to use auxiliary power, particularly for its fast-digitization and memory components.

- What is the envisioned minimum cost per RF station? It would presumably be dominated by the data-acquisition system; the antennae and preamplifiers would probably be cheap by comparison. A preliminary estimate is less than $3K per station [1].

The Southern Hemisphere Auger site is progressing well toward an engineering array of 40 stations, as we have heard at this conference [23]. It is hoped that in a couple of years an investigation at that site of the feasibility of RF pulse detection can be undertaken.

V CONCLUSIONS

No "golden signal" has been seen for an RF transient associated with extensive air showers of cosmic rays at the CASA site. With further processing, the data may permit the setting of useful upper limits on signals relevant to at least some of the previous claims. A number of useful lessons have been learned if a similar technique is to be tried in conjunction with the Auger project.

VI ACKNOWLEDGEMENTS

It is a pleasure to thank Mike Cassidy, Jim Cronin, Brian Fick, Lucy Fortson, Joe Fowler, Rachel Gall, Kevin Green, Brian Newport, Rene Ong, Scott Oser, Daniel F. Sullivan, Fritz Toevs, Kort Travis, Augustine Urbas, and John Wilkerson for collaboration and support on various aspects of this experiment. Thanks are also due to Bruce Allen, Dave Besson, Maurice Givens, Peter Gorham, Kenny Gross, Dick Gustafson, Gerard Jendraszkiewicz, Larry Jones, Dave Peterson, John Ralston, Leslie Rosenberg, David Saltzberg, Dave Smith, M. Teshima, and Stephan Wegerich for useful discussions. This work was supported in part by the Enrico Fermi Institute, the Louis Block Fund, and the Physics Department of the University of Chicago and in part by the U. S. Department of Energy under Grant No. DE FG02 90ER40560.

REFERENCES

1. K. Green, *et al.*, Enrico Fermi Institute Report No. EFI-00-14, article in preparation for submission to Nucl. Instr. Meth.
2. H. R. Allan, in *Progress in Elementary Particles and Cosmic Ray Physics*, v. 10, edited by J. G. Wilson and S. G. Wouthuysen (North-Holland, Amsterdam, 1971), p. 171, and references therein.

3. R. P. Feynman, R. B. Leighton, and M. Sands, *The Feynman Lectures in Physics,* Addison-Wesley, Reading, Mass., 1963, Sec. I-28.

4. J. W. Cronin, Rev. Mod. Phys. **71**, S165 (1998); Nucl. Phys. B Proc. Suppl. **80**, 33 (2000); D. Zavrtanik, Nucl. Phys. B Proc. Suppl. **85**, 324 (2000). For the Pierre Auger Project Design Report see http://www.ses-ng.si/public/pao/design.html.

5. R. R. Wilson, Phys. Rev. **108**, 155 (1967).

6. G. A. Askar'yan, Zh. Eksp. Teor. Fiz. **41**, 616 (1961) [Sov. Phys.–JETP **14**, 441 (1962)]; Zh. Eksp. Teor. Fiz. **48**, 988 (1965) [Sov. Phys.–JETP **21**, 658 (1965)].

7. F. D. Kahn and I. Lerche, Proc. Roy. Soc. A **289**, 206 (1966).

8. E. Zas, F. Halzen, and T. Stanev, Phys. Rev. D **45**, 362 (1992).

9. J. V. Jelley *et al.*, Nature **205**, 327 (1965); Nuovo Cimento **A46**, 649 (1966); N. A. Porter *et al.*, Phys. Lett. **19**, 415 (1965).

10. T. Weekes, this conference.

11. S. N. Vernov *et al.*, Pis'ma v ZhETF **5**, 157 (1967) [Sov. Phys.–JETP Letters **5**, 126 (1967)]; Can. J. Phys. **46**, S241 (1968).

12. P. R. Barker, W. E. Hazen, and A. Z. Hendel, Phys. Rev. Lett. **18**, 51 (1967); W. E. Hazen, *et al.*, *ibid.* **22**, 35 (1969); **24**, 476 (1970).

13. V. B. Atrashkevich et al., Yad. Fiz. **28**, 366 (1978).

14. A. Borione *et al.*, Nucl. Instrum. Meth. A **346**, 329 (1994).

15. K. Kadota *et al.*, Proc. 23rd International Conference on Cosmic Rays (ICRC-23), Calgary, 1993, v. 4, p. 262; Tokyo Workshop on Techniques for the Study of Extremely High Energy Cosmic Rays, Tanashi, Tokyo, 27 – 30 Sept. 1993.

16. P. I. Golubnichii, A. D. Filonenko, and V. I. Yakovlev, Izv. Akad. Nauk **58**, 45 (1994).

17. C. Castagnoli *et al.*, Proc. ICRC-23, Calgary, 1993, v. 4, p. 258.

18. R. Baishya *et al.*, Proc. ICRC-23, Calgary, 1993, V. 4, p. 266; Gauhati University Collaboration, paper submitted to this conference.

19. M. A. Lawrence, R. J. O. Reid, and A. A. Watson, J. Phys. G **17**, 733 (1991).

20. R. Gall and K. D. Green, UMC-CASA note, Aug. 23, 1996 (unpublished).

21. D. F. Sullivan, Master's Thesis, University of Chicago, 1999 (unpublished).

22. P. W. Gorham, "On the possibility of radar echo detection of ultra-high energy cosmic ray- and neutrino-induced extensive air showers," hep-ex/0001041, January, 2000 (unpublished).

23. E. Zas, this conference.

Studies on Radioemission from EAS by the GUCR group

Datta (Mrs) Pranayee [1], Baishya (Mrs) Runima [2] & Roy Sinha (Mrs) Kalpana [3]
[1]   Reader, Deptt of Physics & i/c, Deptt. of Electronics Science; Gauhati University, Guwahati- 781014,Assam ,India.
[2]   Lecturer, Deptt of Physics, Pragjyotish College , Guwahati, Assam ,India
[3]   Lecturer, Deptt of Physics, Assam Engineering Institute, Guwahati, Assam ,India
       Email : krs@postmark.net

Abstract: Different characteristics of LF-MF-HF-VHF Radioemission(RE) from EAS have been studied theoretically as well as experimentally by the Gauhati University Cosmic Ray (GUCR) group since 1970.

Experimental works of GUCR group : Experimental arrangement for RE investigation used by GUCR group is given in fig 1. The general features of the EAS observed by this array viz lateral distribution of the electron component,age parameter distribution,shower size distribution and zenith angle distribution   are found to be consistent with the well known characteristics of EAS .Fieldstrengths of RE obtained at different frequencies are given in table 1. Correlation between simultaneously produced radiopulses were studied for different frequency pairs during the period 1979-1992. Results of this series of experiment are given in table 2.

TABLE : 1. Fieldstrength of radio pulses at different frequencies

Frequency in MHz	Normalised fieldstrength in $\mu Vm^{-1}MHz^{-1}$ (experimental)	
	Datta (1987)	Barthakur (1978)
2	493.6 ± 188.5	-
9	138.07 ± 38.9	-
44	-	3.70 ± 0.2
60	-	3.40 ± 0.2
80	-	2.20
110	1.96 ± 0.28	-
220	0.403 ± 0.124	-

Theoretical works of GUCR group:

Using Castagnoli et al's (1) modifications of Kahn & Larche's (2) model, fieldstrengths were calculated for primary energy $E_p = {\sim}10^{16}$ eV at a point inside the shower disc. The procedure is detailed below.

For a point lying inside the disc of charges of radius R_0 at a distance R from the axis, the point will be an outer one with respect to the disc of radius R and an inner one with respect to the annular ring of inner radius R and outer radius R_0. Accordingly, effects of the disc of radius R and the annular ring were taken into account.

Current field at a distance R from the Axis ($R<R_0$)

a)   Current field due to the disc radius R is given by

$$E'_c(R) = -\frac{ke10^5}{2c(29.4)^2} 2\pi \times 0.4NH_0^{(1)}(k\alpha R)\int_0^R J_0(k\alpha r) r^2 (r/29.4)^{-0.75}(r/29.4+1)^{-3.25}(1+r/29.4\times11.4)dr \qquad -(1)$$

b)   Current field due to the annular ring of inner radius R and outer radius R_0 is given by.

$$E''_c(R) = -\frac{ke10^5}{2c(29.4)^2} 2\pi \times 0.4NJ_0(k\alpha R)\int_R^{R0} H_0^{(1)}(k\alpha r) r^2 (r/29.4)^{-0.75}(r/29.4+1)^{-3.25}(1+r/29.4\times11.4)dr \qquad -(2)$$

CP579, *Radio Detection of High Energy Particles,* edited by D. Saltzberg and P. Gorham
© 2001 American Institute of Physics 0-7354-0018-0/01/$18.00

TABLE : 2. Correlation coefficient 'x' for different frequency pairs.

Experiment	Frequency pair in MHz	Correlation Coefficient 'x'	t-test of significance at 1% & 5% levels.	Remarks regarding production mechanism (PM)
Barthakur(1979)	(60-80)	+ 0.68	Significant	PM are similar
Sarma(1981)	(60-110)	- 0.35	- do -	PM are different
Sarma(1981)	(60-220)	- 0.31	- do -	PM are different
Borah(1983)	(60-9)	+0.38	- do -	PM are similar
Datta (1987)	(2-110)	- 0.69	- do -	PM are different
Datta (1987)	(9-220)	- 0.28	- do -	PM are different
Kalita(1992)	(2-220)	- 0.45	- do -	PM are different
Kalita(1992)	(9-110)	- 0.41	- do -	PM are different

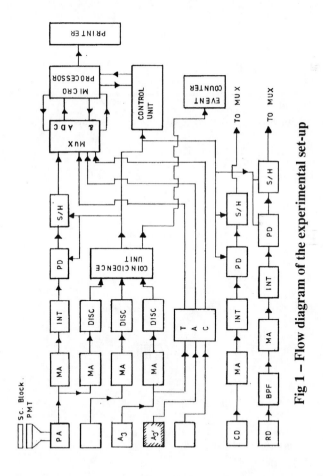

Fig 1 – Flow diagram of the experimental set-up

Dipole field at a distance R from axis ($R < R_0$).
a)   Dipole field due to the disc of radius R is given by

$$E'_p(R) = ik\alpha e \tau \pi \frac{0.4N10^5}{R(29.4)^2} H_0^{(1)}(k\alpha R) \int_0^R J_0(k\alpha r) r^2 (r/29.4)^{-0.75} (r/29.4+1)^{-3.25} (1+r/29.4 \times 11.4) dr \qquad -(3)$$

b)   Dipole field due to the annular ring of inner radius R and outer radius R_0 is given by.

$$E''_p(R) = ik\alpha e \tau \pi \frac{0.4N10^5}{R(29.4)^2} J_0'(k\alpha R) \int_R^{R_0} H_0^{(1)}(k\alpha r) r^2 (r/29.4)^{-0.75} (r/29.4+1)^{-3.25} (1+r/29.4 \times 11.4) dr \qquad -(4)$$

The vector sum of Ec', Ec'', Ep' & Ep'' is the resultant field due to the geomagnetic mechanism.
        In order to solve the problem of RE in the LF- MF- HF band, an encouraging suggestion was put forward by Nishimura in 1985(3).He proposed transition radiation (TR) mechanism as a possible source for LF-MF RE. A RE model for EAS on the basis of TR mechanism is developed by the GUCR group (4)
        When shower particles hit the ground, the phenomenon of transition radiation must occur. The radiation field produced by the positive and negative charges of the moving dipole has a phase difference of (π + an additional term), the additional term is a function of Δ/λ where Δ is the lateral separation of the charges and λ is the wavelength of the radiation being considered. This additional phase change due to charge separation can be neglected for the LF-MF-HF region. Hence, all the charged particles may be assumed to be concentrated at a point instead of distribution over a disc for mathematical convenience, and only the excess negative charge, in effect, will contribute to the TR.
For a vertical air shower, the magnitude of the horizontal component of the field is

$$\vec{E}_{\omega H} = \frac{\varepsilon N e \lambda_1 \eta_1 \vec{k}}{2\pi^2 v\varsigma} \text{Cos}\,\theta \qquad -(5)$$

and the magnitude of the vertical component of the field is

$$\vec{E}_{\omega V} = \frac{\varepsilon N e \eta_1 k^2}{2\pi^2 v\varsigma} \text{Cos}^2\theta \qquad -(6)$$

Where N = total number of shower particles at ground level,
        ε N e = excess negative charge,
        k = ω/c = $2\pi v/c$ = wave number,

$$\lambda_1^2 = \frac{\omega^2}{c^2}\chi_1 - k^2; \chi_1 = \in_1 \mu_1, \qquad \lambda_2^2 = \frac{\omega^2}{c^2}\chi_2 - k^2; \chi_2 = \in_2 \mu_2$$

$$\eta_1 = \frac{\in_2/\in_1 - (v/\omega)\lambda_2}{k^2 - (\omega^2/c^2)\chi_1} + \frac{-1 + (v/\omega)\lambda_2}{k^2 - (\omega^2/c^2)\chi_2}, \qquad \varsigma = \lambda_2 \in_1 + \lambda_1 \in_2$$

tan θ = Z / R,
        Z = height of the antenna from the ground.
        R = distance of the antenna from the shower axis.

Different characteristics of RE studied:
        (i)   Frequency spectrum: Fig 2&3
        (ii)   Variation of fieldstrength with age parameter & lateral distribution of fieldstrength :
               Fig. 4,5 & 6

(iii) TR model for inclined shower:

From the available experimental data of GU CR Research Laboratory at 120 KHz for inclined showers, the following important characteristics of inclined showers are studied.

      (a) Variation of fieldstrength with zenith angle (φ) &

      (b) Lateral distribution of fieldstrength.

These are shown in figures 7,8 & 9 for $\langle E_p \rangle$ = 5.6 x 10^{15} eV, 4.9 x 10^{16} eV and 3.7 x 10^{17} eV respectively.

The transition radiation equation for vertical showers (eq^n 6) is modified on the basis of experimental observations presented in figures 7a, 7b, 8a & 8b. The modified equations are –

For $\langle E_p \rangle$ = 5.6 x 10^{15} eV

$$\left. \begin{aligned} \vec{E}_{\omega v} &= \left(\varepsilon N e \eta_1 k^2 / 4\pi^2 v\varsigma\right) \cos^2\theta - 0.09\phi, 70\,\text{m} \leq R \langle 100\,\text{m} \\ \vec{E}_{\omega v} &= \left(\varepsilon N e \eta_1 k^2 / 4\pi^2 v\varsigma\right) \cos^2\theta - 0.11\phi, 30\,\text{m} \langle R \langle 70\,\text{m} \end{aligned} \right\} \quad (7)$$

For $\langle E_p \rangle$ = 4.9 x 10^{16} eV

$$\left. \begin{aligned} \vec{E}_{\omega v} &= \left(\varepsilon N e \eta_1 k^2 / 5.6\pi^2 v\varsigma\right) \cos^2\theta - 0.11\phi, 70\,\text{m} \leq R \langle 100\,\text{m} \\ \vec{E}_{\omega v} &= \left(\varepsilon N e \eta_1 k^2 / 4.6\pi^2 v\varsigma\right) \cos^2\theta - 0.11\phi, 30\,\text{m} \langle R \langle 70\,\text{m} \end{aligned} \right\} \quad (8)$$

For $\langle E_p \rangle$ = 3.7 x 10^{17} eV

$$\left. \begin{aligned} \vec{E}_{\omega v} &= \left(\varepsilon N e \eta_1 k^2 / 5.2\pi^2 v\varsigma\right) \cos^2\theta - 0.11\phi, 70\,\text{m} \langle R \langle 100\,\text{m} \\ \vec{E}_{\omega v} &= \left(\varepsilon N e \eta_1 k^2 / 4\pi^2 v\varsigma\right) \cos^2\theta - 0.10\phi, 30\,\text{m} \langle R \leq 70\,\text{m} \end{aligned} \right\} \quad (9)$$

Where ϕ = zenith angle, R = Core distance

(iv)   Comparison of Kahn & Lerche's model with Allan's model

(iv)   Variation of Fieldstrength with Height of Shower Maximum (H_m)

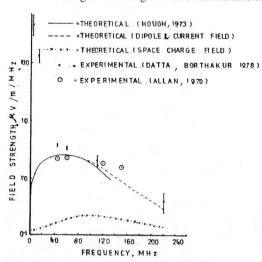

Fig 2 – Variation of fieldstrength with
frequency (Datta & Pathak, 1987)

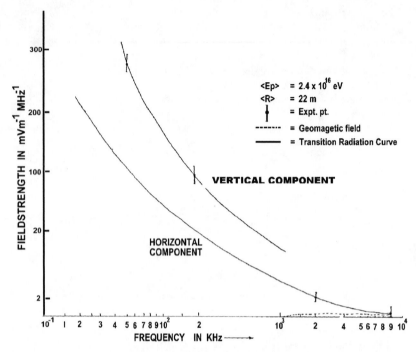

Fig 3 – Frequency Spectra of LF-MF-HF radioemission

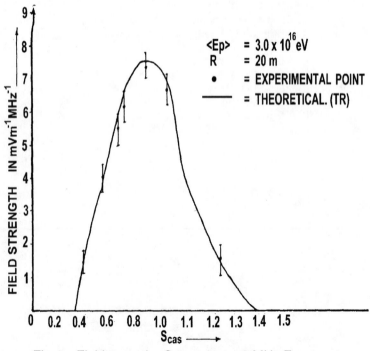

Fig 4 – Fieldstrength –S_{cas} curve at 2 MHz Frequency

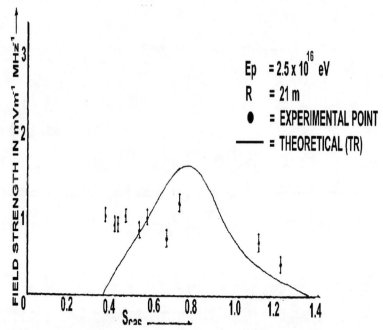

Fig 5 – Fieldstrength –S_{cas} curve at 9 MHz Frequency

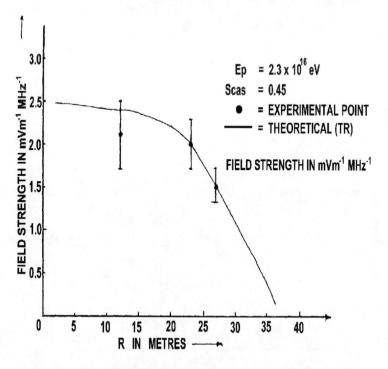

Fig 6 – Lateral distribution of fieldstrength at 2 MHz Frequency

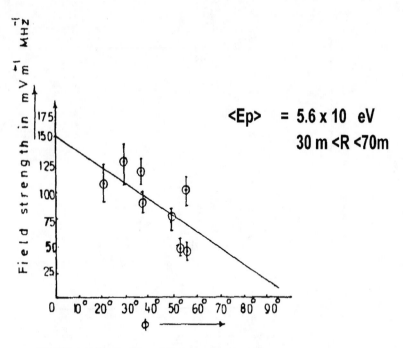

Fig 7(a) – Variation of fieldstrength with φ

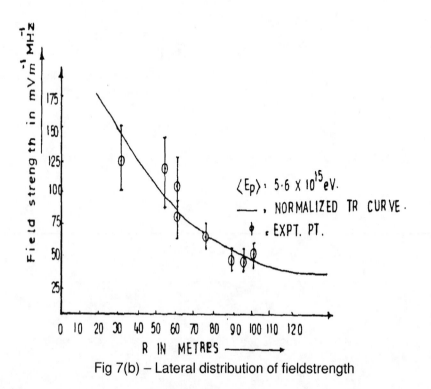

Fig 7(b) – Lateral distribution of fieldstrength

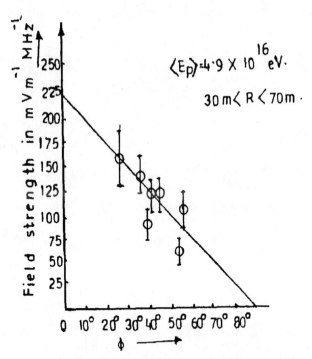

Fig 8(a) — Variation of fieldstrength with ϕ

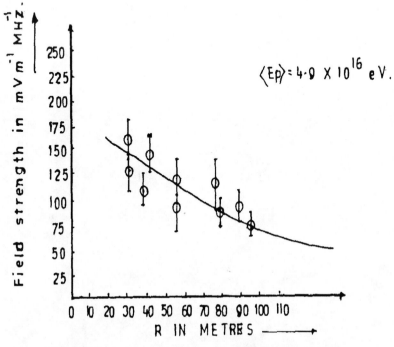

Fig 8(b) — Lateral distribution of fieldstrength

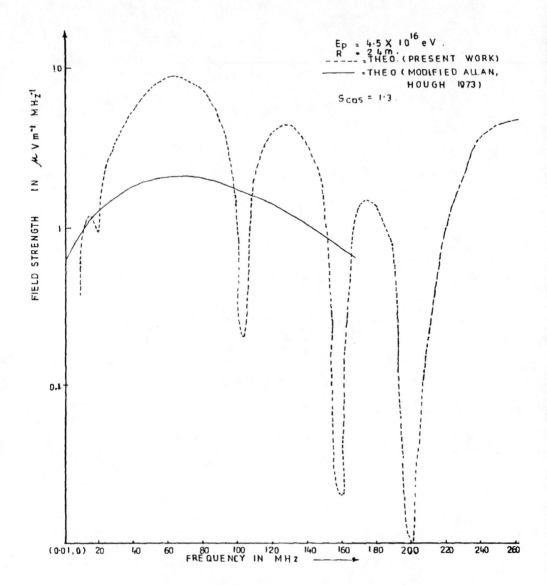

Fig 9 – Frequency spectra of radioemission

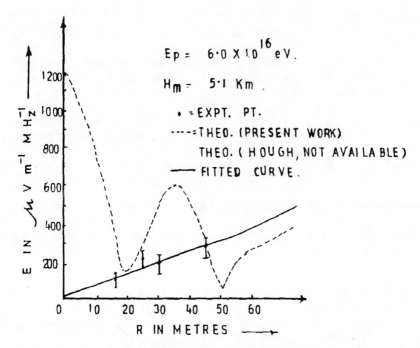

Fig 10 – Lateral distribution of fieldstrength at 110 MHz

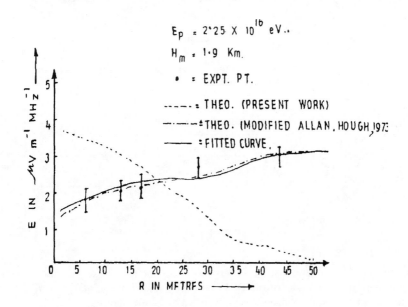

Fig 11 – Lateral distribution of fieldstrength at 220 MHz Frequency

Discussion on LF-MF-HF RE:

For a near vertical shower: The TR model exhibits a sharp increase of fieldstrengths with lowering of frequency whereas geomagnetic mechanism shows a slow decrease of fieldstrength with lowering of frequency (fig 3). The trend of the experimental frequency spectrum is in support of the TR model. Moreover, the data are found to be in good agreement, within experimental error, with the TR model.

The fieldstrengths at 2 MHz given by the TR model increase with s_{cas} reach a maximum and then decrease. These theoretical curves fit well with the experimental fieldstrengths within experimental error (fig.4).

The experimental lateral distribution of fieldstrength at 2 MHz is also in good agreement with the TR model (Fig. 6).

At 9 MHz, experimental fieldstrengths are almost independent of s_{cas} and hence these are not in agreement with the (Fig. 5) TR model.

For Inclined Showers: TR model (eqns. 7, 8 and 9) for inclined showers developed on the basis of experimental observations satisfactorily explains the observed lateral distribution of fieldstrength at 120 KHz.

Hence, in the LF-MF band experimental observations can be explained by TR model. But in the HF band observed fieldstrengths can not be explained by the TR model only, it appears to have contribution from other sources as well.

Investigations made by other workers [5,6,7,8] are summarised in Table 3.

TABLE: 3. Investigations made by other workers in the VLF-LF-MF-HF band.

Group	Frequency	Primary Energy	Results	Conclusions regarding the production mechanism	Inferences
Suga & Nishi (1987) [5]	26 KHz to 300 Khz	-10^{16}eVto -10^{19}eV	Pulses : Negative & unipolar	TR is the most probable mechanism.	
Kakimoto et al. (1990) [6]	LF-MF	$N > 10^7$	Fieldstrength higher than that from TR	Polarity in support but fieldstrength not in support of TR	
Castagnoli et al. (1991) [7]	350-500 KHz & 1.8-5 KHz	5.5×10^{15}eV - 5.0×10^{17}eV	Fieldstrength increases towards lower frequencies.	Using multi-antenna PM can be tested (e.g. TR)	PM Not clearly known
Kadota et al. (1993) [8]	LF-MF	$10^{17.5}$ eV to $10^{18.5}$ eV	Mainly negative unipolar but positive unipolar & bipolar signals are detected.	Deceleration of excess negative charge. Observed pulse widths cannot be explained.	

Table 3 shows that although these observations do not lead to any definite conclusion regarding the production mechanism of LF-MF-HF radioemission, yet most of the groups point out towards transition radiation as the most probable mechanism.

It is worth mentioning that in the present work different important characteristics of LF-MF-HF radioemission viz. Frequency spectrum, dependence of fieldstrength on age parameter and lateral distribution of fieldstrength are studied at four different frequencies with the same experimental setup, but in none of the experiments listed in Table 3 this type of multicharacteristic studies are made.

Discussion on VHF RE :
The theoretical frequency spectrum obtained by Hough (modified Allan's model) gives a smooth variation of fieldstrength with frequency – fieldstrength slowly increases, reaches a maximum and then decreases (fig 9). But the frequency spectrum obtained from the present theoretical work (modified Kahn & Lerche's model) seem to be quite unphysical, because it shows rapid fluctuation of fieldstrength with frequency which is not supported by experimental observationis of any group.

At 110 MHz, the lateral distribution of fieldstrength given by modified Allan's model has quite opposite trend to that obtained from modified Kahn & Lerche's model. The experimental data is in good agreement with modified Allan's model (fig 10)

The theoretical lateral distribution at 220 MHz, based on modified Kahn & Lerche's model is unphysical because it shows rapid fluctuation of fieldstrength with core distance. Experimental lateral distribution shows a smooth variation of fieldstrength with core distance (fig 11)

The experimental observations on variation of fieldstrength with height of shower maximum (H_m) at 110 MHz agree, within experimental error, with theoretical calculations of modified Allan's model. On the other hand, calculations done on the basis of modified Kahn & Lerche's model is in total disagreement.
The experimental observation on variation of fieldstrength with height of shower maximum (H_m) at 220 MHz also do not agree with modified Kahn & Lerche's model .
Hence, it is obvious that the two models of radioemission from EAS cannot be regarded as equivalent to each other. Allan's model. Which is microscopic in concept, is more realistic.

`t`- test for correlation co-efficient at 110 and 220 MHz indicate that for showers having s_{cas} <1, no correlation exist between fieldstrength and s_{cas} at both the frequencies in the VHF band. (Fieldstrength -H_m) curves at 110 MHz & 220 MHz indicate that for old showers (s_{cas} >1) there exist a correlation between fieldstrength and H_m. Thus, for old showers the resultant effect of both the growth and decay must be taken into account in calculating the fieldstrength of radioemission.

Conclusion: This investigation can be summarized in the form of the following conclusions :

(i)   In the LF-MF band experimental observations for near vertical showers can be explained by the TR model developed by the group.
(ii)   In the HF band , TR is not solely responsible for radioemission.
(iii)   TR model developed for inclined showers on the basis of experimental observations at 120 KHz is found to be adequate.
(iv)   Modified Kahn & Lerche's model and modified Allan's model are not equivalent to each other. Modified Allan's model is more realistic than Kahn & Lerche's model.
(v)   In the VHF band, no correlation exists between fieldstrength and s_{cas} for s_{cas}<1.
(vi)   Showers with age parameter s_{cas}>1 can be employed to estimate primary mass composition by radio method in the VHF band.
(vii)   Since observed characteristics of LF-MF Radioemission from EAS indicates very high fieldstrengths,for neutrino detection better results may be expected if frequency of observationis shifted to LF-MF from VHF band. Moreover, in major projects like Pierre-Auger observatory, VLF-LF radio detection may be considered as another technique for Giant Shower studies, which was suggested by Prof. Suga about two decades ago.

Acknowledgement:
This piece of work is the outcome of the contributions towards CR research of all the members, present and past, of the GUCR group. They are, therefore, thankfully acknowledged.

REFERENCES:
1. Castagnoli C et al., Nuovo Cimento (Italy), 63 (1969), 373.
2. Kahn F.D. & Lerche I., Proc. R Soc. London Ser A (GB), 289(1966),206.
3. Nishimura J., Proc.19th ICRC,La Jolla, USA,Vol. 7, (1985),308.
4. Baishya R. et al., II Nuovo Cimento. Vol 16 C,N. 1. (1992),17.
5. Suga K. & Nishi K, Proc. 20th ICRC, 6, Moscow.
6. Kakimoto F., Umezawa T.,Nishiyama T.,and Nishi K.(1990).Proc, 21st ICRC Adelaide,9,213.
7. Castagnoli C., Ghia p.l.,Gomez F. and TriveroP.(1991),Proc,22nd ICRC, Dublin,4,363.
8. Kadota K, Suzuki Y., Nishi K. and Kakimoto F, (1993), Proc. 23[rd]ICRC, Calgary, 4 262.
9. Datta, P. and Pathak, K. M., 20th ICRC (1987) Moscow 6.

Coherent Transition Radiation Produced in Clouds, Ice and Salt by Shower Particles

Edmond D. Gazazian, Karo A. Ispirian and Ashot S. Vardanyan

Yerevan Physics Institute, Alikhanian Brothers' St. 2, 375036 Yerevan, Armenia

Abstract. Using the values of the dielectric constant, ε, determined with the help of various models the intensity of the coherent transition radiation (CTR) produced by extended air shower (EAS) particles is calculated in the meter wave length region. CTR produced at the inhomogeneities of Antarctic ice and mine salt dielectric constant is also considered. The possibilities of detection of such CTR and its influence on cosmic ray particle radio detection is discussed.

INTRODUCTION

Great number of experimental and theoretical works carried out after the predictions of G.A.Askarian [1] on the existence of negative charge excess in EAS and possible detection of coherent Cherenkov and transition radiation produced by this excess, show that the radio detection of very high energy particles can be applied for the investigation of cosmic ray particles and their interactions. In particular, the detection of decimetre wave length Cherenkov radiation produced by super high energy neutrinos in Antarctic ice after penetrating the Earth will open the neutrino astronomy (see the reports presented in the Proceedings of this Workshop). Nevertheless, despite the achievements and confirming results of accelerator experiments [2,3] there are many problems [4] to be solved before considering the radio method completely matured. It is necessary to take into account the contribution of other possible radiation production mechanism, especially, the contribution of coherent transition radiation (CTR). As an alternative mechanism to explain the additional radio signals observed during thunder weather [5] in the work [6] it has been shown that the CTR produced by the charge excess of high energy showers in clouds can be intense and give radio signal of the same amplitudes as the expected Cherenkov radiation. CTR can be produced also in parts of Antarctic ice where a variation of ε takes place, for instance, at the ice surface or in ice volume due to mixtures particles, mechanical inhomogeneities produced during the formation of ice. Such inhomogeneities in mine salt which is also sufficiently transparent to certain ra-

CP579, *Radio Detection of High Energy Particles*, edited by D. Saltzberg and P. Gorham
© 2001 American Institute of Physics 0-7354-0018-0/01/$18.00

dio wave length interval also can be sources of CTR. These types of CTR can be detected by the antennas directly or after reflections.

The purpose of this short note is to turn attention to the consideration of the CTR production in air, ice and salt inhomogeneities of ε. The presented consideration and the estimates carried out using the calculated [6] and directly or indirectly measured (see [4] and this proceedings) values of ε show the necessity to make new investigation in this direction.

CTR due to Inhomogeneities in air ε (Clouds) by EAS

From the point of view of CTR one can describe the clouds and their properties as follows (see [6]): In dependence on the height upon the sea level there are four types of clouds: i)"Low" clouds which in their turn are divided in three types; "St", layered, which are at heights $H \simeq 0.4$ km and have mean thickness $D \simeq 0.4km$; "Sc", layered-cumulus, with $H \simeq 0.5 - 1$ km and $D \simeq 0.5$km and "Ns", layered-pluvious, with $H \simeq 1.5 - 2$km and $D \simeq 1$km; ii)"Middle" clouds, "Ae", high-cumulus and "As", high-layered. The both types have $H \simeq 3 - 4$km and $D \simeq 0.5$km.iii) "Upper" clouds,"Ci" or fleecy clouds with $H \simeq 6 - 8$km and $D \simeq 2$km, and iv) "Cu" or "clouds of vertical development " with $H \simeq 1.2 - 2$km and $D \simeq 0.5 - 5$km.

When the air temperature is $\geq -10^o$ C more than half of the clouds are partially drop-containing, while the clouds of low heights consist mainly of drops. The radius of these drops is from 2 up to 15 μm with mean radius $r_0 \simeq 5\mu$m. The density of the drops is $N_v = (1 - 6)10^8 m^{-3}$ with mean distance between the drops $l \simeq 3\mu$m. Considering the cloud as a strongly rarefied gas one can write

$$\varepsilon = 1 + \varepsilon_0 + \varepsilon_1, \tag{1}$$

where ε_0 and ε_1 are corrections to the vacuum due to air and water molecules, respectively. Having no experimentally measured values of ε for sufficiently long (meter) waves we have modeled [6] the cloud ε and obtained $\varepsilon_0 = 5 \cdot 10^{-4}$ and $\varepsilon_1 = 10^{-5}$.

As it is well known (see [7]) the EAS particles are in a pancake with disk radius $r_1 \simeq L/4$ and thickness $d \simeq L/32$ where $L = L_0 exp(H/H_0)$ and $L_0 = 310m$ are the air radiation length at heights H and sea altitude, respectively, $H_0 = 7.5km$. The number of the negative excess particles $N_v = N_e(N_e - N_{e+})/(N_e + N_{e+}) = \nu N_e$ with ν decreasing with the increase of excess electron energy, E_e and with radial distribution $\sigma_\nu(r) = N_\nu exp(-r/r_1)/2\pi r r_1$. For $E \sim 25$ MeV $\nu \sim 0.04$ [9]. Therefore, for EAS energy $E_0 \simeq 10^{17}$, $N_e \simeq 5.10^7$ and $N_\nu \simeq 2.10^6$. Note that N_e, and N_ν increase linearly with the increase of E_0. If EAS maximum takes place at the lower boundary of clouds then the effective energy of electrons is $E_{eff} \sim 25 MeV$ and $\gamma_{eff} = E_{eff}/mc^2 \sim 50$.

In the meter wave region considering the clouds as a homogeneous medium, i.e neglecting the layered cloud structure resulting in the enhancing the CTR yield, with more or less sharp lower boundary, one can write the following approximate expression for TR intensity frequency distribution from single excess electron [6]

$$\frac{dW}{d\omega} \simeq \frac{\alpha\hbar}{\pi}F(\gamma,\varepsilon_o,\varepsilon_1) = \frac{\alpha\hbar}{\pi}\left[(1 + \frac{2\varepsilon_0}{\varepsilon_1})ln(1 + \frac{\varepsilon_1}{\varepsilon_0 - \gamma^{-2}}) - 2\right]. \tag{2}$$

When $\varepsilon_0\gamma^2 \gg 1$ with an accuracy up to the terms $(\varepsilon_1/\varepsilon_0)^2$:

$$\frac{dW}{d\omega} \simeq \frac{\alpha\hbar}{6\pi}\left[\frac{\varepsilon_1}{\varepsilon_0}\right]^2, \tag{3}$$

which is convenient for estimates when the electron energies are higher than 200 MeV and for media for which the radiation length is sufficiently short. In the case of clouds when $\gamma_{eff} \sim 50$ after integration (2) over frequency interval $\Delta\omega = 0.1\omega$ for wave length $\lambda = 50$m one obtains

$$W \simeq \frac{\alpha\hbar\omega}{\pi}F(\gamma,\varepsilon_0,\varepsilon_1)\Delta\omega = 7.1.10^{-29}J. \tag{4}$$

Integrating the distribution $\sigma_\nu(r)$ for an antenna with diameter R one obtains $N_{\nu,eff} \simeq N_\nu(1 - exp(-R/r_1))$, which for $E_0 = 10^{17}$ eV and R = 50 m gives $N_{\nu,eff} \simeq 9.5.10^5$. If the shower thickness at its maximum is less than the radiation length, then the excess electrons radiate coherently, and the radiation energy increases $N_{\nu,eff}^2$ times giving $W_{N_{\nu,eff}} \simeq 6.4.10^{-17}$ J. Taking the pancake thickness equal to 10 m and assuming the detector integration time equal to the time of the shower passage, we obtain the detected power $P \simeq 2.10^{-9}$ W, providing $\simeq 0.6mV$ signal.

CTR due to Inhomogeneities in Ice and Salt

A nice review of the physical properties of Antarctic ice from the point of view of particle radio detection technique is given in [4]. For the radio wave region in the average the ε values vary form 1 up to 3.18 when the depth from the surface increases. This is connected with the formation of glacial ice from upper snow through firn. However the temperature gradients, mechanical fluctuations, volcanic ash and other reasons can result in the formation of local and layer ε inhomogeneities with thicknesses from centimeters up to kilometers. Similar ε inhomogeneities are observed in mine salt according to the measurements of Japanese physicists (see this proceedings and [8]). Considering the ice or salt as inhomogeneous media with $\varepsilon(\omega,\vec{r}) = \varepsilon_0(\omega) + \Delta(\omega,\vec{r})$ one obtains the following formula for the single excess particle TR frequency spectrum for wave lengths much larger than the mean square sizes l of the inhomogeneities $\lambda \gg l$

$$\frac{dW}{d\omega} = \frac{3\alpha\hbar Dh(\omega)}{16\pi\varepsilon^{3/2}}\left[ln\frac{(\lambda/l)^2}{|1 - \beta^2\varepsilon_0|} - 1\right]\frac{\lambda^2}{l^2}, \tag{5}$$

where $h(\omega) = (\omega^4/18\pi c^4) \int \Delta(\omega, \vec{r_1})\Delta(\omega, \vec{r_2})dV_{\vec{r_1}-\vec{r_2}}$ is the extinction coefficient, D is the thickness on which CTR is produced, i.e. the length when the shower charge excess is large (see the curves of Figs.1 of [2,10]) and $\lambda = \lambda/2\pi$.

Unfortunately, it is impossible to make estimates using (5) because there are no data on $\Delta(r, \omega)$ and l, and experimental investigations for their determination are desirable. Nevertheless, the results obtained for foam and other inhomogeneous media shows that TR intensity is higher than that from single interface which in its turn is higher then the Cherenkov radiation from showers (see [2,10]). Therefore, one can not neglect the radiation given by (5), and additional theoretical as well as experimental studies are necessary.

Discussion

As it follows from the above discussion the CTR production mechanism in clouds, ice and salt is similar to that of x-ray TR. In both cases the differences between ε of the two neighboring media is small, while in case of clouds the εs themselves are very close to unity. For relativistic particles in all cases the emission angles are small, $\theta \sim 1/\gamma$. This later fact makes more difficult (easier) the detection and measurement of the above discussed CTR of the showers produced by high energy cosmic ray (accelerator) particles since detectors with larger (smaller) acceptance are necessary.

REFERENCES

1. Askarian, G.A., *Zh. Eksp. Teor. Fiz.*, **41**, 317 (1961); **48**, 988 (1965).
2. Gorham, P., Saltzberg, D., Schoessow, P., Gai, W., Power, J.G., Konecny, R., and Conde, M.E., *e-preprint* **hep-ex/0004007** (2000).
3. Saltzberg, D., Gorham, P., Walz, D., Field, C., Iverson, R., Odian, A., Resch, G., Schoessow, P., and Williams, D., *e-preprint* **hep-ex/0011001** (2000).
4. Jelley, J.V.,*Astroparticle Physics* **5**, 255 (1996).
5. Allan, H.R., Sun, M.P., and Jones, J.K., *Proc. 14th Intern. Cosmic Ray Conf.*, Munchen, 1975, vol. 8 N. 8, p.3082.
6. Gazazian, E.D., Ispirian, K.A., and Kazarian, A.G., *Proc. of the 2-nd Intern. Symp. on Transition Radiation of High Energy Particles*, Yerevan, 1983, p. 413.
7. Khristiansen, G.,B., Kulikov, G.,V., and Fomin, Yu.,A., Kosmicheskoe Izluchenie Sverkhvisokikh energii, Moscow, Atomizdat, 1975.
8. Odian, A., private communication.
9. Avakian, R., O., Ispirian, K., A., Ispirian, R., K., and Khachatrian, V., A., e-preprint hep-ex/0005039, 2000; *Nucl. Instr. and Meth.* to be published in 2001.
10. Elbakian, S., S., Gazazian, E., D., Ispirian, K., A., Ispirian, R., K., and, Sanosyan, Kh., N., to be published in this Proceedings.

CALCULATIONS OF RADIO EMISSION FROM SHOWERS IN DENSE MEDIA

Calculations of radio pulses from High Energy Showers

Jaime Alvarez-Muñiz

Bartol Research Institute, University of Delaware, Newark, DE 19711.
alvarez@bartol.udel.edu

Enrique Zas

Departamento de Física de Partículas,
Universidad de Santiago de Compostela, E-15706 Santiago, Spain.
zas@fpaxp1.usc.es

Abstract. In this article we review the progress made in understanding the main characteristics of coherent Čerenkov radiation induced by high energy showers in dense media. A specific code developed for this purpose is described because it took a significant part in this process. Subsequent approximations developed for the calculation of radio pulses from EeV showers are reviewed. Emphasis is given to the relation between the shower characteristics and different features of the corresponding radio emission.

I INTRODUCTION

It was about 40 years ago that Askar'yan proposed the detection of very high energy particles through the coherent radio emission from all the particles in a high energy shower [1]. Although radio emission from showers had been discussed before, Askar'yan noted that showers would develop an excess of electrons of order 10% independently of shower energy and that such excess would lead to coherent radiation. Indeed if the emission is coherent the electric field scales as the excess charge and as the shower energy rises, the relative contribution of radio emission increases with respect to other wavelengths (say optical Čerenkov). Since the number of particles in a shower is proportional to shower energy, the power in the coherent signal scales with the square of the primary energy. This fact together with the technical simplicity of detectors makes the technique attractive for detecting very high energy showers. It is remarkable that many detection possibilities addressed in this meeting were already in Askar'yan original work.

Clearly coherence demands that the wavelength of the radiation exceeds the size of the radiating region. As electromagnetic showers develop through successive pair

CP579, *Radio Detection of High Energy Particles,* edited by D. Saltzberg and P. Gorham
© 2001 American Institute of Physics 0-7354-0018-0/01/$18.00

production and bremsstrahlung interactions, the particle distributions of showers of a given energy are, to a very rough approximation, similar when described in radiation length units [1]. The radiation length of different materials in units of matter depth basically scales with the inverse of the atomic number, being 37 g cm^{-2} for air and not too different for many other abundant materials that have been proposed for radio experiments such as water, salt and sand. Because of the medium density the scale of the showers can however be completely different. While air showers have sizes in the km range and coherence effects can be obtained at frequencies of order 10 MHz, for showers in ice the scale is a meter and coherence is kept to frequencies of order 10 GHz.

Since the power associated to the Čerenkov emission increases with frequency Askar'yan suggested detecting showers in dense medium, as those that could be induced by deeply penetrating particles underground or by cosmic rays on the Moon surface. Attempts were made to detect radio pulses in coincidence with other air shower detectors [2] and, although the results were positive, it became clear that there were many difficulties. Calculations revealed that several additional mechanisms competed with excess charge in the generation of radio pulses, including dipole radiation and transverse currents because of the magnetic field of the Earth, and transition radiation when the shower reached the ground [3]. The interpretation of radio signals was obscured by these and other difficulties and the technique was almost completely abandoned but for a few isolated efforts that have been reviewed in this conference.

In the mid 1980's the efforts to build high energy neutrino telescopes renewed the activity in the radio technique [4,5] as a possible alternative to the projects to detect the Čerenkov light from muons under water or ice [7]. As a result early in the 1990's a Monte Carlo program was developed for the calculation of distant radio signals from TeV electromagnetic showers in ice [8,9]. These led to several efforts to assess the possibility to detect neutrino events with arrays of antennas in Antarctica [10–12]. Later on a series of approximations were made to extend the results of these calculations to both electromagnetic and hadronic showers of energies in the EeV and ZeV ranges [13–16]. In this article we will concentrate on the efforts made towards understanding Čerenkov radio emission in dense media from the excess charge that develops in high energy showers. The recent experimental confirmation of this effect in a SLAC experiment [17], in agreement with expectations, leaves now little doubt about the enormous potential of this technique. We believe that many developments will follow this conference and it is our hope that these calculational efforts will help in the task.

---

[1] Hadronic showers can be regarded as a jet of hadrons along shower axis which generates a chain of electromagnetic subshowers through neutral pion decays and have similar scales.

II ZHS: A SHOWER GENERATOR FOR RADIO EMISSION.

The calculation of radio pulses from showers is a complex problem in which the emission from all particles in a high energy shower have to be carefully added in a coherent fashion, taking into account the velocity and direction, position and time of each of the particles involved. For a medium such as ice, having a refraction index $n = 1.78$ for radio waves, electrons of energy exceeding 107 keV emit Čerenkov radiation. The simulation of the shower cascade from this perspective was tackled by Zas, Halzen and Stanev in [9] with a result that has become known as the *ZHS* Monte Carlo. Since the largest contribution to the excess charge in a shower is from electrons, only electromagnetic showers were considered. This program has been a reference point for much progress made during the 1990's in radio detection and deserves a brief description. Efforts have been made by other groups to test the code against other shower codes [18].

The longitudinal development of an electromagnetic shower mostly follows from successive pair production and bremsstrahlung interactions with the electric field of atomic nuclei that are the heart of the showering process. For both these processes the effective interaction distance increases as the energy of the interacting photon or electron rises and atomic screening of the atoms becomes progressively more important. In this respect the Landau-Pomeranchuck-Migdal (LPM) effect can be regarded as the higher energy extreme of this situation in which the atomic potential of the individual atoms has to be modified because of collective effects of several nuclei. The lateral structure of the shower is mostly controlled by elastic scattering. Because of the long range of the Coulomb interaction it turns out that most angular deviations arise because of multiple elastic scattering which is treated in this Monte Carlo keeping the first two terms in Molière's expansion [19].

Besides these interactions that are common to all simulation programs, the ZHS code has special features for radio calculations. It is a three dimensional code that takes timing of the particles in great detail. Time delays are measured with respect to a plane perpendicular to shower axis that starts in phase with the primary particle and moves parallel to the axis at speed c. Most of such delays are due to angular deviations, to multiple elastic scattering and to the subluminal velocities of the shower particles all of which are taken into account. The program includes electron positron annihilation, Möller, Bhabha and Compton scattering, which only become relevant at low energies but are responsible for the excess charge in the shower. These routines were developed following the EGS4 code [20]. In the first version [8] it did not include the LPM effect but this was soon remedied [9]. It can be run with threshold energies as low as 100 keV, but its accuracy is expected to decrease for thresholds below this value because it does not include the photoelectric effect. Finally it is designed to be fast using, when possible, numerical parameterizations instead of actual functions to minimize calculating time.

The results obtained for the longitudinal development of the showers are consistent with the Greisen parameterization. The lateral distributions are consistent with three dimensional theoretical shower results provided the Molière radius is divided by a factor of about 2, in agreement with other simulations [21]. The excess charge obtained is of order 20% depending on particle threshold, it is only slightly dependent on shower depth, see Fig. 1a. The extra track due to electrons is 21% of the total (the Čerenkov emission is proportional to the tracklengths of the charged particles) as shown in Fig. 1b. Compton scattering of atomic electrons that *get accelerated* into the shower account for most of the effect (12%). It is remarkable that close to half (14%) of the excess tracklength is due to particles below 5 MeV (1 MeV). The calculation of coherent signals requires a lower threshold than implied by these numbers because the lower energy particles are expected to interfere more destructively as they get longer delays and larger angular deviations. This is already important because it sets an upper bound to the energy threshold necessary for an accurate calculation.

III ČERENKOV RADIATION

The detection of radio pulses from air showers is clearly a very different problem from detecting them in a dense medium such as ice. Firstly in ice the *coherence*

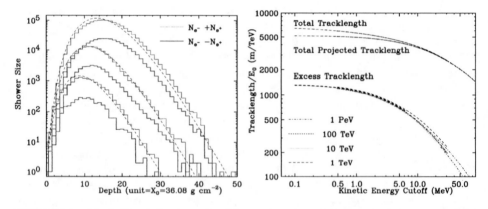

FIGURE 1. a) Left: Depth development of the total number of electrons and positrons and the excess number of electrons for single 1, 10 and 100 TeV electron initiated showers. The Monte Carlo threshold energy is 1 MeV. Depth is measured in radiation lengths of ice. The dashed lines are the results in approximation-B of shower theory in the limit of $E_{th} \to m_e$. **b)** Right: Shower track lengths for all charged particles in 1, 10, 100 TeV and 1 PeV electron showers as a function of the calculational cutoff (E_{th}). The three sets of curves correspond to i) the sum of all track lengths (Total Tracklength), ii) the sum of all track lengths projected onto the shower axis (Total Projected Tracklength) and iii) the difference of electron and positron track lengths projected onto the shower axis (Excess Tracklength). Several showers are plotted for each energy to illustrate the small effect of fluctuations. The tracklengths have been normalized to the shower energy in TeV.

band extends to the GHz regime. Secondly most shower particles travel very small paths of tens of centimeters, and they are hardly deviated because of the Earth's magnetic field. As a result the excess charge mechanism is expected to dominate the radiation and the problem becomes simpler from the calculation point of view. Moreover there is a substantial difference in the relation between the typical observation distances and the shower dimensions. While for air showers the observation distance is of the some order or even less than the shower dimensions, prospects of detecting radio pulses under the Earth surface or from the Moon are attractive precisely because they can be detected from very large distances compared with the shower dimensions which are of order one meter. For most interesting cases the Fraunhofer approximation turns out to be accurate what is useful for calculational purposes and simplifies both the interpretation and the presentation of results.

When a particle of charge z moves through a medium of refractive index n with velocity $|\vec{v}| = \beta c > c/n$ Čerenkov light is emitted at the Čerenkov angle θ_C, verifying $\cos \theta_C = (\beta n)^{-1}$, with a power spectrum given by the well known Frank-Tamm result [22] :

$$\frac{d^2W}{d\nu dl} = \left[\frac{4\pi^2 \hbar}{c} \alpha \right] z^2 \nu \left[1 - \frac{1}{\beta^2 n^2} \right] , \tag{1}$$

with ν the frequency, c the speed of light, $dl = c\beta dt$ a small element of particle track length, and α the fine structure constant. This is the standard approximation used for most Čerenkov applications for wavelengths orders of magnitude smaller than the tracks. This expression corresponds to an infinite track and in the case of a finite track there are diffraction effects.

Working with the time Fourier transform of the electric field, $\vec{E}(\omega)$, Maxwell's equations can be solved in the transverse gauge for an arbitrary current [9]:

$$\vec{E}(\omega, \vec{x}) = \frac{e\mu_r}{2\pi\epsilon_0 c^2} i\omega \int \int \int \int dt' \, d^3\vec{x'} \, e^{i\omega t' + i|\vec{k}||\vec{x} - \vec{x'}|} \frac{\vec{J}_\perp(t', \vec{x'})}{|\vec{x} - \vec{x'}|} \tag{2}$$

where $|\vec{k}| = k = n\omega/c$ and $\vec{J}_\perp$ is a divergenceless component of the current transverse to the observation direction, μ_r is the relative permeability of the medium and ϵ_0 the permittivity of the vacuum. Eq. 1 can be obtained from this expression. For a single particle moving uniformly with velocity $\vec{v}$ between times t_1 and $t_1 + \delta t$ in the Fraunhofer limit it simplifies to:

$$\vec{E}(\omega, \vec{x}) = \frac{e\mu_r}{2\pi\epsilon_0 c^2} \frac{i\omega}{R} \frac{e^{ikR}}{R} e^{i(\omega - \vec{k} \cdot \vec{v})t_1} \vec{v}_\perp \left[\frac{e^{i(\omega - \vec{k} \cdot \vec{v})\delta t} - 1}{i(\omega - \vec{k} \cdot \vec{v})} \right] \tag{3}$$

The Fraunhofer limit corresponds to an observation distance $R \gg [v\delta t \sin\theta]^2/\lambda$. For a characteristic track of one radiation length (~ 40 cm) and λ of 30 cm corresponding to 1 GHz frequencies, the condition implies $R \gg 50$ cm which is well satisfied for detection distances of order one km. It should be remarked that the inequality can be always satisfied provided a sufficiently small time interval is chosen. The ZHS program uses this expression. For an infinitesimal track δl it becomes:

$$R\vec{E}(\omega, \vec{x}) = \frac{e\mu_r}{2\pi\epsilon_0 c^2} \frac{i\omega}{} \, \vec{\delta l}_\perp \, e^{i(\omega - \vec{k}\vec{v}_1)t_1} \, e^{ikR}, \tag{4}$$

This expression shows that the electric field amplitude is proportional to the tracklength and that radiation is polarized in the direction of $\vec{l}_\perp$, the apparent direction of the track as seen by the observer.

For a high energy shower the calculation of the radio pulse is now a matter of book-keeping. Results of the electric field amplitude in the Čerenkov direction display a characteristic frequency spectrum: a linear rise with frequency until the GHz region where destructive interference sets in and the spectrum gradually levels off as shown in Fig. 2a. Although in the Čerenkov direction all points of the trajectory of a particle moving at the speed of light along the shower axis are emitting in phase, lateral deviations of the particle trajectories and time delays destroy the coherent interference at sufficiently high frequencies. The effect of the lateral distribution is dominant. The amplitude at a given frequency is proportional to the excess tracklength which in turn accurately scales with shower energy as shown in Fig. 1b.

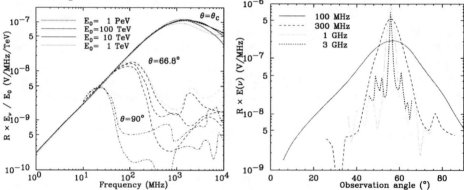

FIGURE 2. a) Left: Frequency spectrum of the electromagnetic pulse. The figure displays 1, 10 and 100 TeV showers for different thresholds. The amplitudes have been divided by the energy of the primary in TeV and renormalized depending on the threshold. The full curves correspond to observation at the Čerenkov angle (θ_C) to the shower axis. The dashed (dot-dashed) curve corresponds to observation at 66.8° (at 90°) to the same axis. **b)** Right: Angular distribution of the electric field generated by a 10 TeV shower ($E_{th} = 611$ keV). The observation angle is the polar angle of the radiation with respect to the shower axis.

The polar diagram of the emitted radiation displays a characteristic peak in the Čerenkov direction of a width which is inversely proportional to the frequency. This is plotted in Fig. 2b and corresponds to a diffraction-like pattern. The width of the diffraction peak about the Čerenkov angle depends on the longitudinal distribution of the charge in the shower. As the observation angle gets away from the Čerenkov direction different longitudinal sections of the shower development start to interfere destructively and the amplitude of the electric field drops.

Long tracks in a shower can be subdivided in arbitrary subintervals. Provided the Fraunhofer approximation is valid for the shower as a whole, the algorithm of Eq. 3 for a particle that moves at constant speed displays cancellations between all the end points of the intermediate subintervals of the track. Only the end points of the total track need to be considered in this case what simplifies the computational task enormously. In approximation "a" (default) each electron gives rise to a single track to which Eq. 3 is applied. As particles do suffer some deceleration along the track the average velocity is used. Different criteria for this subdivision have given rise to approximations "b" and "c" [23]. The approximation "a" has been shown to provide sufficiently good results when compared to longer calculations in which each electron track is subdivided into subtracks each of which contributing with Eq. 3 as shown in Fig. 3. Note that approximation "a" is conservative in the sense that it underpredicts the amplitude of the pulse at very high frequencies.

IV   1-D APPROXIMATION

With the help of scaling it is possible to raise the shower simulation threshold and to generate showers of about 1 PeV but at higher energies the calculations become too lengthy to be handled. The calculation of radioemission from EeV showers is however of great interest because the technique is expected to be most competitive for these energies. Because of the LPM suppression of bremstrahlung and pair production, these showers have very different elongated depth distributions. As a result the angular width of the Čerenkov peaks is expected to be reduced with respect to showers of energy below 1 PeV. As the lateral distributions of these showers are not expected to be very different from the lower energy showers, not many effects can be anticipated in the frequency spectrum of the signal in the Čerenkov direction.

The one dimensional approximation has been developed to study the effects of the longitudinal elongation [13,14]. In this approximation the particles in the shower are all assumed to be moving at the speed of light and the lateral distribution is neglected. The current density simply becomes [16]:

$$\vec{J}_\perp(\vec{x'}, t') = Q(z') \, \vec{c}_\perp \, \delta^3(\vec{x'} - \hat{n}_z ct') \tag{5}$$

where the shower develops along the z axis, $\hat{n}_z$ is a unit vector in this direction and $Q(z')$ is the excess charge in the shower. The substitution of this current into Eq. 2 leads to:

$$\vec{E}(\omega, \vec{x}) = \frac{e\mu_{\rm r}}{2\pi\epsilon_0 c^2} \, i\frac{\omega}{c} \, \sin\theta \, \hat{n}_\perp \int dz' \, Q(z') \, \frac{e^{i\frac{\omega}{c}z' + ik|\vec{x} - z'\hat{n}_z|}}{|\vec{x} - z'\hat{n}_z|} \tag{6}$$

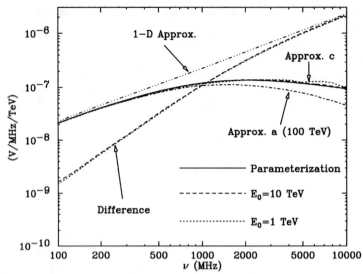

FIGURE 3. Comparison of complete simulation results for the frequency spectrum in the Čerenkov direction for 1 and 10 TeV electromagnetic showers in approximation c, for a 100 TeV in the standard approximation (a) and with the 1D-approximation (top curve). The lower curves represent the difference between the 1D approximation and the full simulation results using in the c approximation. Note that both the spectrum and the difference have the same behavior for different shower energies. All radio pulses scale with shower energy and are normalized to 1 TeV.

where θ is the angle between the shower axis and the direction of observation $\vec{x}$ and $\hat{n}_\perp$ is a unitary vector perpendicular to $\vec{x}$. It should be stressed that no further approximations are made at this stage.

The phase factor in Eq. 6 can be approximated by $ik|\vec{x} - \vec{z}\,'| \simeq ikR - i\vec{k}\vec{z}\,'$ in the Fraunhofer limit. Here $R = |\vec{x}|$ is the distance from the center of the shower to the observation point. The electric field amplitude simply becomes the Fourier transform of the longitudinal charge distribution:

$$\vec{E}(\omega, \vec{x}) = \frac{e\mu_r}{2\pi\epsilon_0 c^2}\ i\omega\ \sin\theta\ \frac{e^{ikR}}{R}\ \hat{n}_\perp \int dz'\ Q(z')\ e^{ipz'} \tag{7}$$

where we have introduced for convenience the parameter $p(\theta, \omega) = (1 - n\cos\theta)\,\omega/c$ in Eq. 7. The angular pattern around the Čerenkov direction is precisely the analog of the classical diffraction pattern of an aperture function.

The accuracy of the approximation has been studied comparing it to complete simulations. Excellent agreement is obtained for frequencies below 100 MHz. For frequencies progressively increasing above this value, the electric field becomes overestimated in the Čerenkov direction, but the angular distribution of the pulse is otherwise preserved. This is not surprising since the lateral distribution is ignored and no levelling off in the frequency spectrum can be expected. Moreover the difference between the complete simulation and the approximation scales with shower energy as shown in Fig. 7. Ad hoc corrections can be implemented to make quantitative predictions in the GHz regime accounting in an effective way for the ignored effects [14,16]. This is analogous to the form factor described in [24]. The approximation gives some insight into the complexity of the spectrum and angular distribution of the pulses relating them to features in shower development.

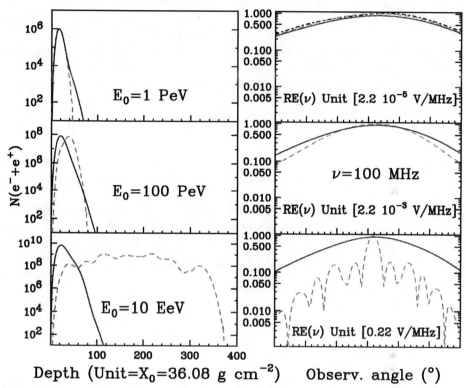

FIGURE 4. Left: Longitudinal development of electromagnetic (solid red lines) and hadronic (dashed lines) showers in ice for different energies. Right: Angular distribution of the electric field amplitude times the distance emitted by the showers shown on the left. Shown is the value $|E(\nu)\,R|$ where R is the distance to the shower, normalized to its maximum at the Cherenkov angle ($\theta_C = 56°$). The units, that is the precise values at maximum are marked in the figure. For the 1 PeV case the electric field amplitude obtained in a complete simulation (dot-dashed line) is also shown to compare with the 1D approximation.

The development of EeV showers of electromagnetic [13] and hadronic [14] character has been studied using hybrid techniques that combine simulation and parameterizations of low energy showers. The LPM suppression leads to a very large elongation of electromagnetic showers which results in a reduction of the angular width of the diffracted pulses. The elongation has been calculated to scale as $E_{sh}^{1/3}$

for shower energies satisfying $E_{sh} > 20$ PeV [13]. For hadronic showers the elongation does not affect shower development significantly until energies in the EeV range. This is because high energy hadronic interactions have large multiplicities and the particles that emerge carry small fractions of the primary energy. Moreover the neutral pions at very high energies interact before decaying into photons, and the transfer of energy to electromagnetic particles is postponed until the pions have energies in the few 10's of PeV. The emerging photons induce ordinary showers. Photons or electrons from very short lived particles are the main source of the occasional shower elongations observed [14]. This is illustrated in Fig. 4.

Eq. 6 can be also used for calculating the radio emission in an approximate fashion to establish Fresnel interference effects. Since this approximation neglects the lateral distribution the results are only expected to be valid for showers which are very elongated because of the LPM effect. For these showers the Fresnel interference effect becomes more important than the lateral distributions. As the observation distance decreases the angular distribution of the pulses becomes wider and the main peak drops. The results show that at the Fresnel distance defined as $R_F = \pi n \nu \ (L_s \sin \theta / 2)^2 / c$ (L_s is the shower length) deviations of the electric field in the Čerenkov direction from the Fraunhofer approximation are below 10%.

For neutrino induced showers the two types of showers become relevant. Neutrinos always induce a hadronic character shower through the nucleon fragments. In the case of charged current electron neutrino interactions, the emerging shower has mixed hadronic and electromagnetic character and the relative weights of each component depend on y, the fraction of energy transferred to the nuclear fragments. For very high energy neutrinos this leads to an interesting combination of two showers of different lengths because of the relative importance of the LPM effect for both types of showers. The resulting angular distributions of the radio pulse displays a complex pattern of the two corresponding angular widths. It has been shown that it is in principle possible to use the angular information of the radio pulse to extract the corresponding value of y for the interaction [15].

V SUMMARY

We have reviewed the development in calculations of the full diffraction patterns and frequency spectra of hadronic and electromagnetic showers developing in ice. The complex emerging patterns are well understood in terms of the shower properties. We have stressed the dependence on the angular width of the Čerenkov pulse on length and its implication for electromagnetic showers with strong LPM elongations.

ACKNOWLEDGEMENTS

We thank D. Saltzberg for helpful comments after carefully reading the manuscript and the organizers P. Gorham and D. Saltzberg for providing this unique oportunity to give a good impulse to this field. This work was supported in part by the European Science Foundation (Neutrino Astrophysics Network N. 86), by the CICYT (AEN99-0589-C02-02) and by Xunta de Galicia (PGIDT00PXI20615PR). The research activities of J.A-M at Bartol Research Institute are supported by the NASA grant NAG5–7009.

REFERENCES

1. G.A. Askar'yan, Zh. Eksp. Teor. Fiz **41**, 616 (1961) [Soviet Physics JETP **14** 441, (1962)]; **48**, 988 (1965) [**21**, 658 (1965)].
2. T. Weekes, in these proceedings and references therein.
3. F.D. Kahn and I. Lerche, *Proc. of the Roy. Soc. A* **289**, 206 (1966).
4. M.A. Markov, I.M. Zheleznykh, Nucl. Instr. and Methods Phys. Res. **A248**, 242 (1986).
5. Ralston, J.P. and McKay, D.M., *Proc. Astrophysics in Antarctica Conference*, ed. Mullan, D.J., Pomerantz, M.A. and Stanev, T., (American Institute of Physics, New York, 1989) Vol. 198, p. 241
6. K. Mannheim, Astropart. Phys. **3**, 295 (1995).
7. F. Halzen, in these proceedings.
8. F. Halzen, E. Zas, and T. Stanev, *Phys. Lett.* **B257** 432 (1991).
9. E. Zas, F. Halzen, and T. Stanev, *Phys. Rev.* **D45**, 362 (1992).
10. A.L. Provorov, I.M. Zheleznykh, Astroparticle Physics **4**, 55 (1995).
11. G.M. Frichter, J.P. Ralston, D.W. Mc Kay, *Phys. Rev.* **D 53**, 1684 (1996).
12. J.V. Jelley, *Astropart. Phys.* **5**, 255 (1996).
13. J. Alvarez-Muñiz and E. Zas, *Phys. Lett.* **B411**, 218 (1997).
14. J. Alvarez-Muñiz and E. Zas, *Phys. Lett.* **B434**, 396 (1998).
15. J. Alvarez-Muñiz, R.A. Vázquez and E. Zas, *Phys. Rev.* **D 61**, 023001 (2000).
16. J. Alvarez-Muñiz, R.A. Vázquez, E. Zas, *Phys. Rev.* **D 61**, 023001 (2000).
17. D. Saltzberg in these proceedings, P. Gorham, *et. al*, e-print archive: hep-ex/0011001.
18. D. Besson *et al.*, in these proceedings.
19. Molière, G., *Z. Naturforsch* **3a** 78 (1948).
20. Nelson, W.R., Hirayama, H. and Rogers, D.W.O., *The EGS4 Code System*, SLAC Rep. SLAC-265, UC-32, Stanford Linear Accelerator Center (1985).
21. M. Hillas, private communication.
22. Frank, I. and Tamm, I., *Dokl. Akad. Nauk SSSR* **14** 109 (1937).
23. J. Alvarez-Muñiz, G. Parente, and E. Zas, *Proc. XXIV Int. Cosmic Ray Conf.*, Rome (1995), vol 1, p. 1023.
24. R. Buniy, J.P. Ralston, astro-ph/0003408.

Prospects for radio detection of extremely high energy cosmic rays and neutrinos in the Moon

Jaime Alvarez-Muñiz* and Enrique Zas[†]

*Bartol Research Institute, University of Delaware, Newark, DE 19711
[†]Departamento de Física de Partículas, Universidade de Santiago de Compostela,
E-15706, Santiago de Compostela, Spain

Abstract. We explore the feasibility of using the Moon as a detector of extremely high energy ($> 10^{19}$ eV) cosmic rays and neutrinos. The idea is to use the existing radiotelescopes on Earth to look for short pulses of Cherenkov radiation in the GHz range emitted by showers induced just below the surface of the Moon when cosmic rays or neutrinos strike it. We estimate the energy threshold of the technique and the effective aperture and volume of the Moon for this detection. We apply our calculation to obtain the expected event rates from the observed cosmic ray flux and several representative theoretical neutrino fluxes.

INTRODUCTION

The observation of atmospheric showers with total energy above the so-called GZK cutoff at energies above $\sim 5 \times 10^{19}$ eV, is one of the most puzzling mysteries of the emerging field of astroparticle physics [1]. The nature of the primaries is still unknown as well as their origin and the mechanisms involved in their acceleration to such energies. The puzzle is even more intriguing because if the primaries are protons, photons or nuclei they should be produced within distances of a few tens of megaparsecs from us, otherwise their interactions with the 2.7 K photons constituting the Cosmic Microwave Background (CMB) as well as with the infrared and radiobackgrounds, should greatly reduce their energy. Since no effective sources capable of accelerating them have been identified within that distance, their very presence is contrary to the expectations. Neutrinos may play a fundamental role in solving this puzzle. They may be produced in interactions of the primary cosmic rays with the matter and radiation in the source or in the matter and radiation they find along their paths to Earth. The so-called GZK neutrinos produced in interactions of cosmic rays with the CMB are almost guaranteed, since both the projectile and the target are known to exist [2,3]. Sources in which ultra high energy neutrinos may be produced include AGN's and GRB's [4] where accelerated

CP579, *Radio Detection of High Energy Particles,* edited by D. Saltzberg and P. Gorham
© 2001 American Institute of Physics 0-7354-0018-0/01/$18.00

particles interact with matter or radiation. Alternative sources are the annihilation of topological defects which can be produced in phase transitions in the early universe where extremely high densities of energy are trapped [5].

Some present and future detectors of ultra high energy cosmic rays [1,6–8] and neutrinos [4,9,10] will soon remedy the lack of statistics [11]. The small observed (predicted) fluxes of ultra high energy cosmic rays (neutrinos) call for huge effective volumes (measured in km^3). Several detection methods and techniques are being employed and/or studied to achieve the required volumes, some of them have been discussed in this workshop [12]. Among them the radio technique is one of the most promising alternatives for neutrino and possibly cosmic ray detection at ultra high energies (10^{15} eV and above). The technique aims at detecting *coherent* Cherenkov radiation in the MHz-GHz range from the excess of electrons in showers initiated by photons or electrons. Several detectors are being planned [12] or they are in the early stages of construction [13]. Due to its excellent radio frequency wave propagation properties, Antarctic ice is being considered as a medium where antennas are being placed to monitor the potential radio signals. Other media such as salt [14] and sand are also being studied. As an interesting alternative, in 1989 Zheleznykh et al. [15] proposed to detect showers initiated by cosmic ray and neutrinos by measuring coherent Cherenkov radiation emitted just below the surface of the Moon when high energy cosmic rays or neutrinos strike it. Two groups have looked for those signals using existing radiotelescopes on Earth with no positive detection during the time they have pointed the instruments to the Moon [16,17].

Here we investigate the feasibility of detecting extremely high energy particles interacting in the Moon. We discuss the characteristics of the radio signals and we stress how they influence the aperture of the Moon as a high energy particle detector. We calculate the energy threshold for the technique as a function of the detector parameters and the signal's main features. Using several representative neutrino fluxes, we estimate the expected event rates and we briefly discuss where, on the surface of the Moon, should the radiotelescopes be pointed at to be able to detect them. We also discuss cosmic ray detection. This work updates previous estimates in which some approximations to extremely high energy shower development in the lunar rock as well as to the associated radiosignals were used [18]. We also take into account the transmissivity of the signals through the Moon's surface and some other geometrical issues which were not fully considered in our previous estimates.

RADIOPULSES IN THE MOON

Neutrino interactions produce different types of showers depending on the neutrino flavor and on whether the interaction is mediated by a W$^\pm$ or a Z boson. In deep inelastic scattering (DIS) charged current (CC) interactions of electron neutrinos (ν_e), the electron produced in the lepton vertex initiates an electromagnetic shower of energy $(1-y)E_\nu$, where E_ν is the neutrino energy in the laboratory frame and y is the fraction of energy transferred to the struck nucleon in the hadronic vertex. The debris of the nucleon initiate a hadronic shower which is superimposed to the electromagnetic one producing a "mixed shower". Hadronic showers are also initiated in both charged and neutral current (NC) DIS neutrino interactions by the hadrons created in the fragmentation of the nuclear debris. At shower energies above $E_{\mathrm{LPM}} \sim 4 \times 10^{14}$ eV (400 TeV) in the lunar regolith, electromagnetic showers are affected by the LPM effect [19]. When the energy of a photon or electron in the shower is above E_{LPM} the interaction distance becomes comparable to the inter-atomic spacing and collective atomic and molecular effects affect the static electric field responsible for the interaction. The result is a reduction in both total cross sections with lab energy (E) which drop like $E^{-0.5}$ above $\sim E_{LPM}$. As a result PeV - EeV photon or electron induced showers are considerably larger than at TeV energies [20]. In hadronic showers this effect is mitigated because the main source of photons is the decay of π^0's which at energies above $E_{\pi^0} = 7 \times 10^{15}$ eV (7 PeV) is suppressed due to the dominance of π^0 interaction [21].

A helpful way of getting insight into the pattern of the electric field emitted by the excess of electrons in the shower, is to think of a shower as a slit illuminated by radio-waves [22]. The diffraction pattern can be obtained as the Fourier transform of the slit, however due to the nature of the Cherenkov radiation the emission has a central peak with its maximum at the Cherenkov angle instead of in the direction perpendicular to the slit. The radiation is coherent when the wavelength is larger than the physical dimensions of the shower. In this situation the radiated electric field becomes proportional to the excess charge and the power in radiowaves scales with the square of the shower energy. This effect was predicted in the 1960's by Askary'an [23] and has been observed recently at SLAC in experiments where an intense beam of GeV photons generates a shower inside a sand target [24]. Given this interpretation it is easy to understand that the LPM effect is going to reduce the angular width of the Cherenkov peak $(\Delta\theta)$ in electromagnetic showers as shower energy increases. Following the same reasoning, the diffraction pattern from hadronic showers is expected to be less affected by the LPM, exhibiting a mild decrease of $\Delta\theta$ with energy [21].

The spectrum of the electric field also increases linearly with frequency up to a maximum frequency which is essentially determined by the lateral structure of the shower at the Cherenkov angle, the wider the shower the smaller the corresponding maximum frequency. At angles away from the Cherenkov angle, the maximum frequency depends only on the longitudinal dimension of the shower. We have performed simulations of electromagnetic showers in the lunar regolith assumed homogeneous (density $\rho \simeq 1.6$ g cm^{-3}). We have adapted the ZHS code [25] originally conceived for simulations in ice, changing the relevant parameters such as density, radiation length, atomic number and refraction index. The frequency spectrum as well as the angular distribution of the pulses emitted by a 100 TeV electromagnetic shower in the Moon are shown in Fig.1. The results of the simulations are in very good agreement with a simple scaling of the simulations from ice to the Moon [18]. With this in mind we have obtained the electric field from electromagnetic showers at the highest energies, by scaling the spectral features in ice. The same scaling has been applied to estimate the radiopulses from hadronic showers. The absolute value of the electric field has been normalized according to the amount of electromagnetic energy present in the hadronic shower, and the width of the Cherenkov peak, which is inversely proportional to the longitudinal dimension of the shower, is scaled with the radiation length and the density of the medium [18]. Simulation work is in progress in this direction and will be published elsewhere. For more details see references [20,21].

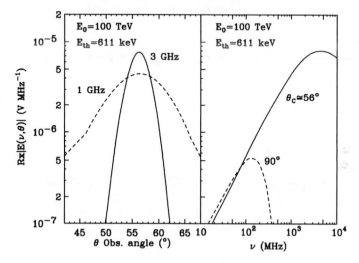

FIGURE 1. Left panel: Results of simulations of the angular behavior of the electric field emitted by a 100 TeV electromagnetic shower in the lunar regolith. θ is the observation angle with respect to the shower axis. Two frequencies are shown. Right panel: Frequency spectrum of the electric field at two observation angles.

APERTURE OF THE MOON FOR NEUTRINO AND COSMIC RAY DETECTION: ENERGY THRESHOLD

One important parameter for assessing the possibilities of using the Moon as a high energy particle detector is the energy threshold for shower observation by radiotelescopes on Earth. This can be estimated comparing the radio signal emitted by an electromagnetic shower at the Cherenkov peak and the noise expected by other processes in the radiotelescope. Here we only take into account thermal noise for which we use two representative values, namely 1σ and 6σ of the flux density given by the standard expression for a radio antenna:

$$F_{\mathrm{Noise}} = \frac{2k_{\mathrm{B}}T_{\mathrm{sys}}}{\sqrt{\Delta t \Delta \nu A_{\mathrm{eff}}}}. \tag{1}$$

Here k_B is Boltzmann's constant, T_{sys} is the noise temperature of the radio detection system, Δt and $\Delta \nu$ are respectively the duration of the pulse and the bandwidth of the detection system around a central frequency ν_C. A_{eff} is the effective area of the antenna. We have estimated the energy threshold for the NASA/JPL Goldstone Deep Space Station 14 (70 m diameter dish) radio telescope [26] with which the measurements in [17] were performed. The 1σ thermal noise level for this system is ~ 400 Jy [1] and the 6σ is 2,400 Jy. The flux density on Earth emitted by an electromagnetic shower at the Cherenkov peak is obtained integrating the frequency spectrum of the electric field $\mathbf{RE}(\nu, \mathrm{R}, \theta_{\mathbf{Cher}})$, where R is the distance to the shower, around ν_C:

$$F_{\mathrm{Signal}} = \int_{\nu_C + \frac{\Delta \nu}{2}}^{\nu_C - \frac{\Delta \nu}{2}} |\mathbf{RE}(\nu, \mathrm{R}, \theta_{\mathbf{Cher}})|^2 \, d\nu \; / \; (4\pi \mathrm{R}^2_{\mathrm{Earth} \to \mathrm{Moon}}). \tag{2}$$

$\mathrm{R}_{\mathrm{Earth} \to \mathrm{Moon}}$ is the distance from Earth to the Moon. The result of this calculation for electromagnetic showers is shown in Fig.2. Reading from the plot one can estimate the energy threshold of a electromagnetic shower to be between $E_{th} \sim 10^{19} - 10^{20}$ eV depending on the noise level, the central frequency and assuming a bandwidth of $\sim 0.1 \, \nu_C$. To translate the electromagnetic shower energy threshold

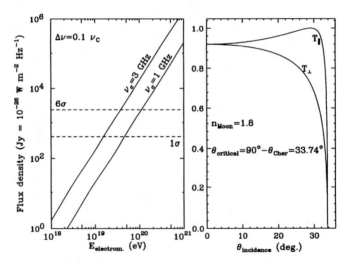

FIGURE 2. Left panel: Flux density at Earth emitted by an electromagnetic shower of energy $E_{\mathrm{electrom.}}$ for two central frequencies and a bandwidth $\Delta \nu = 0.1 \nu_C$. Also shown is the flux density of the thermal noise in NASA/JPL Goldstone DSS 14 radiotelescope. Right panel: Transmissivity (Power$_{\mathrm{transmitted}}$/Power$_{\mathrm{incident}}$) of radio waves at the Moon-vacuum interface as a function of the incidence angle with respect to the perpendicular to the local surface of the Moon.

into a neutrino or cosmic ray energy threshold, one has to keep in mind that at

---

[1] 1 Jy $= 10^{-26}$ W m^{-2} Hz^{-1}

extremely high energy the electromagnetic energy content of a hadronic shower in the lunar regolith is $\sim 90\%$ of the total shower energy, slowly increasing with it [21] and the typical value of y is ~ 0.2 [27]. This implies that the energy threshold for detecting a NC neutrino DIS interaction is roughly $E_\nu \sim 5E_{th}$ whereas it is $\sim E_{th}$ for detection of CR or for a CC ν_e interaction.

Aperture

We have studied in detail the geometry of the problem in order to calculate the aperture (effective area x solid angle) of the Moon for neutrino as well as cosmic ray detection.

For a cosmic ray or a neutrino to be detected it has to interact producing a shower within a distance to the surface of the Moon equal to the typical absorption length of radiowaves inside the lunar regolith: $\lambda_{abs}^{radio} \sim 15\,m(1\,GHz/\nu)$. The emitted radiowave then reaches the surface of the Moon without significant attenuation and has to point to the Earth after refraction in the Moon-vacuum interface. We restrict ourselves to radio emission from the directions between the Cherenkov angle and $\theta_C \pm \Delta\theta$, where most of the power is concentrated. These two considerations allowed us to obtain the potentially detectable incident neutrino and cosmic ray directions. Then we disregarded those directions for which either the cosmic rays or neutrinos are absorbed inside the Moon before reaching its surface[2]. It requires a little bit of thinking and the aid of Fig.3 to realize that this limits the potential interaction points on the surface of the Moon to a fairly narrow rim for cosmic rays and to a wider rim for neutrinos. It also restricts the observed cosmic rays to those hitting the Moon with their tracks almost parallel to its surface. We finally eliminated those directions (and energies) for which the signal, although being "geometrically able" to reach Earth, is below the noise flux density. In this respect the transmissivity properties of the Moon-vaccumm interface play a fundamental role. For a shower propagating inside the Moon parallel to the local surface, the radiation emitted at the Cherenkov angle suffers total internal reflection since the Cherenkov angle is the complementary of the total internal reflection angle. However detection is still possible because, as we mentioned above, the radiation is emitted in a cone of width $\Delta\theta$ and the transmissivity of the interface increases to almost 1 just a few degrees off the total internal reflection angle as can be seen in Fig.2. Our results on the aperture are shown in Fig.4. It is interesting to notice that for hadronic showers once the energy threshold is reached, the aperture varies slowly with shower energy, the reason for this being that the effective solid angle is not restricted by an energy decrease of $\Delta\theta$ caused by the LPM effect. This is not the case for electromagnetic showers where the decrease of the aperture with energy is induced by a corresponding reduction in $\Delta\theta$ due to the LPM. The inner panel in Fig.4 represents the visible face of the Moon and gives an idea of the size of

---

[2] The interaction length L_{int}^{ν} of a neutrino is smaller than the diameter of the Moon when $E_\nu > 10$ PeV

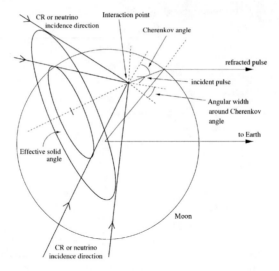

FIGURE 3. Main elements entering in the geometry of radio detection in the Moon. All the cosmic ray and neutrino directions entering in the depicted effective solid angle are "geometrically allowed", although many of them are disregarded due to absorption inside the Moon or threshold effects (see text).

the rim where cosmic ray and neutrino detection might be expected at 10^{21} eV. When combining the aperture with λ_{abs}^{radio} acceptances in excess of 1000 km^3 sr at $E_\nu \sim 10^{20}$ eV might be achieved.

EVENT RATE ESTIMATES

Cosmic ray event rates are calculated in a straightforward manner by convoluting the experimentally observed flux and the aperture of the Moon. For neutrinos the convolution involves the theoretically expected neutrino flux, the neutrino cross section and the effective volume of the detector:

$$N_{\mathrm{CR}} = \int_{E_{\mathrm{CR}}} \frac{d\Phi_{\mathrm{CR}}}{dE_{\mathrm{CR}}} S_\Omega^{\mathrm{CR}} \, dE_{\mathrm{CR}} \quad (\mathrm{yr}^{-1}); \quad \text{for cosmic rays,} \tag{3}$$

$$N_\nu = \int_{E_\nu} \frac{d\Phi_\nu}{dE_\nu} S_\Omega^\nu \frac{\lambda_{abs}^{radio}}{L_{int}^\nu} dE_\nu \quad (\mathrm{yr}^{-1}); \quad \text{for neutrinos,} \tag{4}$$

where $S_\Omega^{CR, \, \nu}$ is the aperture of the detector.

In Fig.5 we show the event rates for the conservative case where the noise level is equal to 6σ of the thermal noise. We have used some neutrino fluxes representative of the different current models and theoretical ideas. For the cosmic ray event rate, we have used a simple parameterization of the experimentally observed flux

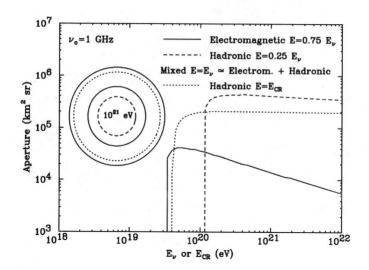

FIGURE 4. Aperture of the Moon for cosmic ray and neutrino detection. For neutrino initiated mixed showers the aperture is the sum of the solid and dashed curves (not drawn). The inset represents the surface of the Moon. The lines correspond to the same lines in the legend. The region between each line and the outermost solid line is the surface where cosmic ray or neutrino detection is expected at $E = 10^{21}$ eV.

which we have extended, perhaps conservatively, to the highest energy observed event $(3 \times 10^{20}$ eV) using a differential slope $\gamma \sim -2.7$. It is clear from the figure that cosmic ray events are going to dominate the event rate near the rim of the Moon outnumbering the neutrino events at least for the representative neutrino fluxes we have chosen. It is then advisable to point the radiotelescopes towards the center of the Moon to be able to observe neutrino events. Due to their high angular resolution, the radiotelescopes such as Goldstone DSS 14 or Parkes in Australia, are able to observe roughly 10% of the surface of the Moon at once, hence one should multiply the numbers in Fig.5 by a factor of 0.1, to get an estimate of the sensitivity of the existing instruments and the observation time required to collect a certain number of events. Our estimates indicate that Goldstone DSS 14 should roughly expect 1 cosmic ray event every 80 hours of observation.

IMPROVEMENTS AND CONCLUSIONS

There are still a few issues that should be explored. The roughness of the surface of the Moon should be included in the calculation. Its wrinkles may facilitate the detection of signals emitted at the Cherenkov angle since, in this case, the particle tracks skimming the regolith surface are in general not parallel to the local surface which creates a problem of total internal reflection. Since showers

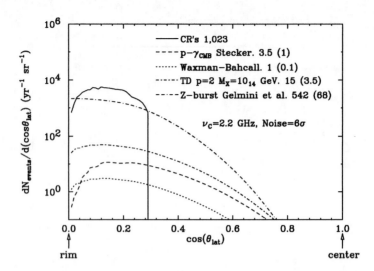

FIGURE 5. Cosmic ray and ν_e CC + ν_e NC + ν_μ CC+NC events in the Moon expected in the NASA/JPL Goldstone DSS 14 radiotelescope. The events are shown as a function of the cosine of the latitude of the points on the surface of the Moon, namely $\cos(\theta_{\rm lat}) = 1$ corresponds to the center of the Moon and $\cos(\theta_{\rm lat}) = 0$ to the outermost rim. One should understand this plot as having azimuthal symmetry around $\cos(\theta_{\rm lat}) = 1$. The numbers accompanying the legends are the total events per year. The numbers in parenthesis represent the neutrino events per year from the latitudes where cosmic ray events are not expected. For cosmic rays the number of events below 3×10^{20} eV is shown.

should be produced near the surface of the Moon, one should worry about near-field effects in the Cherenkov emission [30]. Their main consequences are a decrease of the normalization of the electric field at the Cherenkov peak accompanied by an increase of the width of the Cherenkov cone $\Delta\theta$ [31]. These two effects act in opposite directions with respect to the behavior of the event rate. A decrease in $\Delta\theta$ will increase the effective solid angle while the decrease in the normalization increases the energy threshold. There are also some issues dealing with showers produced at a distance to the surface of the Moon smaller than a radiation length so that Cherenkov radiation doesn't have enough distance to form. Variations of the index of refraction near the surface of the Moon have been measured and should also be included in the calculation of the geometry. Work is in progress to estimate the importance of all these potential issues and the results will be published elsewhere.

Although some room for improvement does exist, we have identified and correctly taken into account many of the elements entering in this difficult problem, updating previous estimates [18]. The actual potential of the Moon as a detector of high energy cosmic rays and neutrinos depends on a comprehensive study of all the elements indicated in this paper and possibly others. In any case, our results are encouraging and should trigger more theoretical work on this exciting possibility.

ACKNOWLEDGEMENTS

We thank Ricardo Vázquez for many discussions on this work. We thank P. Gorham and D. Saltzberg for an excellent and stimulating workshop. We acknowledge the suggestions made by some of the participants: D. Besson, G. Gelmini, P. Gorham, F. Halzen, K. Liewer, D.W. McKay, C. Naudet, J.P. Ralston, D. Saltzberg and I. Zheleznykh among others. The research activities of J.A-M at Bartol Research Institute are supported by the NASA grant NAG5-7009. This work was also supported in part by the European Science Foundation (Neutrino Astrophysics Network N. 86), by the CICYT (AEN99-0589-C02-02) and by Xunta de Galicia (PGIDT00PXI20615PR).

REFERENCES

1. M. Nagano, A.A. Watson, Rev. Mod. Phys. **72** (2000) 689 and refs. therein.
2. F.W. Stecker et al., Phys. Rev. Lett., **66** (1991) 2697.
3. R. Engel and T. Stanev, astro-ph/0101216.
4. F. Halzen, Phys. Rep. **333-334** (2000) 349.
5. P. Bhattacharjee, G. Sigl, Phys. Rep. **327** (2000) 109-247.
6. Design Report of the Pierre Auger Coll., Fermilab Report, Feb. 1997 and refs. therein.
7. D. Cline, F.W. Stecker, contributed to OWL/AW Workshop on Observing Ultrahigh Energy Neutrinos, Los Angeles, California, Nov. 1999, astro-ph/0003459.
8. K. Arisaka for the EUSO collaboration, see these proceedings.
9. T.K. Gaisser, F. Halzen, T. Stanev, Phys. Rep. **258** (1995) 173 and refs. therein.
10. J.G. Learned and K. Mannheim, Ann. Rev. Nucl. Sci. (2000) in press.
11. E. Zas, summary talk given at the *International Workshop on Observing UHECR from Space and Earth*, August 2000, Metepec, Puebla, Mexico, astro-ph/0011414.
12. *Procs. of the 1st International Workshop on Radio Detection of High Energy Particles*, Los Angeles, California, November 2000, to be published by AIP.
13. D. Besson et al. see this proceedings; G.M. Frichter *Proc. of the XXVI Int. Cosmic Ray Conference*, Salt Lake City, Utah, July 1999, Vol. 2, p. 467.
14. M. Chiba, see these proceedings.
15. R.D. Dagkesamansky and I.M. Zheleznykh, Pis'ma Zh. Eksp. Teor. Fiz. [JETP Letters] **50**, 233 (1989).
16. T.H. Hankins, R.D. Ekers, J.D. O'Sullivan, Mon. Not. R. Astr. Soc. **283** (1996) 1027.
17. P.W. Gorham, K.M. Liewer, C.J. Naudet. *Proc. of the XXVI Int. Cosmic Ray Conference*, Salt Lake City, Utah, July 1999, Vol. 2, p. 479, astro-ph/9906504.
18. J. Alvarez-Muñiz, E. Zas, *Proc. of the XXV Int. Cosmic Ray Conference*, Durban (South Africa) August 1997, Vol. 7, p.309.
19. L. Landau and I. Pomeranchuk, *Dokl. Akad. Nauk SSSR* **92** (1953) 535; **92** (1935) 735; A.B. Migdal, Phys. Rev. **103** (1956) 1811; Sov. Phys. JETP **5** (1957) 527.
20. J. Alvarez-Muñiz, E. Zas, Phys. Lett. B **411** (1997) 218.
21. J. Alvarez-Muñiz, E. Zas, Phys. Lett. B **434** (1998) 396.

22. J. Alvarez-Muñiz, R.A. Vázquez and E. Zas, Phys. Rev. D **61** (1999) 023001.

23. G.A. Askar'yan, Soviet Physics JETP **14, 2** (1962) 441; **48** (1965) 988.

24. D. Saltzberg, P. W. Gorham, D. Walz, C. Field, R. Iverson, A. Odian, G. Resch, P. Schoessow and D. Williams, submitted to Phys. Rev. Lett., hep-ex/0011001.

25. E. Zas, F. Halzen, T. Stanev, Phys. Rev. D **45** (1992) 362.

26. For detailed information see *http://gts.gdscc.nasa.gov/*

27. R. Gandhi, C. Quigg, M.H. Reno, I. Sarcevic, Phys. Rev. D **58** (1998) 093009.

28. G. Gelmini and A. Kusenko, Phys. Rev. Lett. **84** (2000) 1378.

29. E. Waxman and J. Bahcall, Phys. Rev. D **59** 023002 (1998) 023002.

30. R. Buniy and J.P. Ralston, astro-ph/0003408.

31. J. Alvarez-Muñiz, R.A. Vázquez and E. Zas, Phys. Rev. D **62** (2000) 063001.

Signal Characteristics from Electromagnetic Cascades in Ice

Soebur Razzaque, Surujhdeo Seunarine, David Z. Besson, and
Douglas W. McKay

Department of Physics and Astronomy
University of Kansas, Lawrence, Kansas 66045-2151

Abstract. We investigate the development of electromagnetic cascades in ice using a
GEANT Monte Carlo simulation. We examine the Cherenkov pulse that is generated
by the charge excess that develops and propagates with the shower. This study is
important for the RICE experiment at the South Pole, as well as any test beam experiment which seeks to measure coherent Cherenkov radiation from an electromagnetic
shower.

I INTRODUCTION

Ultra high energy neutrinos can travel a long distance without scattering. With
these sources of particles, we can probe the physics of standard model and beyond
at energies un-obtainable at current accelerators. An ultra high energy neutrino can
interact via a charge current interaction giving most of its energy to the secondary
electron, which can then initiate an electromagnetic cascade or shower. Askaryan
[1] predicted a negative charge imbalance in the cascade which gives rise to coherent
Cherenkov radiation at radio frequencies. The predicted flux of ultra high energy
neutrinos is small and model dependent [2–4]. An experiment to detect ultra high
energy neutrinos using radio antennas requires a large volume because of the small
flux. A dense, radio-transparent target is needed for the small shower size needed for
coherence and small signal attenuaton. Antarctic ice is suitable for this purpose.
A detailed analysis of such an experiment was done [5] which concluded that a
modest array of antennas can detect many events per year. Radio Ice Cherenkov
Experiment (RICE) at the south pole [6] is a prototype designed to detect neutrinos
with energy $\geq PeV$ using this method. A reliable Monte Carlo simulation tool
is needed to study the shower development, Cherenkov radiation, detector, and
data acquisiton system. One can also test the idea of coherent Cherenkov emission
at accelerator facilities by dumping bunches of electrons or photons in a dense
target like sand or salt or any other suitable medium [7]. Zas, Halzen and Stanev
(ZHS) [8] developed a Monte Carlo simulation to study electromagnetic showers and

CP579, *Radio Detection of High Energy Particles,* edited by D. Saltzberg and P. Gorham
© 2001 American Institute of Physics 0-7354-0018-0/01/$18.00

Cherenkov emission in ice. Buniy and Ralston [9] have also developed a method to estimate Cherenkov signal by parametrizing the cascades.

We have developed a GEANT Monte Carlo simulation primarily to study the coherent Cherenkov emmission in ice but it can easily be modified for other similar studies with different materials. GEANT [10] is a well known and widely used detector simulation package in particle physics. It allows access to all details of the simulation such as controls of various processes, definition of target and detector media, and a complete history of all events simulated. It can be used to simulate all dominant processes in $10 \ keV - 10 \ TeV$ energy range. Cross sections of electromagnetic processes are reproduced in GEANT within a few percent up to a hundred GeV. We use GEANT to simulate electromagnetic cascades inside materials from which we extract track information. We take this track information and determine the resulting radio pulse using standard electrodynamics calculations. We also calculate other shower parameters using the track and energy information from GEANT.

II SHOWER DESCRIPTION

When a high energy electron (e^-) or photon (γ) hits a material target, an electromagnetic cascade is created inside the material. *Bremsstrahlung* and *pair production* are the dominant high energy processes at the beginning of shower development. Due to bremsstrahlung, an e^- loses $1/e$ of its energy on average over a distance X_0, the *radiation length*. The secondary γ can then produce an $e^+ \ e^-$ pair. The number of particles thus grows exponentially. The e^- and e^+ lose energy due to *ionization* as they travel inside the material. After reaching a *critical energy* (E_c), when the energy loss due to bremsstrahlung is equal the energy loss due to ionization, e^-, e^+ lose their energy mostly due to ionization and the cascade eventually stops. A rough estimate of the critical energy is $E_c \approx 605/Z \ MeV$ which can be obtained by equating radiation and ionization loss formulae. Z is the atomic number of the medium.

All models of an electromagnetic cascade, including a simplified model developed by Heitler [11], and a more realistic model developed by Carlson and Oppenheimer [12], show basic features which include linear scaling of the track lengths and logarithmic scaling of the maximum position with primary energy.

The cascade is concentrated near an axis along the direction of the primary. *Multiple Coulomb scattering* is responsible for transverse spread of the cascade. A quantity called *Moliere radius* (R_M) is determined by the average angular deflection per radiation length at the critical energy (E_c) and is used to estimate the transverse spread. One can also look at the fraction of energy that escapes transverse to the shower axis [13,14] to get a good estimate of R_M. About 90% of the primary energy is contained inside a tube of radius R_M along the shower axis. A relationship between the Moliere radius, the critical energy and the radiation length is $R_M = X_0 E_s / E_c$ where $E_s \approx 21.2 \ MeV$ is the scale energy.

Low energy processes like *Compton, Moller* and *Bhabha* scatterings and *positron annihilation* build up a net charge (more e^- than e^+) in the cascade, as atomic electrons in the target medium are swept up into the forward moving shower.

III ELECTROMAGNETIC PULSE

High energy e^- and e^+ in the cascade can travel faster than the speed of light in the medium and give rise to Cherenkov radiation. The electric field due to a single charge moving uniformly from position $\vec{r}_1$ to $\vec{r}_2 = \vec{r}_1 + \vec{v}\delta t$ in *Fraunhoffer* limit, is given by

$$R\vec{E}(\omega) = \frac{\mu_r}{\sqrt{2\pi}} \left(\frac{e}{c^2} \right) e^{i\omega \frac{R}{c}} e^{i\omega(t_1 - n\vec{\beta}\cdot\vec{r}_1)} \vec{v}_T \frac{(e^{i\omega\delta t(1-\hat{n}\cdot\vec{\beta}n)} - 1)}{1 - \hat{n}\cdot\vec{\beta}n} \tag{1}$$

where μ_r is the relative permeability and n is the refractive index of the medium. The condition $1 - \hat{n}\cdot\vec{\beta}n = 0$ defines the Cherenkov angle θ_c as $\cos\theta_c = 1/n\beta$. At or very close to the Cherenkov angle or at low frequency, the equation (1) can be reduced to the form

$$R\vec{E}(\omega) = \frac{\mu_r i\omega}{\sqrt{2\pi}} \left(\frac{e}{c^2} \right) e^{i\omega \frac{R}{c}} e^{i\omega(t_1 - n\vec{\beta}\cdot\vec{r}_1)} \vec{v}_T \delta t. \tag{2}$$

See Fig. 1 for a description of various quantities in equations (1, 2).

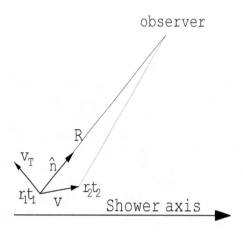

FIGURE 1. Set-up to calculate electromagnetic pulse from a single track.

IV MONTE CARLO RESULTS

Since on average, the electrons lose $1/e$ of their energy in each radiation length due to bremsstrahlung, one can fit an exponential to the energy loss data generated by the Monte Carlo and find a value for the radiation length (X_0). This will serve as an internal consistency check to ensure that we are tracking all the particles along with their energies. We generated 500 showers of 50 GeV electrons and recorded the energy loss due to bremsstrahlung. A fit to this gives a radiation length of 41.5 ± 3.2 cm. Given the molecular composition, GEANT also calculates the medium's radiation length. For ice it is 38.8 cm, which is in agreement with the value we calculated.

Table I GEANT consistency checks

Parameters	Iron	Lead	Ice
X_0 (cm) (from ref [15])	1.76	0.56	39.05
R_M (cm)	2.1	1.6	13
$E_c = X_0 E_s / R_M$ (MeV)	17.77	7.42	63.7
$E_c = 605/Z$ (MeV)	23.3	7.4	83.8

To calculate the Moliere radius (R_M) we construct an imaginary cylinder centered on the shower axis (up to the physical length of the shower). We add the energy (U), of all the tracks that leave the cylinder without re-entering. By varying the radius of the cylinder we arrive at R_M which is the radius of the cylinder when the fraction U/E is equal to 0.1, E being the initial energy of the cascade. We have checked R_M for lead, iron and ice (see table I). One can also calculate the critical energy (E_c) using two different formulae quoted earlier. Shower depth (t) is measured in units of radiation length (X_0) as $t = x/X_0$. We have simulated 30 GeV electron-induced cascades in iron. The longitudinal profile was obtained by adding the number of particles with total energy greater than 1.5 MeV crossing planes at every half radiation length perpendicular to the shower axis as shown in Fig. 2a. The number of particles agree reasonably with EGS4 simulation of the same shower [15]. The Greisen-Rossi distribution for total number of particles (electrons and positrons) is shown by the dashed line. A least squares fit to the GEANT data with Greisen-Rossi distribution gives about 90% confidence level.

We used the same method to study the longitudinal profile of showers in ice. Simulation shows the correct scaling behavior of the number of particles with the initial energy of the shower. The position of the shower maximum also shows the correct logarithmic scaling with the initial energy. The charge excess (ΔQ) is defined as $\Delta Q = \frac{N(e) - N(p)}{N(e) + N(p)}$ where $N(e)$ and $N(p)$ are the number of electrons and positrons respectively as functions of shower depth. According to our simulations, the net charge imbalance is $15\% - 18\%$ at the shower max. Note that there is no direct experimental data on this.

A comparison of 100 GeV shower (averaged over 20 showers) with 611 keV threshold from GEANT to the same from ZHS Monte Carlo (Fig. 2b) shows about a $25\% - 35\%$ discrepancy for the total number of particles at the shower max. The discrepancy between the two simulations remains the same at higher energies.

In Fig. 3, we show the charge excess in a $500\ GeV$ shower from GEANT with kinetic energy threshold $1.5\ MeV$. This is an average of 20 showers and the charge excess is broken down to different energy bins. It shows that most of the contributions to charge excess come from the energy range 1.5 to $30\ MeV$. The Cherenkov pulse from a cascade is proportional to the total track length (energy deposition) of the cascade. Our results show the correct scaling behavior of the track lengths with the initial energy (E_o). If we consider all processes to be elastic except ionization, then an upper bound for the total track length is $L = E_o / \left(\frac{dE}{dx}\right)_{ion}^{avg}$ where $\left(\frac{dE}{dx}\right)_{ion}^{avg}$ is the average ionization energy loss per unit length. To determine $\left(\frac{dE}{dx}\right)$ we generated 500 of $1\ GeV$ tracks and kept a record of the rate at which energy was lost due to ionization. Fig.4a shows the average $\left(\frac{dE}{dx}\right)$ for the 500 tracks. It can be seen that the average value matches theoretical Bethe-Bloch result reasonably well. The average ionization loss in the relativistic rise region is approximately $1.9\ MeV/cm$.

V   PULSE CALCULATION

To calculate the electric pulse from the cascade, we added up the contributions in Eq. (1) from all charged tracks. There are two assumptions we made to evaluate pulse described by Eq. (1) and Eq. (2) from GEANT track information. First, we

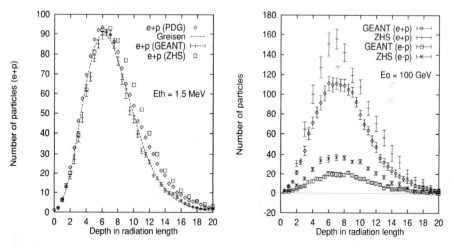

FIGURE 2. (a) The longitudinal profile of a $30\ GeV$ cascade in iron using GEANT (average of 30 showers) and comparison with the same produced by EGS4 (from particle data book) and by ZHS Monte Carlos. (b) Comparisons between the profiles of a $100\ GeV$ cascade (averaged over 20 showers each) from GEANT and from ZHS Monte Carlos. The total energy threshold is 0.611 MeV in both cases.

assume an azimuthal symmetry about the shower axis, which allows us to evaluate pulse equations in 2-dimensions. This is a good approximation as long as we have many tracks in the shower. A typical $500\,GeV$ shower has thousands of tracks and the number goes up as we increase the energy. We checked this approximation on a shower-by-shower basis and did not find any noticeable difference between the pulses evaluated in the $x - z$ plane and in the $y - z$ plane (z - is the shower axis). Second, we evaluate the times t_1 and δt in (1) and (2) assuming that the particles travel at the speed of light. This is a good approximation if the frequency is not too high as can be seen from (1). In Fig. 4b, we calculate pulses[1] from $1\,TeV$ and $500\,GeV$ showers at $1\,GHz$ and $300\,MHz$ frequencies. The pulse height at the Cherenkov peak ($\approx 55.8^{o}$) scales with the energy of the shower, and the Gaussian half width is inversely proportional to the frequency which is analogous to a slit diffraction pattern.

VI CONCLUSION

We find distinctive, coherent, Cherenkov radio frequency emission from simulated electromagnetic cascades in ice with energies in the $100\,GeV$ to $1\,TeV$ range.

---

[1] we use different normalization here for the field, which requiring that one multiply (1) and (2) by $2\sqrt{2\pi}$

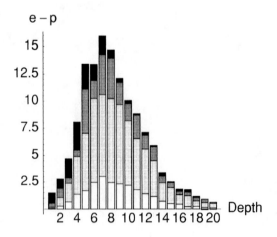

FIGURE 3. Excess charge plot for a $500\,GeV$ cascade (averaged over 20 showers) in ice using GEANT. The excess charge number is shown in energy bins (from bottom to top) 1.5-5 MeV, 5-30 MeV, 30-100 MeV and above 100 MeV.

The multi-purpose, detector simulation package GEANT provides a suitable modelleing of electromagnetic shower details for this purpose, with the flexibility to include hadronic cascades in future studies. This work also serves as an independent treatment of the problem, which has been studied previously [8]. The GEANT generated showers are qualitatively in agreement with those in [8]. The showers differ in several details; however our far-field calculation of the pulse from a shower gives the same result up to a GHz frequency as that of [8] when both are applied to the same shower simulation. This has been checked explicitly by calculating the field using our field code and the track informtions from ZHS code with the direct field output from ZHS code for the same shower. The differences between a GEANT and a ZHS shower of initial energy 100 GeV is summarized in Table II. The origin of differences between the two Monte Carlos is under study.

Our direct calculation of Moliere radius, critical energy and shower population as a function of shower depth using GEANT all agree within a few percent with measurements or with published EGS4 simulations for iron and lead [15]. The extension to ice is straightforward and we expect the same level of reliability here.

Our results agree with the theoretically expected [16] linear scaling of number of particles at shower maximum and logarithimic scaling of depth at shower maximum with energy. These are not straightforward results, since the theoretical predictions are based on inclusion of bremsstrahlung and pair production only, while the physics at the shower maximum is not dominated by these processes. It includes important contributions from Compton, Bhabha and Moller processes as well, which give rise

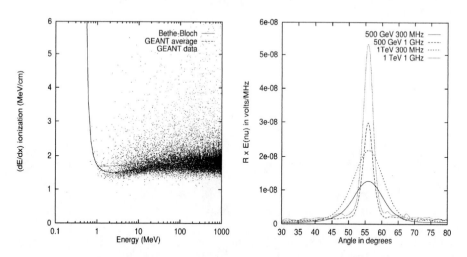

FIGURE 4. (a) $(dE/dx)_{ionization}$ calculated from GEANT output. The average is over 500 electron tracks with 0.1 MeV kinetic energy threshold. We also show theoretical Bethe-Bloch curve. (b) Electromagnetic pulses from cascades in ice generated by GEANT. Pulse height scales linearly and pulse width scales inversely with primary energy. Fraunhoffer limit has been taken to calculate field.

to charge imbalance and, consequently, coherent radio Cherenkov field emission.

Table II Comparisons between GEANT and ZHS Monte Carlos
(averaged over 20 showers each)

Quantity	GEANT	ZHS
Primary Energy (GeV)	100	100
Total Energy Threshold (MeV)	0.611	0.611
Total absolute track length $(meter)$	398.8 ± 4.7	642
Total projected $(e + p)$ track length $(meter)$ (*sum of electron and positron track lengths projected along the shower axis*)	374.4 ± 4.3	518.7
Total projected $(e - p)$ track length $(meter)$ (*difference of electron and positron track lengths projected along the shower axis*)	70.0 ± 8.4	131.2
Position of the shower max. (radiation length)	7	7
Number of particles $(e + p)$ at shower max.	111 ± 26	155 ± 45
Excess electrons $(e - p)$ at shower max.	20 ± 11	37 ± 14
Fractional charge excess at the shower max	$\sim 18\%$	$\sim 24\%$
Cherenkov peak at $1\,GHz$ $(Volts/MHz)$	7.46×10^{-9}	1.08×10^{-8}

Acknowledgement Thanks to J. Alvarez-Muñiz, T. Bolton, R. Buniy, G. Frichter, F. Halzen, J. Ralston, D. Seckel and E. Zas for help and advice at various stages. This work is supported in part by the NSF, the DOE, the University of Kansas General Research Fund, the RESEARCH CORPORATION and the facilities of the Kansas Institute for Theoretical and Computational Science.

REFERENCES

1. G. A. Askar'yan. *Sov. Phys.*, JETP14:441, 1962.

2. S. W. Stecker and M. H. Solomon. *Space Sci. Rev.*, 75:341, 1996.

3. F. Halzen and E. Zas. *astrophys. J.*, 488:669, 1997.

4. J. Rachen, R. Protheroe, and K. Mannheim. astro-ph/99080301, 1999.

5. G. Frichter, D. W. McKay, and J. P. Ralston. *Phys. Rev.*, D53:1684, 1996.

6. RICE Collaboration. ICRC proceedings, 1997.

7. D. Saltzberg et al. hep-ex/0011001, 2000.

8. E. Zas, F. Halzen, and T. Stanev. *Phys. Rev.*, D45:362, 1992.

9. R. V. Buniy and J. P. Ralston. astro-ph/0003408, 2000.

10. CERN. *GEANT-Detector Description and Simulation Tool*, 1994.

11. W. Heitler. *The Quantum Theory of Radiation*. Clarendon, 1954.

12. J. F. Carlson and J. R. Oppenheimer. *Phys. Rev.*, 51:220, 1937.

13. W. R. Nelson et al. *Phys. Rev.*, 149:201, 1966.

14. G. Bathow et al. *Nucl. Phys.*, B20:592, 1970.

15. Particle Data Group. *Phys. Rev.*, D54:136, 1996.

16. B. Rossi. *High Energy Particles*. Prentice Hall, 1952.

Concepts in the Coherence of Radio Cherenkov Emission from Ultra-High Energy Electromagnetic and Hadronic Showers

Roman V. Buniy and John P. Ralston

Department of Physics and Astronomy, University of Kansas, Lawrence, KS 66045

Abstract. Radio Cherenkov emission underlies detection of high energy particles via a signal growing like the particle energy-squared. Many physical scales are involved in calculating the radio emission. We find that simple scaling laws exist which determine much of the physics in different regimes. The classic issues of Fraunhofer and Fresnel zones play a crucial role. The calculated electric fields decompose into a "factorized" product of a form factor, characterizing a moving charge distribution, multiplying a general integral depending on the charge evolution. The resulting expressions can be evaluated explicitly in terms of a few parameters obtainable from shower codes, while retaining exact constraints of analyticity.

I   A MULTIPLE-SCALE PROBLEM

The most effective known method of detecting ultra-high energy particles, via coherent radio emission [1], exploits a wide range of physical phenomena. The process starts with a primary cosmic ray interaction at (say) 100 TeV or higher energy, leading to electromagnetic and hadronic showers. Shower evolution proceeds over many decades of energy and involves complicated strong interaction and electromagnetic physics. A charge imbalance develops in the shower, creating a net current which radiates in the radio regime, until finally the shower and radio emission die away. High energy physics is needed for the primary interaction, which might come from a neutrino in the Standard Model [2] or some other scenario such as Large Extra Dimensions [3]. Relativistic quantum electrodynamics and strong interaction physics predicts the shower evolution [4], atomic physics and low-energy strong interactions are needed for the charge imbalance, and classical electrodynamics is needed for the radio emission. Coherence of the radio emission involves multiple length scales: these include a distance scale for the shower to evolve, a scale length over which the shower is near maximum, the distance from the shower to the radio detector, the transverse and longitudinal spatial extension

CP579, *Radio Detection of High Energy Particles,* edited by D. Saltzberg and P. Gorham

of the moving charge packet, and of course the wavelength of the radio emission.

Despite this complexity, Askaryan's original style of estimating the radio emission [1] works remarkably well. Assume that about 10% of the shower particles contribute to the charge excess, and use the "classic" Cherenkov formulas for power radiated—more on this below. Cut off the radio emission above a frequency scale set by the length of the moving charge packet, which is about 10 cm in ice. Then analytic calculations of km-scale radio-frequency fields [5,6] are surprisingly close to the subsequent outputs of immense Monte Carlos [7].

Yet estimating the last several dB of radio power correctly has proven to be amazingly complicated by any method. Accurate calculations are needed for the purposes of quantifying the RICE project [9], for comparison with test-beam experiments [10–12], and for planning future projects. Among the milestones along the way was the important paper of Zas, Stanev and Halzen [7] (ZHS), which predicted a broad oscillating "diffraction pattern" in the angular distribution of the radiation due to the finite length of the evolving shower. This feature does not seem to have been appreciated before. The ZHS Monte Carlo also predicted a charge excess exceeding 20% in some cases, and a frequency dependence for the electric field that is *linear in* ω up to the coherence cutoff. The conceptual development of RICE [13] accepted the Monte Carlo over the previous analytic estimates, while Price [19] helpfully compared radio and optical methods for the Antarctic ice medium. We intensified our own studies to support RICE starting about 1996-1997. It became clear to us that there was more than one way to do the electrodynamics, yielding different answers, apparently in contradiction to the exact formulas cited in the ZHS text. This created a few puzzles which we resolved, so that by now there is general agreement on certain conceptual issues to be briefly described here. Our longer paper [8] gives a great deal of detail, while the recent papers of Alvarez-Muñiz and Zas can also be recommended, and appear to be in accord.

II THE COHERENCE ZONE

Consider a charge moving on an arbitrary straight line not pointing toward the observation point. Let $R(t)$ be the instantaneous distance from the charge to the observation point. Information propagating at speed c_m will arrive simultaneously from a segment of the track if $\partial R/\partial t = c_m$. This is the Cherenkov condition: $\partial R/\partial t = v\cos\theta = c_m$ for velocity $\vec{v}$ oriented at angle θ relative to the direction $\hat{R}$.

Uniform motion of the charge produces acceleration of $R(t)$. In a time interval Δt near the Cherenkov point, this produces an extra radial displacement of order $\Delta R = 1/2(\partial^2 R/\partial t^2)(\Delta t)^2$. Coherence of modes of wavelength λ is then maintained only over a finite region of $\Delta R < \lambda$. Since $\partial^2 R/\partial t^2 = v^2\sin^2\theta/R$, we solve to find the condition $\Delta t_{coh} < \sqrt{R\lambda}/(v\sin\theta)$. Equivalently, there is a finite spatial coherence region Δz_{coh} for given R, given wavenumber $k = 2\pi/\lambda$, namely

$$\Delta z_{coh} < \sqrt{R/(k\sin^2\theta)}$$

over which the "sonic boom" of radiation is built coherently. Since $\Delta z_{coh} \sim \sqrt{R}$, the coherence zone grows to infinite size as $R \to \infty$: but this limit cannot be taken prematurely. If one is at *finite distance* (as in, say test-beam experiments or RICE for $R < 300 - 1000m$) then the "bite" of radiating source is *smaller* than the same source observed at infinite radius.

As we show momentarily, knowledge of the coherence zone plus basic dimensional analysis brings a good understanding of the radio emission problem. The coherence zone argument was original with us–we discovered it by working a powerful saddle-point calculation [8] backwards for pedagogical purposes–but we subsequently learned that it is closely related to a classic construction. Our coherence zone is the finite R, Cherenkov-regime generalization of the *coherence length* invoked in ultra-high energy arguments for the LPM effect. None of the Landau, Pomeranchuk, or Migdal papers can be recommended to see this: but we eventually found Galitsky and Guerevich [21], and also benefited from the review by Boris Bolotovsky [22] at the meeting.

III   FRAUNHOFER VERSUS FRESNEL

Textbooks routinely (and with little explanation [14]) replace the Helmholz Green function for radiation of frequency ω by the expansion

$$\frac{e^{i\omega|\vec{x} - \vec{x}'|}}{|\vec{x} - \vec{x}'|} \to \frac{e^{i\omega R - i\vec{k} \cdot \vec{x}'}}{R},$$

where $|\vec{x}| = R$ and $\vec{k} = \omega \hat{x}$; we use $c = 1$ except for emphasis. The expansion of the exponent is valid for $R \to \infty$ and sets up the *Fraunhofer approximation*, as it is termed in optics. Because this approximation is so common, it leads to the notion that it is general, along with the misconception that radiation fields always fall like $1/R$. An early Allen [16] review, as well successors [7,20] cited certain Feynman formulas to calculate the Fraunhofer limit of the radio emission from showers in air and ice, respectively.

The difficulty is that the series expansion of the exponent may fail spectacularly when a source, such as a cosmic ray shower, has finite spatial extent. The evolving shower has a length scale a over which it is near its maximum, and radiating copiously. This length scale is known as the "longitudinal spread" in cosmic ray physics [4]. Then the dimensionless ratio

$$\eta = (a/\Delta z_{coh})^2 = \frac{ka^2}{R} \sin^2 \theta$$

controls how the limit $R \to \infty$ is taken. (Indeed the term "large R" is undefined until the limit parameter η is specified.) The Fraunhofer approximation needs $\eta << 1$ to be reliable. In the RICE experiment one optimistically has $a \sim$ meters, $\omega \sim 10cm^{-1}$, and $R \sim 10^3 m$, so $\eta < 1$ holds. Pushing the parameters to $R \sim$

$100 - 300m$, $\omega \sim 300MHz$, and a much larger a, makes $\eta \gg 1$ possible. Corrections "of fractional order unity" occur in the "Fresnel zone", namely the far-field region where Fraunhofer fails, and these are of some concern. The problem may be severe in the super-high energy regime where LPM corrections may affect the shower length [25].

IV "FACTORIZATION"

Under broad conditions, we find [8] that the radio emission amplitudes allow a "factorization" into a product of a form factor describing the spatial extent of the net charge, and a term representing the evolution of the shower as a whole. This is a non-trivial result, reminiscent of the scale-separation between hard and soft physics of the parton model, and due to a separation between the scale of the form factor and the scale of shower evolution. While we cannot expand around $\vec{x}' = 0$ (the Fraunhofer approximation), the conditions of the problem do permit an expansion around $\vec{\rho}' = 0$ (the transverse distance from the shower axis), namely for $R(z') = [(z - z')^2 + \rho^2]^{1/2}$, so that

$$|\vec{x} - \vec{x}'| = [(z - z')^2 + (\vec{\rho} - \vec{\rho}')^2]^{1/2},$$

$$= R(z') - \frac{\vec{\rho} \cdot \vec{\rho}'}{R^2} + \mathcal{O}(\frac{\rho'^2}{R^2}).$$

For typical values in applications, the second term is ~ 10 times smaller than the first, and the third is $\sim 10^3$ smaller than the second. For the expansion of the exponent, the third term does not contribute if $k\Delta\rho'^2/R \ll 1$, that is $\omega \ll 250GHz$.

The shower evolution generates a net charge $en(z)$ whose evolution in z must be taken into account. The vector potential $\vec{A}_\omega$ in (medium-adapted) Lorentz gauge then takes the form [8]

$$cR\vec{A}_\omega = \vec{v} \int dz' \, n(z') \exp\left[i(\frac{\omega}{v}z' + kR)\right]$$

$$\times \int \int dt' \, d^2\rho \, \exp\left\{-i[\frac{\omega}{v}(z' - vt') + \vec{q} \cdot \vec{\rho}']\right\} f(z' - vt', \vec{\rho}').$$

We have shifted the t' integral which produces the translational phase in the z' integral. This gives the factorization:

$$\vec{A}_\omega \approx F(\vec{q})\vec{A}_\omega^{FF}(\eta),$$

where

$$F(\vec{q}) = \int d^3x' \, e^{-i\vec{q} \cdot \vec{x}'} f(\vec{x}'),$$

and $\vec{A}_\omega^{FF}$ is a certain integral [8] that does not depend on the details of the charge distribution. $F(\vec{q})$ is the form-factor of the charge distribution, which happens to be defined, just as in the rest of physics, in terms of the Fourier transform of the snapshot of the distribution. From our definitions $F(0) = 1$. Dependence on $\vec{q}$ may be observable in Giant Air Showers [16,17], where charge separation might cause an azimuthal asymmetry about the shower axis. As a consequence of separating out the form-factor, the integrations have become effectively one-dimensional.

V SCALING LAWS

The disagreement of the Monte Carlo frequency dependence with the classic Cherenkov formulas can be traced to use of the Fraunhofer approximation. The Fraunhofer field assumes complete coherence over the region of radiation, namely $\Delta z_{coh} >> a$. It follows that $E_{Fraunhofer} \sim Qa/R$, where we inserted the $1/R$ of the approximation explicitly. But the dimensions of the electric field $E(\omega) \sim \int dt \epsilon^{i\omega t} E(t) \sim m^1$, where m is a mass scale. (The basic field dimension follows from the action $\sim \int d^4x E^2$ being dimensionless in our units.) There being no other dimensionful scale besides ω, dimensional analysis requires

$$E_{Fraunhofer}(\omega) \sim Q\omega a/R.$$

This is a very general rule, and reliable up to the frequency where another scale (the coherence cutoff) can modify the argument.

Compare the opposite extreme of classic Cherenkov radiation, by which we mean the Frank-Tamm formulas, as cited by textbooks [14,15] and universally applied in particle physics. In this regime the track of the moving charge is assumed infinite in length. Now one cannot take $a \to \infty$, $R \to \infty$ and calculate this in the Fraunhofer approximation at all, leading to great confusion. We may obtain the correct scaling law by replacing $a \to \Delta z_{coh} \sim \sqrt{R/\omega}$, taking into account the finite-sized coherence zone cited earlier. Immediately the Cherenkov electric field scales like

$$E_{Cherenkov}(\omega) \sim Q\Delta\omega z_{coh}/R \sim Q\sqrt{\frac{\omega}{R}}.$$

It is rather unfamiliar to have a radiation field falling like $1/\sqrt{R}$. Yet this behavior also follows from geometry and dimensional analysis: observe that the *radiated power* has *cylindrical symmetry* concentric with the track. In this "cylindrical region", the power $P(\omega) \sim 2\pi R |E_{Cherenkov}(\omega)|^2 \sim const$, which requires $|E_{Cherenkov}(\omega)| \sim 1/\sqrt{R}$ after all. The proportionality of the field to $\sqrt{\omega}$ is also the origin of the well-know rule that the classic Cherenkov power spectrum (not the electric field!) is linear in ω.

The accomplishment of our longer paper [8] was to understand this and to develop a systematic procedure for *interpolating between* the different regimes from cylindrical waves (classic Cherenkov behavior) to spherical waves (Fraunhofer approximation). We exploited the factorization cited earlier, and introduced an elegant saddle-point expansion which automatically takes into account the important coherence regions. Taking the Fraunhofer limit of the interpolating formulas, we still needed a parameter for the shower longitudinal spread a. Running the ZHS code many times and fitting the output of a $1TeV$ shower with a particle threshold of $611MeV$ gives $a \sim 1.5m$. It is rather beautiful that the complicated numerical calculations can be summarized [8] by the parameter a, the charge at shower maximum, and a single length scale for the form factor. After extracting these quantities, we were then able to extend the prediction throughout the entire Fresnel-zone, revealing an intricate complexity in the interplay of frequency, distance, and angular distribution. The analytic formulas are very accurate, and the differences of parameters between different Monte Carlo calculations currently dominates the discussion: see the report by Razzaque at the meeting [26].

VI   ANALYTIC STRUCTURE IN THE COMPLEX PLANE

Experiments on Cherenkov radiation generally see a sharp pulse in an oscilloscope (say) and are done in the time domain. Meanwhile, calculations are done in the frequency domain. The relation between the two is just a Fourier transform, but again there are subtleties. Causality requires [14,15] that any singularities of the electric field $E(\omega)$ lie in the lower half of the complex plane. Among the different methods to parametrize the field, at least two ($E(\omega) \sim 1/(1 + (\omega/\omega_0)^2$; $E(\omega) \sim (\omega/\omega_0)^\delta$) violate the basic requirements of causality. This has been another source of confusion, because the fitting of a Monte Carlo output appears to be a straightforward numerical task, where any naive function that fits "well enough" in a region was previously thought ok. However the time-domain pulses, which are the observable trigger of experiments, depend quite sensitively on proper analytic structure. Wrong-analytic structure lacks the causal front-edge that experiments look for. The explanation is that the Fourier transform is highly *non-local*, so that the correct time-structure demands attention to the niceties of the complex plane.

We resolved this by choosing a set of analytic functions to describe the form factor. A Gaussian for the form-factor works just fine [8]. The saddlepoint calculation also generates proper analytic structure, so that its time-domain dependence is physical: we found there is an effect distinguishing the arrival of the "front" and "back" ends of the evolving shower.

VII   NUMERICAL STUDIES

We compared [8] the saddle-point approximation to direct numerical integration for numerous regions of frequency, angle, radius R, and shower energy for radio emission in circumstances of interest to RICE. In all cases the agreement was extremely good, at the level of 1% or less relative error. Our group has been in the process of evaluating Fresnel-zone changes in the effective volumes of radio receivers for neutrino detection. We believe that the next step will be the resolution of discrepancies among the different Monte Carlos that generate the net current: after the current is known, the electrodynamics seems under good control for the entire far-field regime relevant to RICE and similar experiments.

Acknowledgments: This work was supported in part by the Department of Energy, the University of Kansas General Research Fund, the K*STAR programs and the *Kansas Institute for Theoretical and Computational Science*. We thank Jaime Alvarez-Muñiz, Enrique Zas, Doug McKay, Soeb Razzaque and Suruj Seunarine. We especially thank Enrique for writing and sharing the ZHS code, Jaime for many patient discussions, and Soeb and Suruj for generously sharing their results from shower codes.

REFERENCES

1. G. A. Askaryan, *Zh. Eksp. Teor. Fiz.* **41**,616 (1961) [Soviet Physics JETP **14**, 441 (1962)].
2. Yu. Andreev, V. Berezinsky, and A. Smirnov, *Phys. Lett. B* **84**, 247 (1979); D. W. McKay and J. P. Ralston, *Phys. Lett. B* **167** 103 , (1986); M. H. Reno, and C. Quigg, *Phys. Rev. D* **37**, 657 (1987); G. M. Frichter, D. M. McKay, and J. P. Ralston, *Phys. Rev. Lett.* **74**, 1508-1511 (1995); J. P. Ralston, D. W. McKay, and G. M. Frichter, in *International Workshop on Neutrino Telescopes*, (Venice, Italy, 1996), edited by Baldo-Ceolin, M.; astro-ph/9606007; R. Gandhi, C. Quigg, M. H. Reno, and I. Sarcevic, *Phys. Rev. D* **58** (1998) 093009.
3. P. Jain, D. McKay, S. Panda, and J. P. Ralston, *Phys. Lett.* **B484**, 267 (2000),hep-ph/0001031; in *Intersections of Particle and Nuclear Physics* (Quebec City, 2000), edited by Z. Parseh and W. Marciano (AIP, in press); M. Kachelriess and M. Pluemacher, astro-ph/0005309.
4. B. Rossi, *High Energy Particles* (Prentice Hall, New York, 1952).
5. M. A. Markov and I. M. Zheleznykh, *Nucl. Instr. Meth. Phys. Res.* **A248**, 242 (1986); A. L. Provorov and I. M. Zheleznykh, Astropart. Phys. **4** , 55 (1995).
6. J. P. Ralston and D. W. McKay, in *Arkansas Gamma Ray and Neutrino Workshop: 1989*, edited by G. B. Yodh, D. C. Wold, and W. R. Kropp, Nuc. Phys. B (Proc. Suppl.) **14A** 356 (1990); reprinted in *Proceedings of the Bartol Workshop on Cosmic Rays and Astrophysics at the South Pole* (AIP, NY 1990) Proceedings #198.
7. E. Zas, F. Halzen, and T. Stanev. *Phys. Rev D* **45**, 362 (1992).
8. R. V. Buniy and J. P. Ralston, astro-ph/0003408.
9. D. Besson, these proceedings; G. Frichter, for the RICE collaboration, in *26th International Cosmic Ray Conference* (Salt lake City 1999), edited by D. Kieda and B. L. Dingus (IUPAP 1999).
10. P. Gorham *et al. Phys. Rev.*, **E62**, 8590 (2000).
11. D. Saltzberg, *et al.*, hep-ex/0004007.
12. D. Saltzberg, these proceedings.
13. J. P. Ralston, D. W. McKay, G. M. Frichter. *Phys. Rev.* **D53**, 1684 (1996).
14. See, e.g. J. D. Jackson, *Classical Electrodynamics*, 2nd Edition (Wiley, 1975).

15. Francis Low, *Classical Field Theory: Electrodynamics and Gravitation*, (Addison Wesley, 1999).
16. H. R. Allan, in *Progress in Elementary Particles and Cosmic Ray Physics*, (North-Holland Publishing Company, Amsterdam, 1971).
17. J. L. Rosner and J. F. Wilkerson, EFI-97-10 (1997), hep-ex/9702008; J. L. Rosner, DOE-ER-40561-221 (1995), hep-ex/9508011, and these Proceedings.
18. T. K. Gaisser, *Cosmic Rays and Particle Physics*, (Cambridge University Press, 1990).
19. P. B. Price, Astropart. Phys. **5**, 43 (1996), astro-ph/9510119.
20. J. Alvarez-Muñiz, and E. Zas, Phys. Lett. **B 411**, 218 (1997), and these Proceedings.
21. V. M. Galitsky and I.I. Gurevich, *Nu. Cim.* 32, 396 (1960); see also E. L. Feinberg and I. Ia. Pomeranchuk, *Suppl. Nu. Cim.* **9**, 652 (1956).
22. B. Bolotovsky, this meeting.
23. J. Alvarez-Muñiz, R. A. Vázquez and E. Zas, Phys. Rev. **D 61**, 023001 (1999), and these Proceedings
24. T. K. Gaisser, *Cosmic Rays and Particle Physics*, (Cambridge University Press, 1990).
25. J. P. Ralston and D. W. McKay, D. W., in *High Energy Gamma Ray Astronomy* (Ann Arbor 1990), *APS Conference Proceedings No. 220* (AIP, NY, 1991) J. Matthews, Ed; J. Alvarez-Muñiz and E. Zas, Phys. Lett. **B411**, 218 (1997); *ibid* **B434**, 396 (1998), and these Proceedings.
26. S. Razzaque, D. Besson, and D. W. McKay, (in preparation), and in these Proceedings.

CURRENT AND FUTURE EXPERIMENTS USING RADIO EMISSION IN DENSE MEDIA

Status of the RICE Experiment

Dave Besson, KU Physics Dept., Lawrence, KS 66045-2151

Abstract. The RICE experiment (Radio Ice Čerenkov Experiment) at the South Pole, currently one of the projects included in the overall AMANDA effort, seeks to measure ultra-high energy electron neutrinos ($E_\nu > 10^{15}$) by detection of the long-wavelength Čerenkov Radiation (CR) signal resulting from neutrino-induced showers: $\nu_e + N \rightarrow e + N'$. We present upper limits on the UHE ν_e flux based on analysis of August, 2000 data.

RICE concept and current experimental configuration: The RICE[1] experiment is similar to the primary AMANDA effort in that it seeks to measure ultra-high energy (UHE) neutrinos by detection of Čerenkov radiation resulting from the collision of a UHE neutrino with the target ice.

The RICE concept is illustrated in Figure 1, depicting a Čerenkov cone fit to a set of "struck" dipole receivers in one particular event, along with the extracted neutrino direction. High amplitude/early time receivers are indicated by the shading/icon-size scheme. In the actual array geometry, dipole receivers are spread over a 200 m × 200 m × 200 m cube beneath and around the Martin A. Pomerantz Observatory (MAPO), approximately 1 km. from the geographic South Pole.

Whereas AMANDA is optimized for detection of penetrating muons compact resulting from $\nu_\mu + N \rightarrow \mu + N'$, RICE seeks to measure electromagnetic cascades initiated by electrons or positrons:

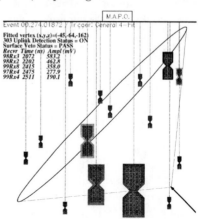

FIGURE 1.: Sample RICE event.

[1] RICE includes: Dave Seckel and Glenn Spiczak (Bartol Research Institute, Newark DE), Jenni Adams (U. of Canterbury, Christchurch, NZ), George M. Frichter (Florida State University, Tallahassee FL) Chris Allen, Alice Bean, Dave Besson, DJ Box, Roman Buniy, Eben Copple, Ryan Dyer, Doug McKay, John Ralston, Soeb Razzaque, Surujhedo Seunarine, Dave Schmitz (University of Kansas, Lawrence KS), and Ilya Kravchenko (M.I.T., Cambridge, MA), and is supported by The Research Corporation, NSF EPSCoR, the University of Kansas, The Marsden Foundation, and the NSF OPP and EPP Divisions.

CP579, *Radio Detection of High Energy Particles,* edited by D. Saltzberg and P. Gorham
© 2001 American Institute of Physics 0-7354-0018-0/01/$18.00

$\nu_e + N \to e + N'$. As the cascade develops, atomic electrons in the target medium are swept into the forward-moving shower, resulting in a net charge on the shower front of $Q_{tot} \sim 0.25 E_s (GeV) e$.

Such cascades produce broadband Čerenkov radiation – for wavelengths larger than the transverse dimensions of the shower front ($2r_{Moliere}$, or ≈ 10 cm in ice), the emitting region increasingly approximates a line charge of magnitude Q_{tot}. At these (RF) wavelengths, the net Čerenkov radiation produced by the shower front can therefore be considered coherent. By contrast, the resultant short wavelength (optical frequency) Čerenkov power is obtained from an incoherent sum over the electric field vectors associated with each short-wavelength Čerenkov wavefront. The experimental sensitivity in the RF regime is further enhanced by the high transparency of cold polar ice ($\lambda_{atten}^{1\ GHz} \sim 1$ km). One analysis [1] indicates that a radio receiver offers greater sensitive volume per module than photomultiplier tubes for 1 PeV$< E_{\nu_e}$.

The RICE experiment presently includes 18 half-wave dipole receivers underice, deployed over the period 1996-1999. Receivers have good reception characteristics over the range 0-1 GHz. Also deployed are three large TEM surface horn antennas which are used as a veto of surface-generated noise. Whenever one of the surface horns registers a signal that exceeds threshold, data-taking is inhibited for the subsequent 3.5 μsec. The signal from each underice antenna is boosted by a 36-dB in-ice amplifier, then carried by coaxial cable to the surface observatory, where the signal is filtered (suppressing noise below 200 MHz), re-amplified (either 52 dB or 60 dB), and split into two copies. One copy is fed into a CAMAC crate; after initial discrimination (using a LeCroy 3412E discriminator), the signal is routed into a NIM crate where the trigger logic resides. The other signal copy is passed to the input of an HP54542 digital oscilloscope where waveform information is buffered and recorded.

The status of the current array deployment is summarized in Figure 2. Indicated in the figure are the AMANDA holes (1-19, drilled using the hot-water technique) containing RICE receivers, and also the four holes drilled specifically for RICE in 1998-99 (using a mechanical drill) containing RICE-only equipment. All channels (receiver "Rx" or transmitter "Tx") are indicated by a 5-character alphanumeric string corresponding to the year of deployment, the type of dipole antenna deployed, and a numerical identifier. Also indicated on the Figure is the MAPO building which houses the RICE (and AMANDA) surface electronics. The AMANDA array is located approximately 600 m (AMANDA-A) to 1400 m (AMANDA-B) below the RICE array in the ice; the SPASE experiment is located approximately at $x \sim -400$m on the Figure.

A valid event trigger is defined when any one of three criteria is satisfied in a time window of 1 μs: i) ≥ 4 underice antennas register signals exceeding threshold, ii) $\geq$ one underice antenna registers a signal above threshold in coincidence with a high-amplitude SPASE event, iii) $\geq$ one underice antenna registers a signal above threshold in coincidence with a 30-fold PMT AMANDA trigger. There are two

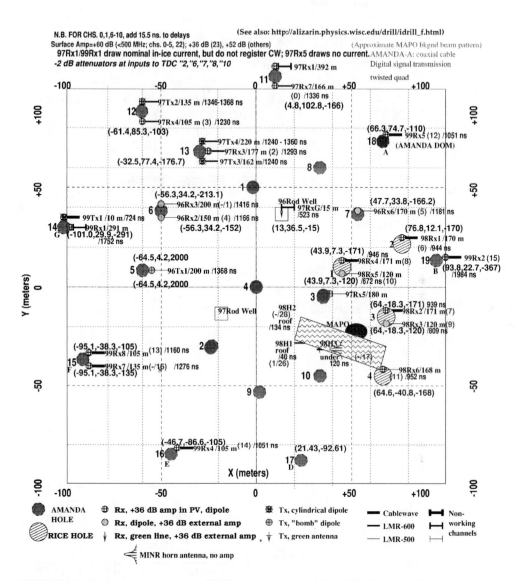

FIGURE 2. Present geometry of the RICE array, relative to AMANDA hole 4. "Tx" designate transmitters; "Rx" desginate receivers. (x,y,z) of receivers and transmitters are shown.

primary ways that surface-generated background transient events can be vetoed – either: a) one of the surface horn antennas registers a signal in coincidence with any of the above three trigger criteria being satisfied, or b) the timing sequence of hits in the underice antennas is determined to be consistent (in software) with the sequence expected from surface generated backgrounds. If any of the above trigger criteria are satisfied and there is no veto signal, the time of each hit above threshold (as recorded by a LeCroy 3377 TDC), and also an 8 μs buffer of data stored in an HP54542 digital oscilloscope at 1 GSa/s (for each channel) is written to disk. Each event is also given a GPS time stamp for synchronization with other South Pole (and more global) experiments. Data is subsequently ftp'ed back to the US for further analysis. Data runs are initiated remotely from the home institutions in the States during the time when there is a satellite connection to the RICE DAQ PC. Raw trigger rates (before veto) are typically 30 Hz; typical data-taking rates after the veto are 0.01 – 0.1 Hz. Experimental sensitivity and data-taking is limited by: a) the time required to write information from the digital scopes to disk (~10 sec/event), b) the time required to perform the surface-background veto in software (~10 msec/event), and c) our inability to take data at those times when the South Pole satellite uplink (broadcast at 303 MHz) is active, due to the high amplitude of the uplink signal. When this uplink is not active, the discriminator thresholds correspond to typical livetimes of ~80%.

A typical surface-generated background event is shown in Figure 3. In the event display, each of the traces corresponds to one of the under-ice antennas. Vertical bars indicate the TDC time recorded for each particular channel; the rise time and timing resolution of each receiver as determined by waveform information is evident from the Figure. This particular event shows receiver -to- receiver timing delays characteristic of surface-generated background noise. As can be seen in the event, leading edge resolution is typically 5-10 nsec; the ring time for each antenna is typically 50 nsec.

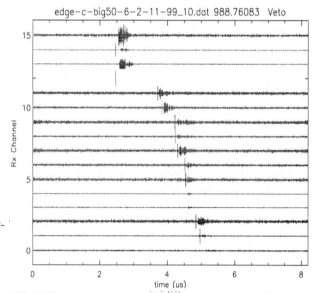

FIGURE 3.: Typical data event recorded by the RICE array.

Expected Detector Performance and Calibration Status:

Expected Angular Resolution: The angular resolution expected for a real neutrino-induced event is determined from Monte Carlo simulations. A ν_e-induced cascade is simulated directly below the radio array, and the Cerenkov cone propagated vertically upwards to the array. Antennas in the simulation are located at 40 m, 80 m, 120 m and 160 m from the event vertex, to correspond to the close-packing of the current array geometry. For the best case in which the Čerenkov cone directly intercepts these four antennas, the corresonding hit times are calculated for each struck antenna, then these times are smeared by 25 ns, to match our experimental timing resolution. Using these smeared times, the

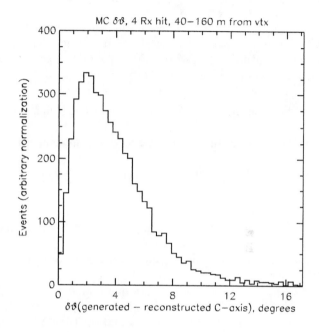

FIGURE 4.: Deviation between reconstructed incident neutrino direction vs. calculated incident neutrino direction, using Monte Carlo simulations.

cone orientation is numerically calculated and then compared with the true input Čerenkov cone orientation. The deviation in cone orientation corresponds directly to the deviation between the true incident neutrino direction and the calculated incident neutrino direction. The results of the simuation are shown in Figure 4.

Typical deviations are of order 4-5 degrees. The simulation is optimistic in that the struck antennas are all considered to be directly on the Čerenkov cone; in reality, the finite width of the Čerenkov cone will smear each hit by an additional degree or so. However, this smearing can be compensated for, to a large extent, by: a) use of amplitude information to deduce how far off the Čerenkov cone angle of 57 degrees a struck antenna is actually located, and b) inclusion of more struck antennas to provide redundancy for the event vertex/neutrino momentum vector calculation.

Timing calibration: Event and source reconstruction is performed by a χ^2 minimization process derived from a knowledge of the relative delays δt_{ij} in the signal arrival times for a given event, in each underice receiver. Given four or more hit antennas (i.e., 3 δt_{ij} values), an event vertex and source direction can be determined numerically. Contributions to δt_{ij} come from several sources, including differences in signal propagation velocity in the ice due to variations in the dielectric constant with depth, differences in signal propagation speed within the different analog cables being used, differences in cable lengths, and channel-to-channel jitter within the surface readout and data acquisition electronics.

Buried transmitters are used to calibrate the channel-to-channel timing delays. A 1 ns duration pulse (comparable to the duration of an expected neutrino-induced event) is sent to one of five buried transmitters, which subsequently broadcasts the signal to the receiver array. An event vertex is reconstructed exclusively from the measured time delays; comparison with the actual transmitter location allows a calculation of the timing residual χ^2 for each channel $(\frac{\delta t_{ij}^{measured} - \delta t_{ij}^{expected}}{\sigma_t})^2$. An iterative procedure is used to calibrate out the observed channel-to-channel timing delays and minimize the timing residuals for an ensemble of transmitter events. Typical corrections, per channel, are of order 25 ns. A calibration event is shown in Figure 5. Here the χ^2 probability is calculated over a 3 dimensional grid (grid size = 20m) of possible vertex points 500 meters on a side. Three orthogonal 2 dimensional slices each passing though the minimum value of χ^2 are shown. The shading indicate probabilities ranging from near zero (black) to near unity. Small squares indicate the positions of all RICE receivers and the the white diamond indicates the known position of the RICE transmitter 96Tx1 which emitted the reconstructed pulse. In this case the difference between known and reconstructed event vertex is about 13m. The spatial residual, at this very early level of calibration, is typically 10 meters, which is consistent with the intrinsic resolution of our calibration technique (with more CPU, this can be improved somewhat).

Amplitude Calibration: Amplitude calibration is achieved as follows: a 1 milliwatt (0 dBm) continuous wave (CW) signal is broadcast through the transmit port of an HP8713C network analyzer. The NWA scans through the frequency range 0→1000 MHz in 200 bins, producing a 0 dBm CW signal in each frequency bin. The signal is transmitted through ∼1000 feet of coaxial cable to one of the five under-ice dipole transmitting antennas. The transmitters subsequently broadcast to the underice receiver array, and the return signal power (after amplification, passing through cable and being fed back into the return port of the network analyzer) is then measured. Then, using laboratory measurements made at KU of: a) the effective area of the dipole antennas, as a function of frequency, and b) the dipole Tx/Rx efficiency as a function of polar angle and azimuth, c) cable losses and dispersion effects for the various cables used in the RICE experiment, d) the gain of the two stages of amplification as determined from RICE data acquired *in situ* by normalizing to thermal noise, and also correcting for $1/r^2$ spherical spreading of the signal, one can model the receiver array and calculate the expected signal

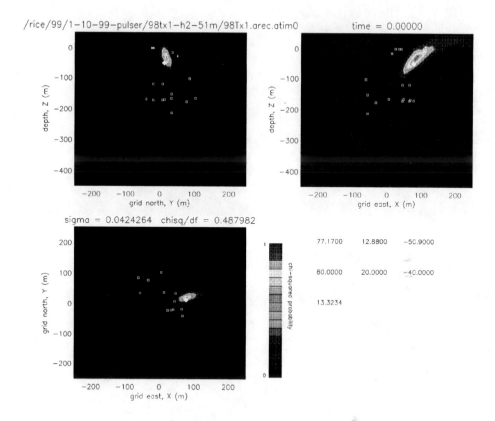

FIGURE 5. Reconstruction of Tx pulse, based exclusively on measured receiver times, shown in the three possible planar slices. Diamond indicates known location of transmitter; shading intensity code indicates χ^2 of reconstruction. Open white squares indicate receivers.

strength returning to the return port of the network analyzer. This can then be directly compared with actual measurement.

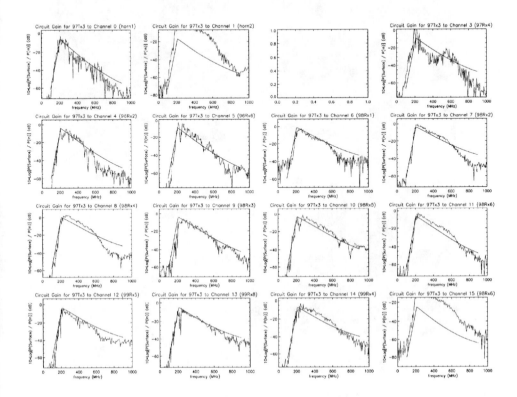

FIGURE 6.: Comparison of expected (solid curve) vs. measured (histogram) Tx→Rx signal strength for one transmitter broadcasting to 16 receivers.

Such a comparison, as a function of frequency, is shown in Figure 6 for one transmitter (97Tx3). With the exception of channels "1 and "15, for which the antenna response is anomalously high, within the "analysis" frequency band of our experiment (200 MHz - 500 MHz), our level of uncertainty in the total circuit power is ±6 dB. Note that, for the case of a neutrino event, this uncertainty can be considered to be conservative, as it folds in effects of both the receiver and the transmitter.

Figure 7 shows the distribution of sources recorded in 15 days of livetime accumulated during August, 2000. The majority of our triggers to date are consistent with either surface-generated noise backgrounds (Z≥0, as shown in the Figure) or random thermal noise hits; no clear neutrino candidates have been observed. Unfortunately, the signal rate expected for other physics processes, given the very high energy experimental threshold and the current configuration, is almost negligible. (For our current array, we expect to observe less than 0.1 hard bremsstrahlung from a UHE muon, e.g., in one year.) Given the known discriminator thresholds, and folding in known experimental effects, an effective volume, as a function of incident neutrino energy, can be calculated (Figure 8).

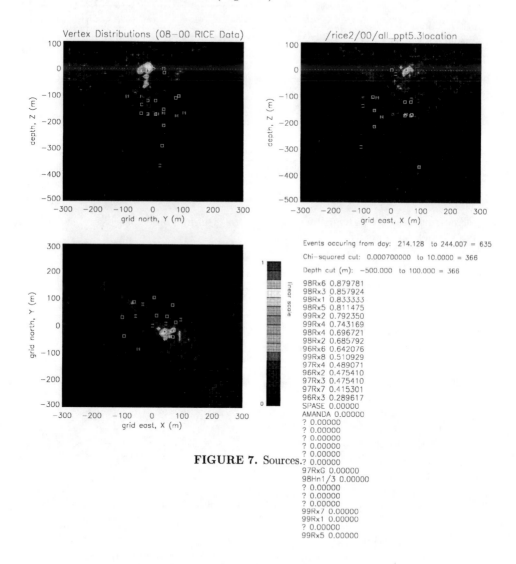

FIGURE 7. Sources.

In the figure, the uppermost black curve corresponds to the effective volume in which only one receiver must be hit, with a discriminator threshold corresponding to thermal noise. The closely spaced curves correspond to more realistic cases requiring n=4, 5, 6... antennas hit, with our current experimental thresholds. The up-

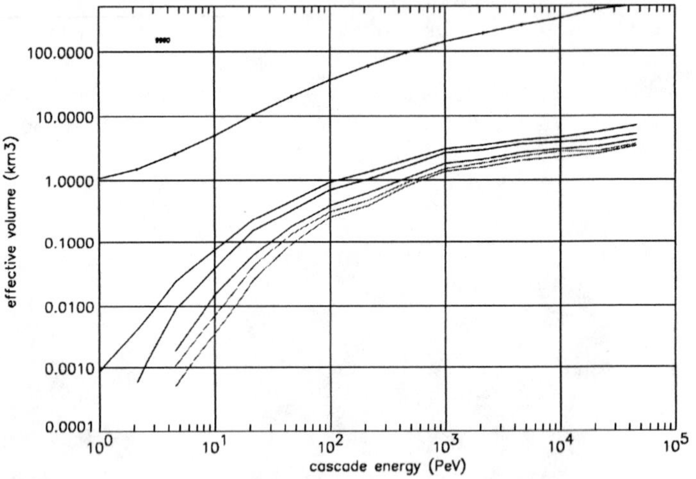

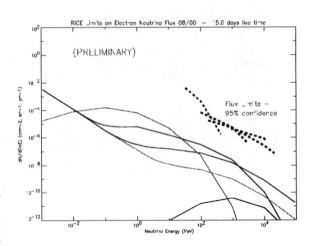

FIGURE 8.: Calculated effective volume, as a function of energy. The curves shown correspond to various assumptions about the experimental thresholds.

permost of the closely-spaced curves (n=4) corresponds to our current experimental configuration.

Knowing the total livetime, the effective volume, and based on two candidates which have not yet been eliminated, one can then calculate an upper limit on the incident ν_e flux, as a function of incident energy (Figure 9). In the figure, the predictions, shown as solid lines, correspond to (in order of descending flux, as measured at 1 PeV): Stecker & Salamon [2], Protheroe [3], Mannheim (B) [5], Mannheim (A) [4], Yoshida et al. GZK-model [6]. Dashed lines represent our current upper limits for the Stecker & Salamon, Protheroe, and Mannheim (B) models (from left to right), based on analysis of the August, 2000 data only.

FIGURE 9.: Neutrino flux model predictions (solid) and corresponding RICE calculated upper limits (dashed), as a function of neutrino energy.

Alternately, we can plot our upper limits as a function of the spectral index

γ of the assumed incident neutrino energy spectrum: $dN/dE_\gamma \sim E^{-\gamma}$. These are shown in Figure 10, for a range of γ values. Note that the dashed lines cover the region for which our sensitivity is maximal given the assumed incident neutrino spectrum; the integral under the dashed line corresponds to our upper limit. As the spectral index increases, our sensitivity is pushed to lower and lower energy regions. Again, we have used only August, 2000 data for this analysis. Data taken from Jan., 1999 - Dec., 2000 has yet to be fully analyzed.

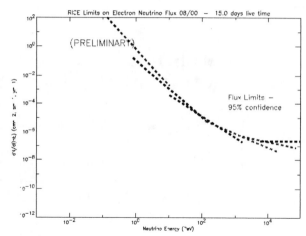

FIGURE 10.: RICE calculated upper limits on neutrino flux assuming $dN/dE \sim E^{-\gamma}$ input spectrum, with the spectral index varying from $\gamma=3.5$ to $\gamma=1.0$, in increments of $\Delta\gamma=0.5$ (from left to right).

Future Plans Further improvement in the current array can be achieved by: a) deployment of additional receiver modules, b) improvement of signal transmission technology (optical fiber, e.g., over coaxial cable), and c) stronger rejection of surface backgrounds.

Acknowledgments The RICE experiment is supported by the National Science Foundation, the University of Kansas, and the Research Corporation. Special thanks go to the AMANDA Collaboration, Matt Peters of the U. of Texas SOAR group, and the winterovers at South Pole Station.

REFERENCES

1. P.B. Price (UC, Berkeley). **astro-ph/9510119**, 1995, published in Astropart. Phys. 5, 43, 1996.
2. F. W. Stecker and M. H. Salamon, **astro-ph/9501064**
3. A. P. Szabo and R. J. Protheroe, Astropart. Phys. **2**, 375 (1994); R. J. Protheroe and A. P. Szabo, Phys. Rev. Lett. **69**, 2285 (1992).
4. K. Mannheim, astro-ph/9306005, Phys. Rev. D**48** (1993) 2408-2414
5. K. Mannheim, R. Protheroe, and J. Rachen, astro-ph/9812398, and Phys. Rev. D**63** (2001) 023003
6. S. Yoshida, G. Sigl, and S. Lee, hep-ph/9808324, Phys.Rev.Lett. 81 (1998) 5505-5508, and G. Sigl, S. Lee, P. Bhattacharjee, and S. Yoshida, hep-ph/9809242, Phys.Rev. D59 (1999) 043504

A Search for Lunar Radio Čerenkov Emission from High-energy Neutrinos

T. H. Hankins[1], R. D. Ekers[2], and J. D. O'Sullivan[3]

[1] *New Mexico Institute of Mining and Technology, Socorro, NM 87801*
[2] *Australia National Telescope Facility, Epping, NSW 2121 Australia*
[3] *News Limited, Surrey Hills, NSW, 2010, Australia*

Abstract.
   We describe a 10-h search for the radio signature of high-energy neutrinos interacting with the near surface of the Moon. The upper limit of the amplitude of the 1-ns duration expected pulses in the 1.5-GHz frequency range, dispersed by one passage through the Earth's ionosphere, is about 400 Jy. We describe here some details of the observations and analysis.

INTRODUCTION

The search for the highest energy particles in the Universe requires enormous detector volumes, partly because their flux is extremely low, and because, in the case of neutrinos, their interaction cross-sections are very small. To increase the detection probability, then, large volumes of matter are required. To this end detectors have been placed in the clear, deep ocean, and in the South polar ice cap to detect Čerenkov radiation from the particle cascade resulting from a neutrino interaction in water. Dagkesamansky and Zheleznykh [1] suggested that if a neutrino has an interaction in the near surface of the Moon, then the radio-frequency Čerenkov emission could escape through ≈10-m of lunar rock, and might be detectable by a sensitive radio telescope on the Earth. The Moon offers a surface area of $10^7 \, \text{km}^2$ and a volume of $10^{14} \, \text{m}^3$, which is more than 1000 times larger than the Antarctic ice detector volume.

Our experience with high-time resolution pulsar studies and the removal of dispersion effects from pulsar signals due to propagation through the interstellar medium led us to design an experiment to test the hypothesis of Dagkesamansky and Zheleznykh, that $E > 10^{20} \, \text{eV}$ neutrino interaction cascades in the near 10-m of the Moon produce 1-ns radio pulses from Čerenkov emission. Our strategy was to look for short, dispersed pulses from the Moon, using existing equipment wherever possible. We have described our results in a previous paper [2]. Here we expand on some of the observational details of our experiment.

CP579, *Radio Detection of High Energy Particles,* edited by D. Saltzberg and P. Gorham
© 2001 American Institute of Physics 0-7354-0018-0/01/$18.00

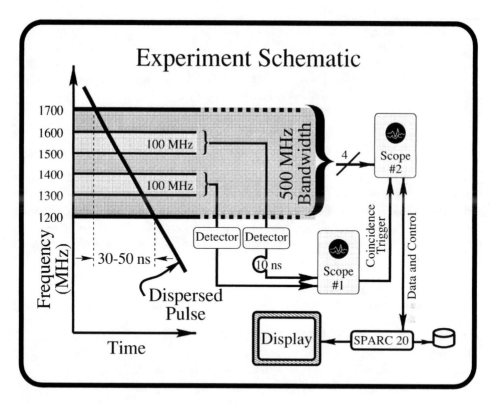

FIGURE 1. Schematic diagram of the data acquisition system.

EXPERIMENT

We used the 64-m telescope at Parkes, NSW, Australia, which had recently been outfitted with a wide-band, dual-polarization receiver constructed for the SETI program. The receiver covers 1.2 to 1.9 GHz; we used a relatively interference-free bandwidth of 500 MHz centered on 1425 MHz to maximize the bandwidth of the signals and to allow 2-ns time resolution of the signals. When pointed at the Moon the receiver's noise temperature is dominated by the 250 K thermal noise from the Moon, yielding a limiting sensitivity (one standard deviation fluctuation of the radiometer square-law detector output with time-bandwidth product equal to 1) of about 450 Jy. The telescope beam-width is 13 arcmin (FWHM), so when the telescope was pointed at the Moon's center we had somewhat reduced sensitivity near the limb.

A schematic diagram of the experimental setup is shown in Figure 1. The received signal was amplified and mixed to a center frequency of 475 MHz; in Figure 1 the signal is shown spanning 1200–1700 MHz. The full 500 MHz from each of two polarizations was fed to two sets of quadrature mixers. These mixers split each polarization into two parts which were then multiplied by $\cos(2\pi f_m t)$ and $\sin(2\pi f_m t)$ respectively, where $f_m = 475\,\text{MHz}$. The mixer outputs are then low-pass filtered to form what we call the in-phase and quadrature parts of each polarization. The mixer oscillator signals at 475 MHz and the mixer output pairs were carefully checked for orthogonality. The four resulting 0 to 250 MHz signals were simultaneously sampled by the LeCroy 9354L oscilloscope.

Triggering

To form trigger channels the full bandwidth was also passed to two 100-MHz wide filters whose center frequencies were separated by 200 MHz. The filter outputs were square-law detected using special detectors with output rise-times <10 ns. We expected the dispersed Čerenkov pulse to be smeared by dispersion across the bandpass by 30 to 50 ns by one pass through the ionosphere. It should therefore reach the upper frequency detector channel 10 to 20 ns before the lower detector channel. By inserting an extra (10-foot) length of coax cable we delayed the upper-frequency channel signal relative to the lower-frequency channel by the expected dispersion delay. We selected the LeCroy 9354L trigger mode which requires two-channel coincidence *and* a pulse width within the (settable) limits 7.5 to 20 ns. The coincidence trigger allows us to minimize the false-alarm probability from random radiometer noise and terrestrial interference. Figure 2 shows a typical detection event. When coincidence was detected, the oscilloscope *output* a trigger signal that was used to "grab" the wide-band, quadrature signals sampled by a separate LeCroy 9354L oscilloscope. The second oscilloscope was sampling continuously at 500 MS/s to a circular memory; it uses the coincidence trigger as a "stop" command, so that data before and after the trigger can be stored. We saved $1\,\mu s$ data (500 2-ns samples per channel) centered on the trigger epoch. The data from the second oscilloscope were then immediately transferred to a Sun Sparc 20 computer, where they were archived and analyzed at a number of trial dispersion values and displayed in quasi-real time.

Detection Criteria

To be considered a real lunar event, we required three criteria to be met.

1. A dedispersed event must be coincident with the trigger time; otherwise it must be considered a fortuitous realignment of noise by the dedispersion filtering process.

2. Since Čerenkov radiation is linearly polarized, and our receiver channels are orthogonally circularly polarized, an event must appear simultaneously in both dedispersed channels.

3. The dedispersion reconstruction must produce an impulse with the correct dispersion value.

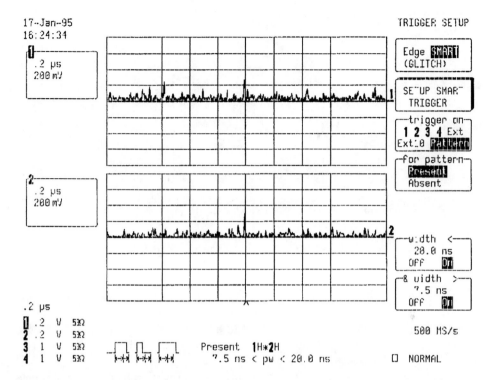

.2 µs
200 mV

.2 µs
200 mV

.2 µs

0	.2	V	5Ω
2	.2	V	5Ω
3	1	V	5Ω
4	1	V	5Ω

Edge **SMART**
(GLITCH)

SE-UP SMART
TRIGGER

─trigger on─
1 2 3 4 Ext
Ext-0 **Pattern**

─for pattern─
Present
Absent

─width <──
20.0 ns
OFF **ON**

─8 width >──
7.5 ns
OFF **ON**

500 MS/s

Present 1H∗2H
7.5 ns < pw < 20.0 ns   □ NORMAL

FIGURE 2. Trigger oscilloscope display of a typical event. The trigger point is set to the center of the traces as indicated by the caret symbol below the lower trace. The upper trace shows the square-law detected signal in the 1300–1400 MHz band, and the lower trace shows the (delayed) signal from the 1500–1600 MHz band. The total trace time shown is 2 µs, and the sample rate is 500 MS/s. The coincidence trigger threshold levels were set to 8 times the *rms* detected signal level.

Ionospheric Dispersion

Since our receiver passband is narrow relative to its center frequency, we may represent the electric field, s(t), of the lunar radio event in narrow-band signal or complex-envelope notation. (The frequency limits set by the receiver bandwidth are implicit in the following discussion. To keep the presentation simple, the effects of quadrature mixing the signal to a center frequency of zero are omitted. See [3] for details.)

$$s_{\text{emitted}}(t) = \Re\left\{a(t)e^{i\omega t}\right\},$$

where $a(t)$ is the narrow-band, complex envelope of the signal, and we assume that the bandwidth of $s(t)$ is centered around ω. Propagation through ionosphere is equivalent to passing the signal through a filter:

$$s_{\text{received}}(t) = s_{\text{emitted}}(t) * h(t)$$

where $h(t) =$ impulse response of the ionosphere and $*$ denotes convolution. We can write this in the frequency domain using $S(\omega) \equiv$ Fourier transform $\{s(t)\}$. Then

$$S_{\text{received}}(\omega) = S_{\text{emitted}}(\omega) \cdot H(\omega).$$

$H(\omega) = e^{ik(\omega)z}$ is (formally) the Transfer Function of the ionosphere of thickness z. But $k(\omega) = n\omega/c$ and $n =$ refractive index of the ionosphere. For a cold plasma $n^2 = 1 - \omega_p^2/\omega^2$, where $\omega_p = \sqrt{N_e e/m_e \epsilon_0}$ is the plasma frequency of the ionosphere. N_e, e, and m_e are the electron density, charge, and mass. Then

$$H(\omega) = \exp\left(i\left(1 - \frac{N_e e}{m_e \epsilon_0}\right)\frac{\omega}{c} \cdot z\right)$$

The factor $N_e z$ is usually written as dispersion measure or *slant total electron content* of the propagation path, $DM = STEC = \int_z N_e \, dz$. Since $H(\omega)$ is frequency dependent, lower radio frequencies have a slower group velocity and hence arrive at Earth later than high frequencies. So the lunar impulsive events received in a finite bandwidth are smeared out by the time it takes the signal to "sweep" from the higher to lower frequency edge of the receiver passband.

The dispersed signal then provides a unique signature for discriminating between events of lunar (dispersed by passage through the ionosphere) and terrestrial origin.

Dispersion Removal Processing

The dispersion distortion suffered by the lunar radio events is linear and can easily be removed by inverse filtering. To recover the signals as they were emitted we need only to multiply the Fourier transform of the received signal by $H^{-1}(\omega) = e^{-ik(\omega)z}$:

$$S_{\text{received}}(\omega)H^{-1}(\omega) = S_{\text{emitted}}(\omega) \cdot H(\omega)H^{-1}(\omega)$$
$$= S_{\text{emitted}}(\omega) \xleftrightarrow{\text{FT}} s_{\text{emitted}}(t)$$

In Figure 3 is shown the result of a simulation of the dedispersion process. A noise-free dispersed pulse with a STEC of 0.56×10^{14} el/cm^3 has been dedispersed using inverse filters with 100 trial values of STEC from 0 to $2(0.56 \times 10^{14})$ el/cm^3. The time axis is plotted so that the nominal arrival time of the dispersed pulse at the center of the receiver passband is at time 0.0. The magnitude of the dispersed pulse with zero dispersion removed appears as a low rectangle centered at $t = 0.0$. Each subsequent trace has been shifted one sample to the right for clarity. The effectiveness of the dedispersion processing gain is readily apparent. When the filter

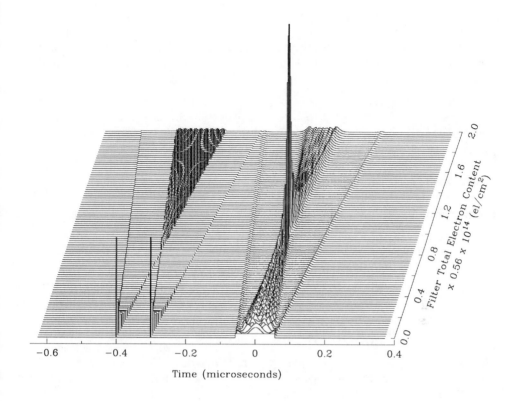

FIGURE 3. Simulation of dispersed and local interfering pulses. The simulated lunar event was dispersed by passage through an ionosphere whose slant total electron content was 0.56×10^{14} el/cm³.

dispersion matches the test signal dispersion the original impulse is reconstructed and nearly all of its power appears in a single time sample. The *dispersed* pulse duration is τ_d, which corresponds to $N_d = \tau_d \Delta \nu$ complex time samples when the receiver bandpass $\Delta \nu$ is properly sampled at the Nyquist rate. The processing gain, g_d, which we define as the ratio of the dedispersed pulse amplitude to the average value of the dispersed signal while it is in the receiver passband, is $g_d \approx N_d$.

When noise is present, and in particular, impulsive terrestrial interference, the effective processing gain is somewhat reduced. Since the dispersion removal filter is an "all-pass" filter, none of the interfering frequency components in the passband are rejected; they are instead rearranged in time. In Figure 3 we show the effect of two noise-free impulses arriving at $t = -0.4$ and $-0.3\,\mu s$. Since they are dispersed by the inverse filter rather than passage through the ionosphere, they become spread out in time as they are filtered with larger STEC values. When several local interference pulses arrive nearby in time, as shown in Figure 3, their filter responses may overlap and superpose imitating a real lunar event. Therefore we scrutinized each captured event to be sure that superposition of local events did not masquerade as lunar events.

For the simulation calculations in Figure 3 a perfectly rectangular receiver pass-band was used. This causes the sidelobes or edge effects seen in the figure. Simulations with Hanning or Hamming weighting of the spectra show considerably reduced edge effects at a cost of about a factor of 2 in processing gain. The actual receiver pass-bands were established by ordinary analog Butterworth filters which do not have sharp band-edge cutoffs, and hence do not produce edge effects as pronounced as in the simulation. Nonetheless, we analyzed all of the data with and without Hanning spectral weighting.

To maximize the processing gain, then, we should use the lowest practical frequency (assuming that the event spectrum is flat), the widest possible bandwidth and the largest STEC. For our experiment the receiver center frequency was chosen to coincide with the radioastronomy protected $\lambda 21$ (1.4 GHz) band, and the bandwidth was determined by the maximum sampling rate available. Despite the fact that the ionospheric STEC is 3–5 times larger in the daytime than at night, we chose night time to avoid interference. Our expected processing gain was therefore about $g_d \approx 10$. We used values of STEC more appropriate for the daytime ionosphere in the simulation of Figure 3, where $g_d \approx 50$. Note that observing at low elevation angles can also increase the actual STEC, and thereby increase the ionospheric signature on lunar events.

For our quasi-realtime display and subsequent re-analysis we plotted the dedispersed signals in a format similar to Figure 4, but with fewer (20) trial values of STEC, and we plotted the two polarizations in different colors and with opposite signs.

In Figure 4 we show a typical event processed with our experimental parameters. The STEC used for the dedispersion processing was obtained from the Australian Ionospheric Prediction Service for the epochs of our observations. The central impulse meets criterion 1, but it does not appear in both polarizations. Since its width does not change with filter STEC, it must be an impulse with a narrower bandwidth than our receiver, and therefore does not meet criterion 3. Furthermore, we see a second weak event about 180 ns after the trigger event. We found this to be typical of the signature of local automobile ignition noise. Therefore we reject this event.

We recorded about 700 events in 10 hours over the course of three nights. At least 30% of these were easily attributable to ignition noise (when ignition interference was clearly present, we disabled the triggering system). We set the detector thresholds at a level that should have produced several hundred random triggers; careful inspection of the data off-line did not reveal any events which we consider significant.

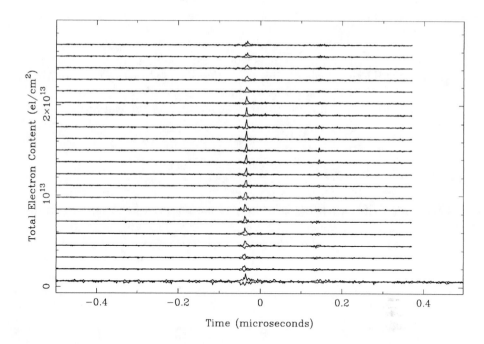

FIGURE 4. A typical event is shown, dedispersed using 20 trial values of STEC from 0 to 2 times the predicted value of 0.14×10^{14} el/cm^3, and plotted with the full 2-ns time resolution. The unfiltered, but square-law detected data are shown as the bottom trace. One circular polarization is plotted upwards, the other downwards.

CONCLUSIONS

Radio detection of lunar Čerenkov events is feasible, but it requires careful rejection of impulsive terrestrial interference. There are several ways to improve the "false-alarm" response of our experiment. Foremost is to use spatial receiver diversity and insist upon temporal coincidence of events that could only come from the Moon's direction. Since the sample rates required to capture the expected pulses are so high, it is only practical to arrange for coincidence tests in real time. Then the communications link bandwidth between spaced receivers must be wide to preserve the time resolution. We feel that the dispersion signature of the ionosphere is essential for identifying any received pulses as extra-terrestrial. Our two-channel coincidence triggering scheme was crude, and could certainly be improved by the addition of more detector/delay triggering channels on both polarizations. In principle one could tailor an antenna or array beam so that it would "illuminate" only the limb of the Moon where events are more likely, but this would require new techniques. Finally, frequency diversity could be employed to take additional advantage of the ionospheric dispersion delay. However, finding adequately wide, interference-free radio spectrum below 1 GHz would be difficult anywhere this side of the Moon!

ACKNOWLEDGMENTS

We thank Mal Sinclair and George Graves for building the fast detectors. We would like to acknowledge Mark Lotter, Scientific Devices, Sydney, Australia, for the generous loan of the second LeCroy 9354L oscilloscope, the National Science Foundation for support through grants AST-9315285 and AST-9618408, and the Australia Telescope National Facility for sabbatical year support of THH. The Parkes Radio Telescope is operated as part of the ATNF, a division of CSIRO. We thank Project Phoenix of SETI Institute for use of the wide-band receiver, and Jiping Wu, Australian Ionospheric Prediction Service for STEC. THH thanks Peter Gorham and David Saltzberg for their kind invitation to RADHEP-2000.

REFERENCES

1. Dagkesamansky R. D., Zheleznykh I. M., 1992, in Nagano, M. and Takahara, F., eds, Astrophysical Aspects of the Most Energetic Cosmic Rays, 1992, World Scientific, Singapore, p. 373
2. Hankins T. H., Ekers R. D., and O'Sullivan J. D. 1996, MNRAS, 283, 1027
3. Hankins T. H., Rickett B. J., 1975, Meth. Comp. Phys., 14, 55

New Limits on a Diffuse Flux of ≥ 100 EeV Cosmic Neutrinos

P. W. Gorham[*], K. M. Liewer[*], C. J. Naudet[*], D. P. Saltzberg[†], and D. Williams[†]

[*]*Jet Propulsion Laboratory, California Institute of Technology*
4800 Oak Grove Drive, Pasadena, CA, 91109
[†]*Dept. of Physics & Astronomy*
UCLA, Los Angeles, CA

Abstract. We report on results from about 30 hours of livetime with the Goldstone Lunar Ultra-high energy neutrino Experiment (GLUE). The experiment searches for ∼ 10 ns microwave pulses from the lunar regolith, appearing in coincidence at two large radio telescopes separated by about 20 km and linked by optical fiber. The pulses can arise from subsurface electromagnetic cascades induced by interactions of up-coming ∼ 100 EeV neutrinos in the lunar regolith. A new triggering method implemented after the first 12 hours of livetime has significantly reduced the terrestrial interference background, and we now operate at the thermal noise level. No strong candidates are yet seen. We report on limits implied by this non-detection, based on new Monte Carlo estimates of the efficiency. We also report on preliminary analysis of smaller pulses, where some indications of non-statistical excess may be present.

I INTRODUCTION

Recent accelerator results [1,2] have confirmed the 1962 prediction of Askaryan [3,4] that cascades in dense media should produce strong coherent pulses of microwave Cherenkov radiation. These confirmations strengthen the motivation to use this effect to search for cascades induced by predicted diffuse backgrounds of high energy neutrinos. At neutrino energies of about 100 EeV (1 EeV = 10^{18} eV), cascades in the upper 10 m of the radio-transparent lunar regolith result in pulses that are detectable by large radio telescopes at earth [5,21]. One prior exploratory experiment has been reported, using the Parkes 64 m telescope [22].

At frequencies above 2 GHz, ionospheric delay smearing is unimportant, and the signal should appear as highly linearly-polarized, band-limited electromagnetic impulses, well above the thermal noise levels. However, since there are many terrestrial anthropic sources of impulsive radio emission, the primary problem in detecting such pulses is eliminating sensitivity to such interference.

CP579, *Radio Detection of High Energy Particles,* edited by D. Saltzberg and P. Gorham
© 2001 American Institute of Physics 0-7354-0018-0/01/$18.00

Since 1999 we have been conducting a series of experiments to establish techniques to measure such pulses, using the JPL/NASA Deep Space Network antennas at Goldstone Tracking Facility near Barstow, California. We employ the 70 m and 34 m telecommunication antennas (designated DSS14 and DSS13 respectively) in a coincidence-type system to solve the problem of terrestrial interference, and this approach has proven very effective. Since mid-2000, the project has moved into a new status as an ongoing experiment, and receives more regularly scheduled observations, subject to the constraints imposed by the spacecraft telecommunications priorities of the Goldstone facility.

Although the total livetime accumulated in such an experiment is a relatively small fraction of what is possible with a dedicated system, the volume of material to which we are sensitive, a significant fraction of the Moon's surface to ~ 10 m depth, is enormous, exceeding 100,000 km^3 at the highest energies. The resulting sensitivity is enough to begin to constrain some models for diffuse neutrino backgrounds at energies near and beyond 10^{20} eV. We report on the status of the experiment, and astrophysical constraints imposed by limits from about 30 hours of livetime. We are also improving our understanding of the emission geometry and detection sensitivity through simulations, and are describe initial results in extending our sensitivity through small-event analysis.

II   DESCRIPTION OF EXPERIMENT

The lunar regolith is an aggregate layer of fine particles and small rocks, thought to be the accumulated ejecta of meteor impacts with the lunar surface. It consists mostly of silicates and related minerals, with meteoritic iron and titanium compounds at an average level of several per cent, and traces of meteoritic carbon. It has a typical depth range of 10 to 20 m in the maria and valleys, but may be hundreds of m deep in portions of the highlands [10]. It has a mean dielectric constant of $\epsilon \simeq 3$ and a density of $\rho \simeq 1.7$, both increasing slowly with depth. Measured values for the loss tangent vary widely depending on iron and titanium content, but a mean value at high frequencies is $\tan \delta \simeq 0.003$, implying a field attenuation length at 2 GHz of $(\alpha)^{-1} = 9$ m [11].

A   Emission geometry & Signal Characteristics

In Fig. 1 we illustrate the signal emission geometry. At 100 EeV the interaction length L_{int} of an electron or muon neutrino for the dominant deep inelastic hadronic scattering interactions (averaging over the charged and neutral current processes) is about 60 km ($R_m = 1740$ km). Upon interaction, a ~ 10 m long cascade then forms as the secondary particles multiply, and compton scattering, positron annihilation, and other scattering processes then lead to a $\sim 20\%$ negative charge excess which radiates a cone of coherent Cherenkov emission at an angle of $\sim 56°$, with a full-width at half-maximum of $\sim 1°$. The radiation propagates in the form of a sub-ns pulse through the regolith to the surface where it is refracted upon transmission.

Because the angle for total internal reflection (TIR) of the radiation emitted from the cascade is to first order the complement of the Cherenkov angle, we consider for the moment only neutrinos that cascade upon emerging from a penetrating chord through the lunar limb. Under these conditions the typical neutrino cascade has an upcoming angle with respect to the local surface of

$$\theta_{up} = \sin^{-1}\left(\frac{L_{int}}{2R_m}\right) \tag{1}$$

which implies a mean of $\theta_{up} \sim 1°$ at 10^{20} eV.

At the regolith surface the resulting microwave Cherenkov radiation is refracted strongly into the forward direction. Scattering from surface irregularities and demagnification from the interface refraction gradient fills in the Cherenkov cone, and results in a larger effective area of the lunar surface over which events can be detected, as well as a greater acceptance solid angle. These effects are discussed in more detail in a later section.

B   Antennas & receivers

The antennas employed in our search are the shaped-Cassegrainian 70 m antenna DSS14, and the beam waveguide 34 m antenna DSS13, both part of the NASA Goldstone Deep Space Network (DSN) Tracking Station. DSS13 is located about 22 km to the SSW of DSS14. The S-band (2.2 GHZ) right-circular-polarization (RCP) signal from DSS13 is filtered to 150 MHz BW, then downconverted with an intermediate frequency (IF) near 300 MHz. The band is then further subdivided

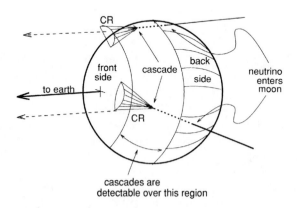

FIGURE 1. Schematic of the geometry for lunar neutrino cascade event detection.

into high and low frequency halves of 75 MHz each, and no overlap. These IF signals are then sent via an analog fiber-optic link to DSS14. At DSS14, the dual polarization S-band signals are downconverted with the same 300 MHz IF, and bandwidths of $\sim$ 100 MHz (RCP) and 40 MHZ (LCP) are used. A third signal is also employed at DSS14: a 1.8 GHz (L-band) feed which is off-pointed by $\sim$ 0.5° is used as a monitor of terrestrial interference signals; the signal is downconverted in the same manner as the other signals and has a 40 MHz bandwidth.

C   Trigger system

The experimental approach in our initial 12 hours of observations was to use a single antenna trigger with dual antenna data recording [7]. This was accomplished by using the local S-band signals as DSS14 to form a 2-fold coincidence with an active veto from the L-band interference monitor. Since any system with an active veto is subject to potential unforeseen impact on the trigger efficiency, we have now developed an approach which utilizes signals from both antennas to form a real-time dual-antenna trigger, with no active veto.

Fig. 2 shows the layout of the trigger. The four triggering signals from the two antennas are converted to unipolar pulses using tunnel-diode square-law detectors. Stanford Research Systems SR400 discriminators are used for the initial threshold level, and these are set to maintain a roughly constant singles rate, typically 0.5-1 KHz/chan for DSS14 and 30 KHz/chan for DSS13. A local coincidence is then formed for each antenna's signals. The DSS14 coincidence between both circular polarizations ensures that the signals are highly linearly polarized, and the DSS13 coincidence helps to ensure that the signal is broadband.

Fig. 3 indicates the timing sequence for a trigger to form (negative logic levels are used here). A local coincidence at DSS14, typically with a 25 ns gate, initiates the trigger sequence. After a 65 μs delay, a 150 μs gate is opened (the delays compensate for the 136 μs fiber delay between the two antennas). This large time window encompasses the possible geometric delay range for the moon throughout the year. Use of a smaller window is possible but would require delay tracking and a thus more stringent need for testing and reliability; use of a large window avoids this and a tighter coincidence can then be required offline.

If a $\sim$ 25 ns local coincidence forms between the DSS13 signals within the allowed window, a trigger is formed. The sampling scopes are then triggered, and a 250 μs record, sampled at 1 Gs/s, is stored. The average trigger rate, due primarily to random coincidences of thermal noise fluctuations, is about 1.6 mHz, or 1 trigger every 5 minutes or so. Terrestrial interference triggers are uncommon (a few percent of the total), but can occasionally increase in number when a large burst of interference occurs at either antenna, with DSS14 more sensitive to this effect. The deadtime per event is about 6 s; thus on average we maintain about 99% livetime during a run.

III ESTIMATED SENSITIVITY

Estimates of the sensitivity of radio telescope observations usually involve systems that integrate total power for some time constant Δt which is in general much longer than the antenna's single temporal mode duration $\tau = (\Delta\nu)^{-1}$. Since the pulses of interest in our experiment are much shorter than the time scale of a single temporal mode, the observed pulse structure of induced voltage in the antenna

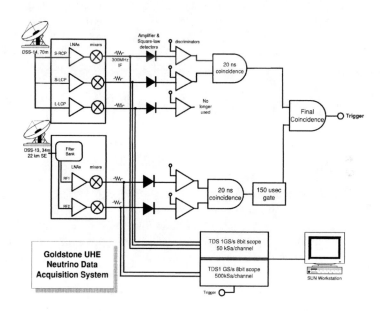

FIGURE 2. The GLUE trigger system used for the lunar neutrino search.

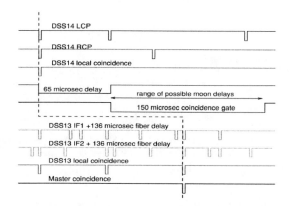

FIGURE 3. A timing diagram for the GLUE trigger system.

receiver is determined only by the bandpass function; that is, the pulses are band-limited. Thus the typical dependence of sensitivity on the factor $(\sqrt{\Delta t \Delta \nu})^{-1}$ does not obtain; this factor is always unity in band-limited pulse detection.

Because much of the theoretical work in describing such pulses has been done in terms of field strength rather than power, we analyze our sensitivity in these terms as well. Such analysis is also compatible with the receiving system, which records antenna voltages proportional to the incident electric field, and leads to a more linear analysis. It also yields signal-to-noise ratio estimates which are consistent with Gaussian statistics, since thermal noise voltages are described by a Gaussian random process.

The expected field strength from a cascade of total energy W_T can be expressed as [13,9,8]:

$$E_0 \text{ (V m}^{-1} \text{ MHz}^{-1}) = \frac{2.53 \times 10^{-7}}{R} \left(\frac{W_T}{1 \text{ TeV}} \right) \frac{\nu}{\nu_0} \left(\frac{1}{1 + 0.4(\nu/\nu_0)^{1.44}} \right) , \quad (2)$$

where R is the distance to the source in m, ν is the radio frequency, and the decoherence frequency $\nu_0 \simeq 2500$ MHz for regolith material (ν_0 scales mainly by radiation length). For typical parameters in our experiment, a 10^{19} eV cascade will result in a peak field strength at earth of $E \simeq 0.5$ μV m^{-1} for a 70 MHZ BW. Equation 2 has now been verified to within factors of ~ 2 through accelerator tests [1,2].

Given that the use of a dual antenna trigger has virtually eliminated the problem of terrestrial interference that was the primary limitation to the sensitivity of the one previous experiment [22], we can now express the minimum detectable field strength E_{min} for each antenna in terms of the induced signal and the thermal noise background.

The expected signal strength E_0 induces a voltage at the antenna receiver given by

$$V_s = h_e E_0 \Delta \nu \quad (3)$$

where the antenna effective height h_e is given by [12]

$$h_e = 2 \sqrt{\frac{Z_a \eta A}{Z_0}} \cos \theta_p \quad (4)$$

where Z_a is the antenna radiation resistance, η and A are the antenna efficiency and area, respectively, $Z_0 = 377$ Ω is the impedance of free space, and θ_p the polarization angle of the antenna with respect to the plane of polarization of the radiation.

The average thermal noise voltage in the system is given by

$$V_n = \sqrt{4kT_{sys}Z_T\Delta \nu} . \quad (5)$$

Here k is Boltzmann's constant, T_{sys} is the system thermal noise temperature, and Z_T the termination impedance of the receiver. If we assume that $Z_a \approx Z_T$ then the resulting SNR $S+$ is

$$S \equiv \frac{V_s}{V_n} = E_0 \cos \theta_p \sqrt{\frac{\eta A \Delta \nu}{k T_{sys} Z_0}} \; . \tag{6}$$

The minimum detectable field strength ($E_0 \to E_{min}$) is then given by

$$E_{min} = S \sqrt{\frac{k T_{sys} Z_0}{\eta A \Delta \nu}} \frac{1}{\cos \theta_p} \; . \tag{7}$$

Combining this with equation 2 above, the threshold energy for pulse detection is

$$W_{thr}(EeV) \simeq 4 E_{min} \left(\frac{R}{1 \text{ m}} \right) \left(\frac{\nu_0}{\nu} \right) \left(1 + 0.4(\nu/\nu_0)^{1.44} \right) \; . \tag{8}$$

For the lunar observations on the limb, which make up about 85% of the data reported here, $T_{sys} \simeq 110$ K, $\nu = 2.2$ GHz, and the average $\Delta \nu \simeq 70$ MHz. For the 70 m antenna, with efficiency $\eta \simeq 0.8$, the minimum detectable field strength is $E_{min} \simeq 1.2 \times 10^{-8}$ V m^{-1} MHz^{-1} for $\cos \theta_p = 0.7$. The estimated threshold energy for these parameters is $W_{thr} = 3.6 \times 10^{19}$ eV, assuming a detection level of 5σ per IF.

A Monte Carlo results

To estimate the effective volume and acceptance solid angle as a function of incoming neutrino energy, events were generated at discrete neutrino energies, including the current best estimates of both charged and neutral current cross sections [14], and the Bjorken-Y distribution. Both electron and muon neutrino interactions were included, and Landau-Pomeranchuk-Migdal effects in the shower formation were estimated [8]. At each neutrino energy, a distribution of cascade angles and depths with respect to the local surface was obtained, and a refraction propagation of the predicted Cherenkov angular distribution was made through the regolith surface, including absorption and reflection losses and a first order roughness model. Antenna thermal noise fluctuations were included in the detection process.

A portion of the simulation is shown in Fig. 4. Here the intensity is shown as it would appear projected on the sky, with (0,0) corresponding to the tangent to the lunar surface in the direction of the original cascade. The units are Jy (1 Jy = 10^{-26} W m^{-2} Hz^{-1}) as measured at earth, and the plot is an average over several hundred events at different depths and a range of θ_{up} consistent with a 10^{20} eV neutrino interaction, averaging over inelasticity effects and a mixture of electron and muon neutrinos consistent with decays from a hadronic $\pi^{\pm}$ source.

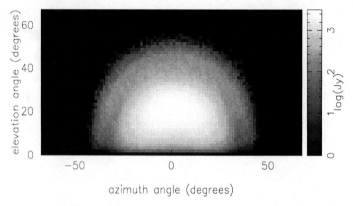

FIGURE 4. The microwave Cherenkov radiation pattern from an event in the lunar regolith.

Although the averaging has broadened the distribution somewhat, a typical cascade still produces an intensity pattern of comparable angular size. The angular width of the pattern directly increases the acceptance solid angle, and the angular height increases the annular band of the lunar surface over which neutrino events can be detected, as indicated in Fig. 1. The net effect is that, although the specific intensity of the events are lowered somewhat by refraction and scattering, the effective volume and acceptance solid angle are significantly increased. The neutrino acceptance solid angle, in particular, is about a factor of 50 larger than the apparent solid angle of the moon itself.

B EHE cosmic rays

We have noted above that the refraction geometry of the regolith favors emission from cascades that are upcoming relative to the local regolith surface. Thus to first order EHE cosmic ray events, which cascade within a few tens of cm as they enter the regolith, will not produce detectable pulses since their emission with be totally internally reflected within the regolith. This effect has been now demonstrated in an accelerator experiment [2].

This conclusion does not account for several effects however. THese effects are illustrated in Fig. 5. In Fig. 5A, the varying surface angles due surface roughness on scales greater than a wavelength will lead to escape of some radiation from cosmic ray cascades. Fig. 5B illustrqates that, because the Cherenkov angular distribution is not infinitely narrow arround the Cherenkov angle (FWHM $\sim 1°$), emission at angles larger then the Cherenkov angle can escape total internal reflectance. In Fig. 5C cascades from cosmic rays that enter along ridgelines can encounter a change in slope of the local surface that results in more efficient transmission of the radiation.

Even if total internal reflection strongly suppresses detection of cosmic rays in cases A and B, the latter case C of favorable surface geometry along ridgelines or hilltops will lead to some background of EHE cosmic ray events. We have not as yet made estimates of this background.[1]

Fig. 5D shows the formation zone aspect of the process of Cherenkov emission from near-surface cosmic ray cascades. This constraint may suppress Cherenkov production even if surface roughness and the width of the Cherenkov distribution would otherwise favor some escape of emission.

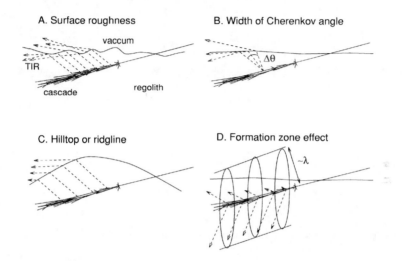

FIGURE 5. Various effects associated with EHE cosmic ray hadron interactions.

It has now been conclusively shown [15] that coherent Cherenkov emission is a process involving the bulk dielectric properties of the radiating material. Cherenkov radiation is induced over a macroscopic region of the dielectric (with respect to the scale of a wavelength), and does not even require that the charges particles enter the dielectric for radiation to be produced—a proximity of several wavelengths or less is sufficient [16]. A corollary to this result is that a cascade travelling along very near a boundary of the dielectric will not radiate (or radiate only weakly) into the hemisphere with the boundary.

Thus in the case of a cosmic ray entering the regolith at near grazing incidence (say within $\sim 1°$) the resulting cascade reaches maximum within ~ 3 cm of the surface, still less than a wavelength for S-band observations. We therefore expect that the suppression of Cherenkov emission in such events significantly reduces our sensitivity to cosmic rays. Such effects have not been included yet in other estimates [9,17] of the cosmic ray detection efficieny of such experiments.

---

[1] These conclusions also apply of course to the fraction of neutrinos which interact on entering the regolith as well as those which interact near their projected exit point. Thus we have nto yet accounted for all possible neutrino events as well as the cosmic ray background events.

IV RESULTS

A Neutrino limits

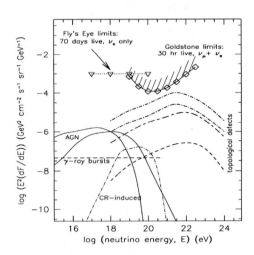

FIGURE 6. Plot of model neutrino fluxes and limits from the Fly's Eye experiment and the present work.

Figure 6 plots the predicted fluxes of EHE neutrinos from a number of models including AGN production [24] gamma-ray bursts [18], EHE cosmic-ray interactions [23], and topological defects [25,20] Also plotted are limits from about 70 days of Fly's Eye livetime [19] (accumulated in several years of runtime), which apply only to electron neutrino events.

Our initial 90% CL limit, for 30 hours of livetime is shown plotted with squares (see also Table 1), based on the observation of no events above an equivalent 5σ level amplitude (referenced to the 70 m antenna) consistent with the direction of the moon. These limits assume a monoenergetic signal at each energy; thus they are independent of source spectral model, and represent the most conservative limits we can apply. Our limits just begin to constrain the highest topological defect model [25] for which we expected a total of order 1–2 events.

TABLE 1. Limits on EHE neutrino fluxes[†]

Energy (eV)	10^{19}	3×10^{19}	10^{20}	3×10^{20}	10^{21}	3×10^{21}	10^{22}	3×10^{22}
$\log_{10}(E^2 dF/dE)$ (GeV cm^{-2} s^{-1} sr^{-1})	-3.14	-3.66	-3.92	-3.91	-3.73	-3.42	-3.03	-2.66

†limits based on assumption of monoenergetic source at each given energy.

B   Small event analysis

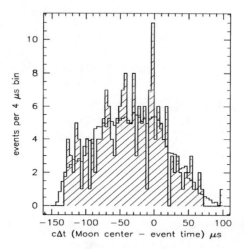

FIGURE 7. Small event histogram of GLUE data in delay with respect to the moon center delay. An excess occurs near zero delay, with a 2 microsecond offset.

In addition to the limits set above from the non-observation of events above, we have also analyzed events which triggered the system, but did not pass our more stringent software amplitude cuts. A sample of events was prepared by applying our standard cuts to remove terrestrial interference events. We required somewhat tighter timing that the hardware trigger, as well as band-limited pulse shape, but allowed smaller amplitudes. The results are shown in Fig. 7, where the passing events have been binned according to their delay timing with respect to the expected delay from an event at the center of the moon. The background level (solid line) has been determined by randomizing the UT of the events and indicates the somewhat non-uniform seasonal coverage of our observations.

An excess is observed in the vicinity of zero delay where the lunar events are expected to cluster. At present there is a $\sim 2\mu s$ offset from zero delay; this is too large to be accounted for by differential delays to the lunar limb, which can produce offsets of several hundred ns. Further study of the small events is in progress.

V   CONCLUSIONS

We have developed a robust system for observing microwave pulses produced in the lunar regolith by electromagnetic particle cascades above $\sim 10^{19}$ eV. We have operated this system to achieve a livetime of 30 hours, with no large apparent signals detected to date. We have set conservative upper limits on the diffuse cosmic neutrino fluxes over the energy range from $10^{19-22.5}$ eV. We have also begun to analyze smaller events and have some preliminary indications that a signal may be present, but requiring further study.

We thank Michael Klein, George Resch, and the staff at Goldstone for their enthusiastic support of our efforts. This work was performed in part at the Jet Propulsion Laboratory, California Institute of Technology, under contract with NASA, and supported in part by the Caltech President's Fund, by DOE contract DE-FG03-91ER40662 at UCLA, the Sloan Foundation, and the National Science Foundation.

REFERENCES

1. P. W. Gorham, D. P. Saltzberg, P. Schoessow, et al., 2000, Phys. Rev. E. 62, 8590.
2. D. Saltzberg, P. Gorham, D. Walz, et al. 2001, Phys. Rev. Lett., in press.
3. Askaryan, G.A.,1962, JETP 14, 441
4. Askaryan, G.A.,1965, JETP 21, 658
5. I. M. Zheleznykh, 1988, Proc. Neutrino 88 ....
6. Dagkesamanskii, R.D., & Zheleznyk, I.M., 1989, JETP 50, 233
7. P. W. Gorham, K. M. Liewer, and C. J. Naudet, Proc. 26th Intl. Cosmic Ray Conf., v 2, 479, (1999); also astro-ph/9906504
8. J. Alvarez–Muñiz, & E. Zas, 1997, Phys. Lett. B, 411, 218
9. J. Alvarez–Muñiz, & E. Zas, 1996, Proc. 25th ICRC, ed. M.S. Potgeiter et al.,7,309.
10. D. Morrison & T. Own, 1987 *The Planetary System*, (Addison-Wesley: Reading,MA)
11. G. R. Olhoeft & D. W. Strangway, 1975, Earth Plan. Sci. Lett. 24, 394
12. J. D. Kraus, 1988, *Antennas*, (McGraw-Hill:New York)
13. E. Zas, F. Halzen,, & T. Stanev, 1992, Phys Rev D 45, 362
14. R. Gandhi et al., 2000...
15. T. Takahashi, Y. Shibata, K. Ishi, et al. 2000, Phys Rev. E, 62, 8606
16. R. Ulrich, 1966, Zeit. Phys., 194, 180.
17. Alvarez–Muñiz, J., & Zas, E., 2001, this proceedings.
18. Bahcall, J.N., and Waxman, E., 1999, LANL preprint astro-ph/9902383 .
19. Baltrusaitas, R.M., Cassiday, G.L., Elbert, J.W., et al 1985, Phys Rev D 31, 2192.
20. Bhattacharjee, P., Hill, C.T., & Schramm, D.N, 1992 PRL 69, 567.
21. Dagkesamanskii, R.D., & Zheleznyk, I.M., 1989, JETP 50, 233.
22. Hankins, T.H., Ekers, R.D. & O'Sullivan, J.D. 1996, MNRAS 283, 1027.
23. Hill, C.T., & Schramm, D.N., 1985, Phys Rev D 31, 564.
24. Mannheim, K., 1996, Astropart. Phys 3, 295.
25. Yoshida, S., Dai, H., Jui, C.C.H., & Sommers, P., 1997, ApJ 479, 547.

Monitoring of Cherenkov Emission Pulses with Kalyazin Radiotelescope: real sensitivity and prospective program

Rustam D. Dagkesamanskii

Pushchino Radio Astronomy Observatory, Astro Space Center of the Lebedev Physical Institute, Moscow region 142290, RUSSIA

Abstract. Possibilities of searching for the Cherenkov emission pulses from super-high-energy particle cascades by monitoring of the Moon with Kalyazin 64-meter radiotelescope are considered. Sensitivity of the available facilities and a level of man-made interference at the radiotelescope site are estimated. Some details of a proposed experiment are described.

INTRODUCTION

In the mid-1980s Igor Zheleznykh and the people of his team from Institute for Nuclear Research of Russian Academy of Sciences actively worked out the RA-MAND (Radio Antarctic Muon and Neutrino Detector) project [1,2]. The project was based on one of the brightest idea by Gurgen Askaryan [3,4] to detect high-energy cosmic rays by detection of Cherenkov emission pulses produced by the negative charge excess of cascades generated in relatively dense dielectric medium when the high energy particle interacts with atoms of the dielectric. The properties of Antarctic ice make it possible to use an effective target of 10^9 or even 10^{11} m^3 [5].

To discuss the possible ways of improving the efficiency of the experiment Igor Zheleznykh and his colleagues contacted me and some other people from Pushchino Radio Astronomy Observatory. It seems that our discussion was rather fruitful and during the discussion Igor formulated his new idea to enlarge the target volume by observing the Cherenkov emission pulses from the Moon. First estimates had shown that the large meter-wavelength radiotelescopes of Pushchino Radio Astronomy Observatory (DKR-1000 and BSA FIAN) were not suitable for such observations because they could track the Moon for only few minutes around its culmination and their frequency ranges were too far from the expected maximum. On the other hand, the estimates revealed that using the already existing fully-steerable radiotelescopes of 70-meter (and more) diameter we could hope to detect at decimeter and centimeter wavelengths the Cherenkov pulses from 10^{20} eV cascades.

CP579, *Radio Detection of High Energy Particles,* edited by D. Saltzberg and P. Gorham
© 2001 American Institute of Physics 0-7354-0018-0/01/$18.00

RAMHAND PROJECT

The following more careful considerations confirmed, in general, the rather optimistic estimates mentioned above. The new project was named RAMHAND — RAdio Moon Hadron And Neutrino Detector [5,6] because Cherenkov emission pulses have to be emitted after interactions of hadrons with the atoms of lunar regolith and the corresponding events could be observed near outer ring of the lunar disk. Sensitivity of this giant neutrino telescope is determined mainly by effective area of the ground-based antenna used as a detector and by the wavelength of observations. The power of the coherent Cherenkov emission pulse should be proportional to E^2, where E is the energy of the cascade. The expected rate of the events depends on the spatial density of particles in corresponding energy range and from the effective total mass of the target.

Using a predicted spectrum and some other parameters of the Cherenkov emission pulses it had been shown [5–8] that the pulses arising from interaction between particles of 10^{20} eV with the Moon's regolith should be about 5 times the r.m.s. noise level when observed with 70-meter ground-based radiotelescope. However, the expected rate of such events is very low and all *a priori* information about the pulses should be used to separate the real signal from noise fluctuations and the interference. For example, predicted short duration (several nanoseconds) of expected pulse could help to extract it from some types of man-made interference. The continuous spectrum of the predicted emission provides efficiency for multi-frequency observations together with coincidence schemes. Taking into account a dispersion delay of the signal in the Earth's ionosphere and using multi-frequency observations it could be possible to reject many sources of narrow-band and wide-band interference. Finally, spatial correlation of the signal enables using the simultaneous monitoring with several radiotelescopes separated by large distances.

A small variation of the RAMHAND project was suggested in the mid-1990s [9,10]. The possibility to use an on-board antenna orbiting the Earth or even a lunar satellite was considered. It was pointed out that neutrinos of somewhat lower ($\approx 10^{18}$ eV) energy could be detected in these cases.

To realize the main RAMHAND project one or several large fully-steerable radiotelescopes equipped with rather good receivers and special terminals of very high (≈ 10 nsec) time resolution were needed. By the end of the 1980s and in beginning of the 1990s, there were no such terminals at our radiotelescopes and we could not start the monitoring of pulses from the Moon.

IMPORTANT RESULTS OF 1990S

Already the first discussions of the RAMHAND project with physicists and astronomers at "Neutrinos-88" [5], "Quarks '88" [6], USSR-USA Seminar on High Energy Astrophysics (Tbilisi, 1999) and other meetings showed that some of our colleagues did not believe the Cherenkov emission pulse monitoring would succeed because of several far extrapolations and rather rough estimates that were used in calculations of the pulse intensities and the rate of the observable events. New and more careful theoretical considerations as well as computer simulations were needed to overcome this scepticism. In the beginning of the 1990s besides Zheleznykh's team [11] the theoretical aspects of the problem were studied by Halzen, Zas and Stanev [12,13] and later by Alvarez-Muniz and Zas [14,15]. The results of computer simulations described in these papers confirmed as a whole preliminary optimistic RAMHAND project estimates and predicted duration, spectrum and some other parameters of Cherenkov pulse emission. For example, calculated the spectrum maximum position $\nu_{max} = 2$ GHz or higher (compare with $\nu_{max} = 1$ GHz used for earlier estimates).

The first attempt to detect the Cherenkov emission pulses from the Moon was made by T. Hankins, R. Ekers and J. O'Sullivan [16]. Using the Parkes 64-meter radiotelescope with a wideband (1187-1662 MHz) receiver and high time-resolution (of few nanoseconds) oscilloscope they observed the Moon for about 10 hours. No pulses of 1 nanosecond duration with flux density above 4000 Jy were detected during the observations.

Another attempt was made by P. Gorham, K. Liewer and C. Naudet [17] using the NASA Goldstone 70 m and 34 m antennas. The observations were made at S-band ($2.2 - 2.3$ GHz) for a total of 12 hours: 8 hrs in dual-antenna and 4 hrs in single-antenna mode. An L-band (1.6 GHz) receiver was used at 70 m antenna simultaneously with S-band receiver. During the observations the 1.6 GHz beam was pointed 0.5° away from the 2.2 GHz beam. Coincidences and anti-coincidences (between S- and L-band receivers) scheme were used during observations and processing so that almost all interference was rejected. So only the system noise of 400 Jy r.m.s. fluctuations limited the final sensitivity. No pulses of expected duration with flux density more than 6 r.m.s., i.e. 2400 Jy, were detected in 12 hours. The corresponding neutrino energy threshold was 5×10^{19} eV.

The negative results of both experiments are not surprising. Indeed, the upper limits to the event rates found in the experiments are above most theoretical estimates, and exclude only very exotic theoretical models like a Universe with super-conducting strings and so on. It was obvious, to test more real models, the observations of hundreds hours with sensitivity which was realized already were needed. Nevertheless, the scepticism mentioned above could be increased after these negative results. Fortunately, two excellent laboratory experiments [18,19] were fulfilled by UCLA/NASA group of physics shortly after the publication of Goldstone results. Using Argonne Wakefield Accelerator and Stanford Linear Accelerator authors of [18,19] have modeled the cascades of 10^{18-19} eV in silica sand target and showed that they had the expected negative excess and measured the radio pulses from the cascades. It turned out that measured energy dependence of the pulse amplitude is very close to that predicted before. On the other hand, the maximum of the pulse emission spectrum is at least at 8 GHz or even higher frequency. So, nothing is left for scepticism after the last two papers and the flux of neutrinos with energy $E \geq E_0$ can be directly found from the measured (or estimated) rate of the pulses with flux densities S above corresponding limit $S \geq S_0$.

PREPARATIONS TO MONITORING WITH KALYAZIN RADIOTELESCOPE

The 64-meter fully-steerable dish of OKB MEI (Special Construction Bureau of Moscow Power Institute) near Kalyazin town, which is about 200 km to north of Moscow, was constructed in 1992 for the deep space network service and for radioastronomical observations. Now this radiotelescope is equipped with receiving and recording complex mainly for pulsars and VLBI observations [20]. (see Figure 1). There is Kalyazin laboratory of Pushchino Radio Astronomy Observatory supporting most of these observations.

Today the Kalyazin 64 m radiotelescope is one of the most suitable tool for searching the Cherenkov emission pulses from the Moon in Russia. The radiotelescope

FIGURE 1. Kalyazin 64-meter radiotelescope.

is of alt-azimuth mount. There is a 6 m secondary reflector and a multi-frequency feed placed at a Cassegrain focus. There are five dual- channel (LCP/RCP) receivers for 0.6, 1.4, 2.3, 4.8 and 8.3 GHz frequency bands. Four of them (except 4.8 GHz receiver) can be used in simultaneous observations. System noises are less than 100 K at all mentioned frequencies, so the Moon brightness temperature will determine the noise fluctuations during proposed observations. The radiotelescope is equipped with Canadian S2 VLBI terminals of 32 MHz bandwidth. There is precise time service at radiotelescope.

There is a good experience in using the Kalyazin radiotelescope in pulsar and VLBI observations. Most observations are free from any strong interference. Figure 2 shows the short pieces of records during observations of two pulsars with Kalyazin radiotelescope.

The observations were made with S2 recording terminals in LCP-channels. Original time resolution was 30 nsec. Total duration of each Fig.2 piece is 10 seconds and the all three are averaged over 10 milliseconds. The spike near the beginning of Fig.2a record is one of giant pulses of Crab nebulae pulsar. Most of spikes at Fig.2c are the pulses of PSR 0329+54. After subtraction of these spikes from the original (non averaged) records the distributions of fluctuations at both frequencies are very close to Gaussian with standard deviations corresponding to expected system noise level.

We are planning to start the monitoring of the Moon with Kalyazin 64m radiotelescope at two or three frequencies before mid-2001. For the beginning we are going to use S2 recording systems as terminals.

CONCLUSIONS

It seems that now radio astronomers are close to getting the real estimate of extremely-high-energy neutrinos flux or at least a very strong upper limit of it. However, the rate of Cherenkov emission pulses from the Moon should be very low, and may be not more than one or two per month. To increase the reliability of such estimates and perhaps to determine the position of the cascade at lunar disk the simultaneous monitoring with two or three radiotelescopes separated by hundreds or thousand kilometers is very desirable. Several European radiotelescopes could be joined in such program. Close collaboration between radio astronomers from different countries should help us to solve this task.

Acknowledgments.   I gratefully acknowledge the Organizing Committee of RADHEP-2000 for invitation to this meeting and financial support.

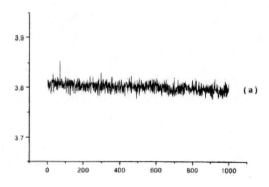

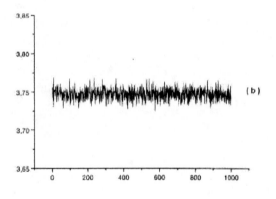

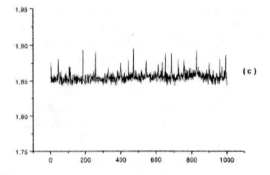

FIGURE 2. Pieces of pulsars records: (a) and (b) - two 10 sec. pieces of simultaneous records of Crab nebulae pulsar at 594 MHz and 2228 MHz, (c) - 10 sec. record of PSR 0329+54 at 2244 MHz

REFERENCES

1. Gusev G.A., and Zheleznykh I.M., *Pis'ma Zh. Eksp. Teor. Fiz.[JETP Lett.]* **38**, 505 (1983).

2. Markov M.A., and Zheleznykh I.M., *Nucl. Instr. and Methods in Phys. Res.* **A248**, 242 (1986).

3. Askaryan G.A., *Zh. Eksp. Teor. Fiz. [JETP]* **41**, 616 (1961).

4. Askaryan G.A., *Zh. Eksp. Teor. Fiz. [JETP]* **48**, 988 (1965).

5. Zheleznykh I.M., *Proc. 13th Intl. Conf. on Neutrino Physics and Astrophysics ("Neutrino 88")*, ed. J.Schreps, World Scientific, Boston, 1988, p. 528.

6. Zheleznykh I.M., *Proc. Int. Seminar "Quarks 88", Tbilisi, USSR*, World Scientific, 1988, p.789.

7. Dagkesamanskii R.D., and Zheleznykh I.M., *Pis'ma Zh. Eksp. Teor. Fiz. [JETP Lett.]* **50**, 233 (1989).

8. Dagkesamanskii R.D., and Zheleznykh I.M., *ICRR Intern. Symp. "Astrophysical Aspects of the Most Energetic Cosmic Rays", Kofu, Japan, 26-29 Nov. 1990, ed. M.Nagano and F.Takahara*, World Scientific, Singapore, 1992, pp. 373-380.

9. Dagkesamanskii R.D. and Zheleznykh I.M., *First International Conference on Cosmoparticle Physics "COSMION-94", Moscow, Dec.5-14, 1994*

10. Dagkesamanskii R.D. and Zheleznykh I.M., *Talk at the Workshop "High Energy Neutrino Astrophysics" (HENA-96)*, Aspen Center, Colorado, USA, 1996 (unpublished).

11. Zheleznykh I.M., *Proc. of XXIst ICRC, Adelaide, Australia, 1989, ed. R.Protheroe* Northfield, South Australia: Graphic Services, 1990, Vol.6, pp.528-533.

12. Halzen F., Zas E. and Stanev T., *Physics Letters* **257**, number 3,4, 432-436 (1991).

13. Zas E., Halzen F. and Stanev T., *Physical Review D* **45**, Number 1, 362-376 (1992).

14. Alvarez-Muniz R.A., and Zas E., *Phys. Lett. B* **411**, 218 (1997).

15. Alvarez-Muniz R.A., and Zas E., *Phys. Lett. B* **434**, 396 (1998).

16. Hankins T.H., Ekers R.D., and O'Sullivan J.D., *MNRAS* **283**, 1027 (1996).

17. Gorham P.W., Liewer K.M., and Naudet C.J., *Proc. 26th ICRC* **2**, 479 (1999).

18. Gorham P.W., Saltzberg D.P., Schoessow P., Gai W., Power J.G., Konecny R., and Conde M.E., Phys. Rev. E **62** 8590 (2000)

19. Saltzberg D.P., Gorham P., Walz D., Field C., Iverson R., Odian A., Resch G., Schoessow P., and Williams D., astro-ph/0011001, to appear in Phys. Rev Lett.

20. Ilyasov Yu.P., Poperechenko B.A., and Oreshko V.V., *"Radioastronomical Tools and Techniques", part 2. "Ground-based radiotelescopes, equipment and techniques"*, Moscow: Trudy FIAN, 2000, V.229, pp.44-60.

In-ice radio detection of GZK neutrinos

David Seckel

Bartol Research Institute
University of Delaware
Newark Delaware 19716

Abstract. Models for the source and propagation of cosmic rays are stressed by observations of cosmic rays with energies $E > 10^{20}$ eV. A key discriminant between different models may be complementary observations of neutrinos with energies $E > 10^{18}$ eV. Independent of the source of the cosmic rays, neutrinos are produced during propagation via the GZK mechanism. Event rates for GZK neutrinos are expected to be in the range of $0.01 - 0.1$ per km^3 yr, suggesting a detector mass in excess of 1 Eg. Detection of radio cherenkov emission from showers produced in Antarctic ice may be an economical way to instrument such a large mass. It is suggested that a 100 km^2 array of antennas centered on Icecube may allow confirmation of the radio technique and also increase the science achievable with Icecube by providing vertex information for events with throughgoing muons.

INTRODUCTION

Observations of cosmic rays with energies greater than 10^{20} eV present a problem. Assuming the particles are baryons, energy losses due to pair production and photoproduction demand that the sources of such particles be located within about 10 Mpc. Since the Universe is fairly lumpy on those length scales, it is somewhat surprising that the events appear to be distributed smoothly on the sky. Further, there are no obvious candidate sources when you look back along the arrival directions. Other particle choices present no better alternatives. Faced with this conundrum, source models generally fall into two classes. a) Astrophysical sources which satisfy the locality and isotropy constraints via a set of just so conditions. b) Models motivated by particle physics, which invoke the injection of high energy particles through the decay of massive particles or topological defects. These models may be distinguished through their associated neutrino fluxes.

Most astrophysical cosmic ray acceleration models involve proton acceleration in a modestly dense environment. Within that source region, interactions may take place of the form $p + X \rightarrow n + \pi^+ + X'$, where X and X' may be anything. Two things happen at this juncture. 1) The neutron may escape from the acceleration region. 2) The π^+ decays, ultimately producing an e^+, ν_e, ν_μ and an $\bar{\nu}_\mu$. Such

CP579, *Radio Detection of High Energy Particles,* edited by D. Saltzberg and P. Gorham
© 2001 American Institute of Physics 0-7354-0018-0/01/$18.00

models produce neutrinos at energy E in comparable abundance to nucleons at energy $10E$. It is also possible to have acceleration in low density environments, with no associated neutrino production. By contrast, the particle physics models result in the production of quark jets, which fragment into mesons, and eventually produce neutrinos, other leptons or photons. Baryon production in the fragmentation process accounts for a few percent of all particles, so the neutrinos outnumber the baryons. Finally, once energetic protons are released into the cosmological medium, they produce more neutrinos through the GZK process (Greisen, Zatsepin, Kuzmin) involving photoproduction of pions from collisions with microwave background photons. Since protons with energies above the GZK theshold have been observed, one cannot avoid the conclusion that neutrinos are produced during cosmic ray propagation.

With these comments, one can distinguish source models by their neutrino fluxes. An absolute minimum flux is the GZK flux inferred from the cosmic ray observations themselves. If this is all, then one is inclined to astrophysical acceleration models where the acceleration takes place is low density environments, such as at the interface of galactic winds, the termination shocks of jets from active AGN, etc. If the neutrino fluxes are modestly in excess of the predicted GZK fluxes, then acceleration in a dense astrophysical environment is favored, such as within AGN jets, around magnetars, or in gamma ray bursts, etc. Finally, if neutrino fluxes are greatly in excess of GZK models, then exotic particle physics models are indicated.

To understand the observations of high energy cosmic rays, it is imperative to make complementary observations of ultra high energy cosmic neutrinos. To ensure that one has reached a correct interpretation, one should be prepared to search for fluxes as low as the GZK fluxes. The rest of this contribution is devoted to estimating the size detector required and comparing the potential of proposed detection techniques. In doing this analysis, a rate of 100 events per yr is suggested to go beyond simple discovery level science, an argument which should be familiar to supporters of the AUGER project. Since total fluxes are at a minimum in astrophysical models, I focus on this possibility.

EVENT RATES

Event rates are estimated by convolving cross-section and neutrino flux. Calculations of the GZK neutrino flux are illustrated by Figure 1, based primarily on results from Yoshida and Teshima [1] (YT). They start with a simple cosmic ray injection model, a homogeneous distribution of sources with proton spectrum $dN/dE = P(t)E^{-2}$. The power law is motivated by shock acceleration models. Their source evolution is described by $P(t) = P_0(1 + z)^m$ for $z < z_{max}$ and zero otherwise, with P_0, m and z_{max} taken as parameters, and $(1 + z)(t)$ described by the cosmological expansion. YT normalize their models by integrating the predicted cosmic ray flux with energies $E > 10^{19.5}$ eV and comparing to the AGASA data. Since photoproduction is efficient at high energies, the normalization only

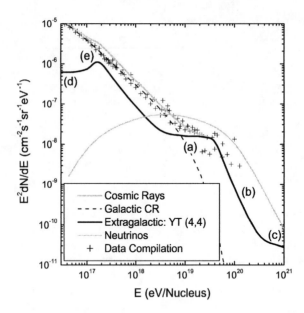

FIGURE 1. Neutrino and cosmic ray flux for model (4,4) of Yoshida and Teshima.

counts protons that originate within a few tens of Mpc. It therefore fixes P_0, but is independent of the evolution model. YT models are labeled by (z_{max}, m).

In a homogeneous source model the GZK neutrinos are primarily produced at high red shift, so their flux depends both on P_0 and the source evolution. It follows that models with strong source evolution will produce more GZK neutrinos. In this regard, the model picked out for Figure 1 is the YT (4,4) model. This is the strongest source evolution model that YT consider and it gives a GZK neutrino flux approximately an order of magnitude stronger than their middle of the road (2,2) model. Even so it is interesting for several reasons.

First, the evolved cosmic ray flux shows clearly several generic features. (a) The shelf at 10^{19} eV is fixed by the normalization and reflects the current injection rate integrated over the age of the universe. It does not, qualitatively, include injection at high redshift ($z > 1$) and so is relatively insensitive to parameters other than P_0. It is also relatively insensitive to the assumption of homogeneity unless magnetic diffusion effects are strong and sources are long lived. (b) The roll off above 10^{19} eV is due to dE/dX from photoproduction. (c) The shelf at 10^{21} eV reflects the current injection rate integrated over the energy loss time. The level of (c) relative to (a) is sensitive to the assumption that the sources are distributed homogeneously. (d) The shelf below 10^{17} eV is due to injection at high red shift of protons at low enough energy that the only energy losses are adiabatic. The level of (d) relative to (a) is sensitive to the evolution parameters. (e) The bump at 10^{17} eV contains protons injected with high energy at high redshift that quickly loose energy due to photoproduction followed by pair-production. This bump is directly related to the flux level of GZK neutrinos. Its height is sensitive to m and its position to z_{max}.

Second, since publication of YT, it has become apparent that stronger source models are favored. For example, Engel and Stanev [2] argue for source evolution at least as strong as a (2,3) model, and they find conservative neutrino fluxes comparable to those from the YT (4,4) model. ES use a more detailed treatment of photoproduction than YT which may alter the evolution somewhat. They also use a different normalization, based on work by Waxman [3], which involves integrating the cosmic ray flux above 10^{19} eV. Comparing YT's integral spectra to Waxman's suggests normalization differences of $\pm 20\%$. Taken together this suggests the YT flux in Figure 1 may not be unreasonable for homogeneous models.

Third, Figure 1 shows that when an extragalactic flux is combined with a reasonable galactic flux model interesting features relevant for consideration of GZK neutrinos may arise, beyond just the presence of a GZK cutoff. Specifically, in models of strong source evolution, the bump (e) should be noticeable as a distortion of the cosmic ray spectrum or its composition around 10^{17} eV. There is no hint of either effect in reviews by Gaisser [4].

The second half of the event rate calculation is the νN cross section. Within the standard model, cross-sections at energies up to 10^{20} eV have been calculated by several authors [5–8]. All such calculations are extrapolations, and should be treated with some caution. Typical variation in the charged current cross-section at 10^{19} eV is 20% with larger variations in the neutral current cross-sections. Here, we use the results of Glück, Kretzer and Reya [7] (GKR), which are a bit above average for charged currents and a bit below for neutral currents.

Figure 2a shows event rates for different reactions assuming the ES flux and the GKR cross-sections. Rates assume 2π angular acceptance for downward going neutrinos, the upward flux being absorbed by the Earth, and are given in terms of neutrino energy, not shower energy. For charged current events total energy may be measurable based on known showering and dE/dX properties of various leptons. Even τ leptons, at an EeV, lose some 7% of their energy per km of path length in ice [9]. For neutral current events only about 20% of the primary neutrino energy [5] goes into the hadronic recoil, so event rates should be estimated using a neutrino threshold energy ~ 5 times the detector threshold.

Figure 2b shows integral event rates for four YT models, as well as the ES flux. To achieve 100 events per yr requires roughly 1000 km^3 instrumented volume (or 1 Eg of mass) with threshold of 1 EeV. Pessimistic models, such as YT (2,2) may require 10 times the mass, but even then discovery could be accomplished on a modest 100 km^3 detector within a year or two of running.

RICE AND OTHER TECHNOLOGIES

A summary of existing and proposed neutrino detectors is shown in Figure 3. The most widely deployed technique is to use optical Cherenkov photons to recon-

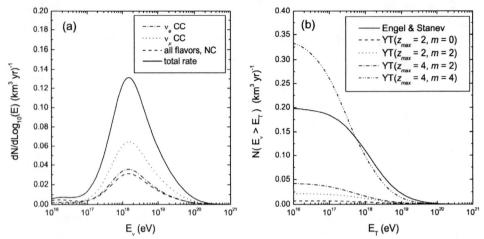

FIGURE 2. (a) Differential GZK neutrino interaction rates in ice, by flavor and event type, for the flux of Engel and Stanev. Rates for each flavor include both neutrinos and antineutrinos. The figure assumes no oscillations. (b) Integrated GZK neutrino event rates as a function of detector threshold. It is assumed that all energy is visible for charged current events and 20% is visible for neutral current events.

struct the tracks of muons produced in charged current reactions. In addition to muons, showers produced by the hadronic recoil in νN scattering or by charged current electrons will produce intense concentrations of Cherenkov photons. Optical Cherenkov detectors are capable of energy determination for contained events and an extended sensitive volume for throughgoing heavy leptons. In the latter case, the energy resolution for the initial neutrino is likely to be quite poor. The largest optical Cherenkov detector constructed to date is AMANDA [10]. Cubic kilometer detectors are planned, such as Icecube, but even these are too small to adequately probe GZK neutrinos. In principle, the effective volume for throughgoing muons may exceed that for contained events by an order of magnitude; however, without energy resolution the technique at best can set upper limits on EeV neutrino fluxes, but claims of GZK neutrino detection will be problematic.

Air shower techniques have also been discussed for neutrino detection. Soon, the largest air shower array will be the AUGER [11] experiment including some 3000 km^2 active area. However, since the column density of air is about 10 m, the instrumented mass corresponds to only about 0.03 Eg. The mass shown in Figure 3 accounts for the efficiency to detect neutrino induced cascades via the ground array [11]. Air showers may also be detected via fluorescence techniques. To increase area, OWL [12]/EUSO [13] have proposed to look down on $\sim 1000 \times 1000$ km^2 of atmosphere from Earth orbit. The effective mass shown in the figure includes a 0.1 efficiency factor to include duty cycle and constraints on interaction depth and zenith angle so that the air shower can both fully develop and be cleanly separated from cosmic ray induced air showers. With these factors, space based fluorescence

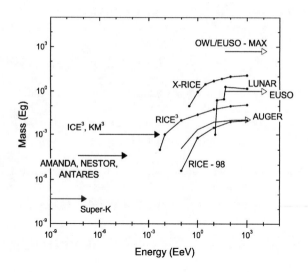

FIGURE 3. Effective mass of experiments proposed for detection of high energy neutrinos, as a function of neutrino energy. Experiments based on optical Cherenkov in water or ice are labeled by filled arrowheads. Open arrowheads denote air shower techniques. Radio Cherenkov experiments have dots overlayed and no arrowhead.

techniques can provide 1 Eg of detector mass. However, if the indicated threshold of 5×10^{19} eV [13] is sharp, then using the ES flux would result in just 1 ν_e charged current event per yr, and less than one of other event types assuming that only the hadronic recoil is visible. For amusement, "OWL/EUSO Max" refers to the total mass of the atmosphere utilized with 10% efficiency.

A third option is to detect neutrinos through the coherent radio Cherenkov emission from the showers produced by neutrino interactions. Hopefully, it is not necessary, at this conference, to summarize the basic ideas behind the radio detection of high energy particles. One radio based technique involves the search for radio flashes from events in the lunar crust [14]. The effective mass shown in the figure is due to Alvarez-Müniz [15]. It falls somewhat short of the mass and energy thresholds required for GZK neutrino detection. Additionally, only those events that occur well away from the limb of the moon will be separable from more frequent cosmic ray events causing similar signals.

The figure also shows sensitivity for experiments based on the technique of deploying radio antennas in ice at the South Pole [16] (RICE). RICE-98 [17] shows the effective mass of the current pioneering effort as configured at the end of the 98 polar season. RICE^3 depicts an array as may be deployed with Icecube. The lower threshold for RICE^3 arises from a relatively dense spacing of over 300 radio antennas. The largest version, X-RICE, is a 10^4 km^2 array suggested for the detection of GZK neutrinos. The result shown is for antennas on a rectangular grid of 1 km spacing [18]. 1 Eg of effective mass is achieved for $E > 1$ EeV. Increasing the antenna density tenfold lowers the threshold, and allows a 1 Eg detector from 1000 km^2 of ice. (This is about 50% efficiency since the ice is 2 km thick.) Allowing for detection of only electromagnetic and hadronic showers, such a detector would obtain approximately 30 ν_e and 25ν_μ charged current, and 15 neutral current events from the Engel and Stanev GZK flux.

To summarize, optical Cherenkov experiments are too small, or too expensive, to ensure detection of GZK neutrinos. Air shower techniques require 10^5 km^2 for particle detectors or 10^6 km^2 for fluorescence detectors due to the low density of air, beyond the scope of current ground based efforts. Space based fluorescence experiments, as proposed, do not have a low enough threshold to detect GZK neutrinos. The lunar radio technique also has too high a threshold. Only the RICE technique seems to offer a suitable combination of mass and threshold to detect GZK neutrinos at the rate of $\sim$ 100 neutrinos per yr.

DEPLOYMENT STRATEGIES WITH ICECUBE

The main difficulties with RICE are a) demonstrating that the physics of the technique is correctly described, b) establishing a calibration system for a deployed experiment, and c) overcoming the technical challenges of deploying on a remote basis up to 100 km from South Pole. (a) Recent experimental work by Saltzberg and collaborators [19], and continuing theoretical efforts on several fronts [20,21] suggests that the basic physics is sound, although there is still room for refinement. (b) One of the main points behind the RICE[3] plan is to provide explicit verification and calibration of a deployed experiment. (c) No serious work.

There are some flaws with the RICE[3] concept. First, Icecube is targeted toward detection of AGN neutrinos in the PeV region. Even with dense deployment, RICE[3] barely reaches below 10 PeV, although more sophisticated designs may improve on this situation. Second, the detectors do not really overlap. Optical transparency (bubbles) demands that Icecube be deployed in the bottom kilometer of ice, while radio transparency (geothermal warming) demands that RICE be deployed in the top kilometer. Since the Earth is opaque, most events in Icecube come from above or the side. The geometry of such events is such that the radio Cherenkov cone does not intersect a RICE array deployed above Icecube for most event vertecies contained in Icecube. The best geometry for coincident detection is charged current ν_μ interactions where the muon passes through Icecube and RICE[3] gets the Cherenkov cone from the hadronic shower. For each zenith angle and impact parameter there is of order a linear km along the line of sight where the event geometry will satisfy this condition. Total effective overlap is about 1 km^3. Third, little progress is taken towards difficulty (c). Fourth, there may, in fact, be no AGN neutrinos, and fifth Icecube still cannot expect to see GZK neutrinos other than straggler external muons.

An alternative deployment strategy is to make RICE[3] look like the central part of X-RICE, i.e. deploy $\sim$ 100 antennas over 100 km^2 centered on Icecube. First off, such a deployment directly addresses difficulty (c). Second, such an array would provide vertex information for throughgoing muons detected by Icecube, allowing energy calibration of those events. Third, such an array would detect ~ 10 GZK neutrinos per yr, which would be a significant discovery. On the down side, the threshold for radio detection would rise, so potential overlap for 10 PeV AGN neutrinos is eliminated. Also, it seems that overlapping detection of GZK neutrinos still seems unlikely. Simple geometry considerations indicate that none of the ten GZK events are likely to be detected by Icecube.

Which deployment strategy is better? In either case one cannot rely on the GZK neutrinos for calibration of RICE, even utilizing throughgoing muons. A bright AGN flux seems somewhat more likely than a bright flux in excess of GZK expectations at EeV energies. If neither bright option is present, then an independent confirmation and calibration of the radio technique is in order. Once achieved, the X-RICE prototype would likely give a first look at GZK neutrinos, while providing a technology development platform for the full X-RICE. On the other hand, the compact RICE[3] lacks the sensitive volume for GZK detection, and would leave unsolved the daunting technical problems of a large area Antarctic deployment.

I thank R. Engel for discussion and for sharing results of the ES model.

REFERENCES

1. Yoshida, S. and M. Teshiima, Prog. Theor. Phys. 89, 833 (1993).
2. Engel, R., and T. Stanev, astro-ph/0101216, (2001).
3. Waxman, E., Astrophys. J., 452, L1 (1995).
4. Gaisser, T.K., astro-ph/0011524, astro-ph/0011525 (2000).
5. Gandhi, R., C. Quigg, H. Reno, I. Sarkevic, Astropart. Phys. 5, 81 (1996).
6. Gandhi, R., C. Quigg, H. Reno, I. Sarkevic, Phys. Rev. D58 (093009) (1998).
7. Glück, M., S. Kretzer, and E. Reya, Astropart. Phys. 11, 327 (1999).
8. Kwiecinski, J., A.D. Martin, and A.M. Staso, Phys. Rev. D59, 093002 (1999).
9. Dutta, S.I., et al., Phys. Rev. D, In press. (astro-ph/0012350) (2001).
10. http://amanda.berkeley.edu/amanda/amanda.html (2001).
11. http://www.auger.org/ (2001).
12. http://owl.gsfc.nasa.gov/ (2001).
13. http://www.ifcai.pa.cnr.it/~EUSO/ (2001).
14. Gorham, P.W., K.M. Liewer and C.J. Naudet, Proceedings 26th Int. Cos. Ray Conf., Salt Lake City (astro-ph/9906504) (1999).
15. Alvarez-Muniz, J. These proceedings, astro-ph/0102173. (2000).
16. Frichter, G., J. Ralston and D. McKay, Phys. Rev. D 53, 1684 (1996).
17. http://kuhep4.phsx.ukans.edu/~iceman/index.html (2001).
18. Seckel, D. and G. Frichter, Proc. 26th Int. Cos. Ray Conf., Salt Lake City (1999).
19. Saltzberg, D., et al., astro-ph/0011001 (2000).
20. Buniy, R.V. and J. Ralston, Phys. Rev. D, in Press, astro-ph/0003408 (2000).
21. Alvarez-Muniz, J. and E. Zas, Phys. Lett. B 411, 218 (1997); Phys. Lett. B 434, 396 (1998); Phys. Rev. D62, 063001 (2000).

Study of Salt Neutrino Detector

Masami Chiba, Toshio Kamijo[*], Miho Kawaki, Athar Husain,
Masahide Inuzuka, Maho Ikeda, Osamu Yasuda

Department of Physics, Tokyo Metropolitan University
1-1 Minami-Ohsawa Hachioji-shi, Tokyo, 192-0397, Japan
[*]Department of Electrical Engineering, Tokyo Metropolitan University
1-1 Minami-Ohsawa Hachioji-shi, Tokyo, 192-0397, Japan

Abstract. Rock salt is studied as a radio wave transmission medium in an ultra high energy (UHE) cosmic neutrino detector. The radio wave would be generated by Askar'yan effect (coherent Chrenkov radiation from negative excess charges in the electromagnetic shower) in the UHE neutrino interaction in the rock salt. We collected the samples of the rock salts from various rock salt mines in order to investigate whether they have a possibility as a Salt Neutrino Detector (SND) sites or not. As a tentative result, the absorption length of the rock salt samples was measured to be between 40m and 400m at 1 GHz

INTRODUCTION

Several cosmologically distant astrophysical systems could produce ultra-high-energy (UHE) cosmic neutrinos [1] over the energy of 10^{15} eV or PeV whose flux probably exceeds that of the atmospheric neutrino [2]. Therefore, in spite of the very low flux, we could detect extraterrestrial neutrinos coming from far distance over 100Mpc (10 million light years) without the atmospheric neutrino background. UHE neutrinos are produced by astrophysical sources in cosmological distance such as Quasars (QSOs) or Active Galactic Nuclei (AGNs) and Gamma Ray Bursts (GRBs). They follow the production and decay of related unstable hadrons: pions, kaons, and charmed mesons. These unstable hadrons could be produced when the accelerated protons interact with the ambient photon or protons present in the environment. The electron and muon neutrinos are mainly produced in the decay chain of charged pions whereas the tau neutrinos are mainly produced in the decay chain of charmed mesons in the same collisions at a suppressed level.

The maximum energy of observed cosmic rays (protons or nuclei) exceeds 10^{20} eV [3]. The energy is over the allowed energy for cosmic rays traveled from the extra galaxy over 50 Mpc, which loses its energy in scattering processes against 3° K cosmic background radiations. (GZK cutoff) [4]. It is one of the centers of attention in astrophysics and high-energy physics [5]. On the other hand, a neutrino scarcely interacts with the 3° K cosmic background radiation, and then it can travel long distance without losing the energy. The aim of the UHE-neutrino detection is to study, (1) the UHE neutrino interaction, which is not afforded by an artificial neutrino beam, generated by an accelerator, (2) the neutrino mass problem through neutrino

CP579, *Radio Detection of High Energy Particles,* edited by D. Saltzberg and P. Gorham
© 2001 American Institute of Physics 0-7354-0018-0/01/$18.00

oscillation effect after the long flight distance, (3) the UHE accelerating mechanism of the proton existing in the universe.

In order to detect the UHE neutrinos, we need a huge mass (at least 10^9 tons) of the detector since neutrinos interact in the detector volume via weak interactions and the flux of UHE neutrinos is very low. However, from the practical point of view, it is difficult to construct such a huge detector. Therefore we decide to use the rock salt mine in nature as the UHE neutrino detector, Salt Neutrino Detector (SND). For such a huge detector we could not give any treatment on it e.g. purification of the medium etc., then we should use it as it is naturally existed.

The interaction cross-section of UHE neutrino increases up to 10^{-33} cm^2 but still it is not so large compared with the other interactions [6]. Due to the increase, the number of upward neutrinos is suppressed due to the absorption by the earth. The differential shadow factor by the earth and the neutrino fluxes are presented in ref. [6]. A moderate event rate over PeV neutrinos is expected to be 10/year by 1km $\times$ 1km $\times$ 1km of 2×10^9 tons of rock salt from QSOs or AGNs and GRBs by Athar, H. [7].

The mass of the detector required to measure the UHE neutrinos should be compared with the existing large neutrino detector, Super Kamiokande (S-K), which consists of 5×10^4 tons of pure water [8]. SND is 4×10^4 times large compared with S-K. They detect visible light generated by Cherenkov effect in the pure water. The transparency (absorption length) in pure water is 100 m at most. On the contrary, rock salt is one of the most transparent materials for radio wave. The transparency of electromagnetic wave in microwave region or longer wavelength is expected to be larger than that of the optical wavelength in pure water. Therefore moderate number of radio wave sensors could detect the neutrino interaction in the massive rock salt.

Coherent Cherenkov radiation

Almost one third of the energy in the hadronic shower generated at UHE neutrino interaction is given to the electromagnetic part by two-photon decays of neutral pions. Askar'yan, G. A. proposed detection of radio emission with coherent amplification from the excess negative charges of an electron-photon shower in a dense material [9]. In the range of wavelengths greater than the dimensions of the excess electron cluster, the intensity of the radiation is proportional to the square of the total excess electron track length. The density of the medium determines the dimensions of the localized region of shower particles. The shower size and the wavelength determine how much the radiation is in coherence. The small attenuation of the radio waves in rock salt, marble, granite, and etc. underground is preferable. Also the absorption of radio waves in the ground of the moon should be small to use such an aim.

The presence of the excess charge in a shower may emit Cherenkov and bremsstrahlung radiation in a uniform medium. But the bremsstrahlung radiation is suppressed by LPM effect [10] and dielectric material effect [11] in the radio wave range. LPM effect is caused by that the coherent formation zone extends to the longitudinal direction due to the small longitudinal momentum transfer by the uncertainty principle. In the process of Coulomb multiple scattering of electrons in the coherent formation zone, radio emission gets destructive interference effect. For 2.5 MeV electrons, emission of radio wave under the frequency ~800 GHz is suppressed. Stronger suppression in the emission frequency comes from the dielectric material effect. Forward Compton scattering amplitude changes in phase giving destructive interference in the radio emission. For 2.5 MeV electrons, emission of photons under the energy ~150 eV is suppressed. Therefore practically only Cherenkov radiation effect should be considered in a uniform medium.

Wavelength λ of radio wave are compared at 1 GHz ($\lambda = 0.3$ m) and visible light at 1 PHz ($\lambda = 0.3$ μm) or 1×10^{15} Hz

$$\frac{\lambda(1\,\text{GHz})}{\lambda(1\,\text{PHz})} = 10^{6}. \tag{1}$$

Emission power of Cherenkov radiation is expressed by Frank-Tamm formula when charge ze travelling a path-length l with the velocity of $c\beta$ in a medium of refractive index n

$$\frac{dW}{dv} = (2\pi h\alpha/c)z^{2}v[1 - \frac{1}{(\beta n)^{2}}]l \tag{2}$$

Radiation power W at the frequency v of 1 GHz and 1 PHz with the photon energies of 4.1×10^{-6} eV and 4.1 eV, respectively, are compared

$$\frac{W(1\,\text{GHz})}{W(1\,\text{PHz})} = 10^{-6}. \tag{3}$$

If we take into account the bandwidth, the power to be detected becomes hopelessly low ~10^{-12} to detect the radio wave. Visible wavelength is much shorter than the shower size, and then the radiation power is proportional to the total track length of the charged particles. The radiation power emitted by visible light $W_{visible}$ is expressed

$$W_{\text{visible}} \sim N < l >, \tag{4}$$

where N is the number of whole charged particles, $N(e^{-})+N(e^{+})$ with the average track length of $< l >$. The notations $N(e^{-})$ and $N(e^{+})$ are the numbers of electrons and positrons in the shower, respectively. On the contrary if the wavelength is larger than the shower size enough, radiated wave from each excess-negative charged track becomes constructively coherent with the same phase. Then the radiation power increases with the assistance of the coherent effect to a detectable level and is proportional to the average track length of charged particles times the square of negative excess charges $\Delta q \bullet N$, which is $N(e^{-})-N(e^{+})$. The radiation power for the radio wave, $W_{radiowave}$ is given

$$W_{\text{radiowave}} \sim (\Delta q \bullet N)^{2} < l >, \tag{5}$$

where Δq is a ratio of negative excess charges to the whole charges. $W_{radiowave}$ is amplified by a factor of $\Delta q^2 \bullet N$ compared with $W_{visible}$. The factor is the negative excess charges times Δq, which is described

$$\Delta q = \frac{N(e^-) - N(e^+)}{N(e^-) + N(e^+)}.$$ (6)

The possibility of using ice as the medium of the neutrino detector to propagate the radio wave is considered [12]. The excess electrons of the electromagnetic shower and coherent Cherenkov radiation by them are studied in detail in the case of ice [13]. The radiation power was calculated without absorption in the ice

$$W \sim 5 \times 10^{-14} [\frac{E}{1\,TeV}]^2 [\frac{\nu_{max}}{1\,GHz}]^2 \; erg \sim 5 \times 10^{-15} \; J.$$ (7)

Where we assume $E = 1$ PeV, $\nu_{max} = 1$ GHz.

They estimated the threshold energy E_{th} for the electromagnetic shower as a function of distance to the observation. At the distance of R (km), the detectable threshold energy with a receiver noise of 300° K thermal equivalent with the bandwidth of 1 GHz becomes,

$$R \; (km) \sim \frac{E_{th} \; (PeV)}{4}.$$ (8)

The excess electrons (Δq) are reported to be 15 % of the total track-length of particles in ice [13]. They are caused by the interaction of photons, positrons and electrons in the shower with the atomic electrons in the medium and generated by the Compton (~60 %) and Bhabha scattering (~30 %), and positron annihilation in flight (~10%). On the contrary, Moller scattering decreases the excess charge (~ -5 %) due to the energy loss by the collision. Radio Ice Cherenkov Experiment (RICE) in Antarctic ice is being executed successfully and gave a neutrino flux limit [14]. Recently, Radio-frequency measurements of coherent transition and Cherenkov radiation is done using a 15.2 MeV pulsed electron beam at the Argonne Wakefield accelerator. They detected coherent radiation, which is attributed both to transition and possibly Cherenkov radiation [15]. The coherent radiation was confirmed that the radiation strength was proportional to the square of the number of electrons in a single pulse.

Comparison with optical Cherenkov radiation

Detection of optical Chrenkov radiation is the well-established method using water (IMB, S-K, DUMAND, NESTOR, Lake Baikal, ANTARES, etc.) and ice (AMANDA, ICECUBE) [14] as the radiator. They could detect Cherenkov radiation emitted by high-energy charged particles faster than the velocity of light in the medium. The detection of the light is done well-developed photo-multiplier tubes, which could detect even one photoelectron converted at the photo cathode from an optical photon. The conversion efficiency (quantum efficiency) is over 20 % in the optical wavelength region. Therefore only one charged particles passing in the detection region could be measured easily. Moreover muons could be detected even if a muon passed the detector region which was produced in muon-neutrino charged-current interaction at a long way off. As a result it could provide a large mass of neutrino reaction volume although the initial energy is not determined so definitely due to being insensitive to the energy loss between the interaction point to the detector.

Meanwhile radio Cherenkov detection method could not detect single muon, and could not utilize the long range of high-energy muons. But it utilizes special characters that the radiation power is roughly proportional to the square of the shower energy due to the coherence effect among radio Cherenkov radiation from the excess electrons in the shower. And the attenuation length of radio wave is long in the medium. Consequently, the radio sensors could be set sparsely in a huge medium. The electron and the tau neutrinos could be detected with the information of the energy and the incident direction utilizing the specific Cherenkov radiation pattern in the same way as the optical Cherenkov detectors. Although for a muon generated by a charged-current interaction of a muon neutrino, it could not be registered after it escaped from the shower generated at the interaction.

Consideration on the medium density

We compare density, radiation length, refractive index, Chrenkov angle and Cherenkov threshold kinetic energy of electrons for air, ice and rock salt in Table 1.

TABLE 1. Comparison between air, ice and rock salt in density, radiation length X_0, refractive index, Chrenkov angle θ_c and Cherenkov threshold kinetic energy for electrons T_{th}.

	Density (g/cm^3)	X_0 (cm)	Refractive Index	θ_c (deg)	T_{th} (keV)
Air (STP)	0.0012	30420	1.000293	1.387	2060407
Ice	0.924	39	1.78	55.8	107
Rock salt	2.22	10	2.43	65.7	50

The density of air is 1/833 times less than rock salt and the radiation length is 3042 times larger than that even on the earth surface. Then the shower length and the diameter are much larger than ice and rock salt. The detection is not easy for such an extended air shower. In addition Cherenkov or fluorescent light detection in air shower depend on weather condition severely as well as the Sunlight, the Moonlight, lightning and artificial light. Direct particle detection of the UHE shower by the particle detector set on the surface is not affected by such the conditions but the detection of horizontal air showers is difficult. Underground detectors neither depend on the weather nor the Sunlight and the Moonlight. In principle they could furnish the same detection ability for all the time and the directions of the neutrino incidence.

In comparison with ice, rock salt has 2.4 times higher density, shorter radiation length (~1/4) and 1.4 times larger refractive index. Then the high-energy particle shower size is small owing to the high density. For electrons, the Cherenkov angle is large and the threshold energy is small due to the high refractive index. Consequently, rock salt is adequate medium to get larger Cherenkov radiation power. The frequency at the maximum electric field strength at θ_c is ~2 GHz without the absorption in ice [13]. Whereas it would increase up to ~5.6 GHz taking into account the short radiation length and the wavelength contraction due to the higher refractive index.

When a UHE proton hits the atmosphere, it could not put the large energy deposit underground in a narrow region. Radio-wave neutrino detector could sense only from many tracks of excess electrons inside the concentrated region with the help of strengthening by the coherence effect. For high sensitivity detector e.g. optical Cherenkov detectors, which could detect single muons, a great number of downward muons become large backgrounds. On the contrary the radio Cherenkov detector is immune from the muons. Therefore, practically only UHE neutrinos could induce the electric field on the radio antennas underground. However it could be a shortcoming when we intend to detect abundant lower-energy neutrinos in high statistics.

Normally, rock salt is covered by thick soil, which absorbs electromagnetic wave completely. Then SND is background free from natural or artificial radio wave coming from the surface on the earth. As a result background is only blackbody radiation corresponding to the temperature of the surrounding rocks. As the remaining potential background, radiation may come from a seismic movement of the surrounding rocks, which may generate radio wave due to the piezoelectric effect by the stress in the rocks. If such a radiation could be detected, the observation contributes to the seismology.

Appearance of tau neutrino due to oscillation

Detection of high-energy cosmic tau neutrinos [16] could be achieved by making use of the characteristic double shower events [17] in the detector medium. The two showers are connected by a μ like track. The size of the second bang is on the average a factor of two larger than the first bang. Electron neutrinos produce a single bang at these energies whereas the muon neutrinos typically produce a single shower along with a zipping μ -like track. Based on these rather distinct event topologies, cosmic neutrino flavor may be conceivable. We assume the standard ratio of the intrinsic cosmic neutrino flux being $F(\nu_e)$: $F(\nu_\mu)$: $F(\nu_\tau) \cong 1:2:0$ at the production. With the constraints of the solar and atmospheric neutrino and the reactor data, we obtain that the ratio of the cosmic high-energy neutrino fluxes in the far distance is 1:1:1, irrespective of which solar solution is chosen.

Possible experimental sites of rock salt mine for SND

We searched a medium which has long attenuation length in a Science chronological table [18]. Then we found that rock salt has one of the lowest $\tan\delta$ of 5-1×10^{-4} in the wide range in the radio frequency of 1 kHz - 10 GHz with the relative real permittivity of 5.9. The loss angle $\tan\delta$ is the ratio of the imaginary divided by the real part in the complex permittivity. The imaginary part presents the absorption in the material. Therefore smaller $\tan\delta$ means longer absorption length. But if water is present in the rock salt, the transparency reduces much. Another idea come from a finding that rock salt is free from water permeation described in a topography dictionary [19]. On the salt dome, petroleum or natural gas are likely to deposit. They are covered with soil, which prevents radio wave to penetrate from the surface. Unfortunately there is no rock salt in Japan. In that era, Japan had no chance to form rock salt since it was under the ocean.

Due to the characters of no water permeation and the long time stability, they are used to maintain radiation wastes, e.g. Asse and Gorleben salt mines in Germany, and WIPP (Waste Isolation Pilot Plant) at USA. Some laboratory measurements of RF complex permittivity had been made on a variety of rocks encountered in mining, tunneling, and engineering works. An RF impedance bridge and a parallel-plate capacitance test cell were employed at frequencies of 1,5, 25, and 100 MHz [20]. In their measurement rock salt was the most transparent for the radio wave.

Rock salt had beam made in old days, around 250 millions yeas ago when a huge ocean was situated in a hot, dry desert climate. As the water in a closed region evaporated, dissolved substances were left behind. Marine sediments later buried these deposits, but since halite is plastic and less dense than the materials that make up the overlying sediments, the salt beds often punched up through the sediments to create dome-like structures. Additional sediments now mostly bury these. Salt consists of the two basic elements, sodium and chlorine, and forms colorless, cube-like crystals. Various additional elements such as iron, potassium, magnesium or calcium, discolor the slat and also alter its taste; magnesium produces a bitter taste and potassium a soapy one. There are many rock salt basins in the world [21].

We had visited five rock salt mines. Among them, at Austria, we went to see Durrenberg salt mine at Hallein near Saltzburg and Altaussee salt mine at Salzkammergut, July 1999. Durrenberg salt mine already closed mining 10 years ago and active only for sight seeing. The Altaussee is still active to produce salt by the borehole brine method. The entrance of the salt mine is located at the mountainside going into horizontally. The temperature inside the tunnel was around 5°C. It is said to be the largest and the cleanest among European Alps rock salt mines. The surfaces of the tunnels got wet and there were lakes underground. Tourists could ride on a boat in the lake. As far as the view from the sightseeing course, it seemed that it was not adequate for the radio wave passage inside the rock salt since the water penetrated rock salt crevices might absorb radio wave much.

We visited Asse research mine on August 1999 and March 2000, which is located near Braunschweig between Hanover and Berlin, Germany. The Asse is a range of hills of approximately 8 km length in the northern Harzforeland. In the course of geological time, the Zechstein salt deposited on the ocean floor migrated into the core of the Asse structure. The mine is used as an underground laboratory to be a service facility of the GSF-National Research Center for Environment and Health. Most of the tunnels would be used to store radiation wastes. We could go down driving a car through a spiral section (average gallery incline 10 %) between 490 m and 930 m levels as well as by a facility of elevators in shafts. The Asse salt mine consists of vertical area between 490 and 850m depth, divided into 16 levels. At the present, the lower area of the salt structure is being mined to the planned level of 930 m. The horizontal tunnels make a lattice extended nearly 750 m at each levels. Air was sent inside to cool down the temperature. The humidity was low. Near the bottom there was a low background radiation laboratory operated by PTB (the Physikalisch-

Technische Bundesanstalt in Braunschweig, Germany). In the rock salt environment, natural radiation background from uranium and thorium is lower than inside of normal rocks. In Asse mine, a group from Hanover studied the salt bodies used as the repositories for nuclear waste using a 50 MHz ground-penetrating radar. They had succeeded to find positions and shapes of cavities and other kind of rocks other than the rock salt in tomographically [22].

The Hockley salt mine is near Houston, Texas. The salt mining is executed for commercial use. There is an underground "The Hockley Seismic Station" operated by the Institute for Geophysics, University of Texas and is a cooperative project between the University of Texas, IRIS, USNSN, and the United Salt Corporation. The Hockley seismometer is located in an excavated chamber within an active salt mine owned by United Salt Corporation. The chamber lies approximately 470 meters below the surface within a large salt dome. Locating the instrument within a salt dome reduces the noise from human activity that generally affects instruments on the surface. The air was dry and the temperature was rather high. Tunnels are developed as a lattice nearly 1 km at the underground level. The salt dome size seems enough for SND.

The Waste Isolation Pilot Plant (WIPP) is the first underground repository licensed to safely and permanently disposes of transuranic radioactive waste left from the research etc. in USA. After more than 20 years of scientific study, public input, and regulatory struggles, WIPP began operations on March, 1999. It is located in the remote Chihuahuan Desert of Southeastern New Mexico; project facilities include disposal rooms mined 540 m underground in a 500m thick salt formation that has been stable for more than 200 million years. The tunnels are excavated nearly 1 km in a lattice style. Over the next 35 years, WIPP is expected to receive shipments. The temperature of the rock salt is 24° C and the air is dry. One of the cavities is planned to use underground experiments such as dark matter search etc. The rock salt extends horizontally in the wide range around WIPP although the salt deposit is not so deep. There is enough land of the governmental property to construct SND near WIPP.

Measurement of microwave absorption length in the laboratory and the noise measurement of a radiometer

We have measured complex permittivities of rock salt samples, shown in Fig. 1.

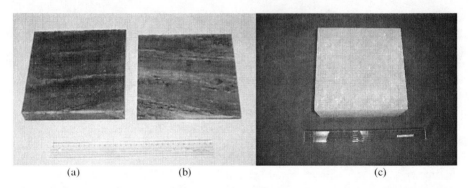

(a)	(b)	(c)

FIGURE 1. Photos of rock salt samples from Hallstadt salt mine: a) 200mm × 200mm × 30mm, b) 200mm × 200mm × 11mm, Austria, and Asse salt mine: c) 200mm × 200mm × 99mm], Germany.

The samples from the Hallstadt salt mine are colored in brown and have a striped pattern. On the contrary, the sample from Asse salt mine is white and has no striped pattern. It seems the rock salt consists of small crystals.

A free space measurement method was employed which had been used in a non-destructive manner for the assessment of microwave materials [23]. The principle of the method is illustrated in Fig. 2. The vector network analyzer HP85107A measures the strength and the phase of the scattering wave with and without a metal-plate reflector on the sample. The feature of the method is the ability to subtract the effect from the background wave scattered from the wall and the floor, etc. other than the area of the sample covered with the metal-plate reflector, namely measuring area. By displacing the position of the metal-plate reflector on the specimen to be tested, the incident wave and the scattered wave from the measuring area were determined without the influence of extraneous waves such as the direct coupling between the transmitting and the receiving antennas and scattered waves from the background objects. The behavior of the metal-plate reflector is similar to that of an optical shutter in optics. The photograph of the equipment is shown in Fig. 3.

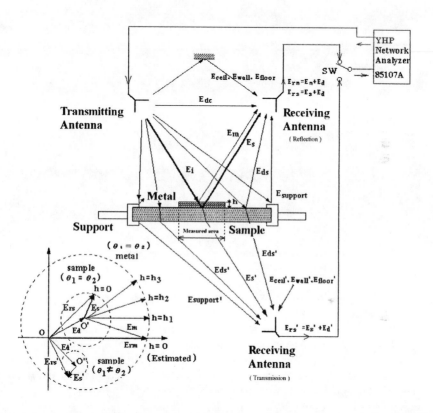

FIGURE 2. Principle of the measurement is illustrated with various wave components. A vector diagram of the received signals is shown with and without the metal-plate reflector on the sample.

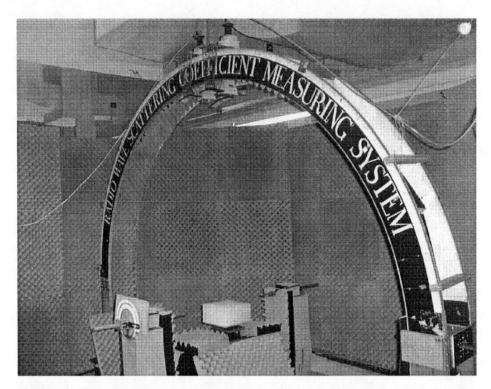

FIGURE 3. Photograph of the measuring apparatus in the X-band region with the sample from Asse salt mine. The transmitting and receiving horn antennas are so set at the upward on the arch that both the incident and scattering angles become 5°. The diameter of the arch is 2.6 m.

In Fig. 2 the illustration shows the measurement with the metal-plate reflector where three metal plates are piled up on the sample to adjust the height of the metal plate surface. The receiving antenna is set at the reflection configuration. Just for instance, the configuration of transmission measurement is also given. The notations E_m, E_s and E_d mean scattered signals from the metal plate, the measuring area of the sample and the background objects, respectively. The received signals of E_{rm} and E_{rs} are the scattered signals from the metal-plate reflector E_m with the background E_d, and the sample within the measuring area E_s with the background E_d, respectively. Changing the height of the metal-plate reflector gives a small difference in the path length between the antennas and the metal-plate reflector, which alters the phase of the scattering wave from the metal-plate reflector only. In the vector diagram the horizontal and the vertical axes express the real and the imaginary coordinates, respectively. The large and the small circles show the cases with and without the metal-plate reflector, respectively. The heights of the metal-plate reflector in the different path lengths are expressed as h = 0, h_1, h_2 and h_3 on the large circle for E_m. At the small circle, E_s is shown at h = 0 without the metal-plate reflector. We could subtract the background effect deducting E_d.

As shown in Fig. 3, the sample was set horizontally on a metal table, which could be lifted up and down to adjust the height of the surface of the metal-plate reflector or the sample. The microwave was irradiated downward with the incident angle of 5° to a sample by a horn antenna having an aperture of 97 mm × 68mm at 8.0-10.0 GHz. The reflected radio wave was detected by another horn antenna of the same aperture with the same angle of 5°. The distance from the horn antennas to the sample was 1.3 m. The distance from the antennas was adjusted to keep the same in the two cases of with and without the metal-plate reflector of 200 mm × 200 mm × 3 mm on the sample.

The measurement was done at 7 heights of the sample or the metal-plate reflector being separated in 2mm, which correspond to $\lambda/8$ changes in the path lengths in the air. The measured complex values of 7 received signals are plotted in Fig. 4 with (solid line) and without (dashed line) metal-plate reflector.

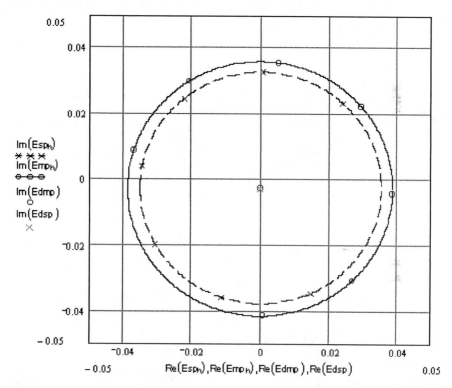

FIGURE 4. The measurement was done at each position of the sample being separated in 2 mm, which corresponds to $\lambda/8$ changes in phase. The measured complex values of received signal are plotted in each position with (solid line) and without (dotted line) metal plate. The figure is in parallel polarization with the scattering plane. The 7 measured points in each circle with (circle) and without (cross) the metal-plate reflector are fitted, and the centers are found to be coincident.

Figure 4 shows the vectors of $\mathbf{E_{rm}}$ from the origin to the large circle and $\mathbf{E_{rs}}$ to the small circle in parallel polarization with the scattering plane, respectively. The notations are the same at Fig. 2. The centers of the two circles coincide so precisely that the background signal E_d is the same in the two cases. Also the centers are so

215

close to the origin that the background signal $\mathbf{E_d}$ is small. Naively, in case of a single reflection at the bottom surface of the sample, the angle and the radius differences in the complex plane correspond to real and imaginary permittivity, respectively. In more precise analysis, we take into account the multiple paths between the upper and the lower surfaces inside the sample. The complex permitivity was deduced from the ratio (scattering coefficient) of the complex measurements of the electric fields as $\mathbf{E_{rs}}$ /$\mathbf{E_{rm}}$. The measured real part of the complex permittivities are tabulated in Table 2 as the linear polarization of parallel R_p and perpendicular R_s with the scattering plane. The values are well consistent each other and with the value of 5.9 tabulated in the handbook [18]. While the measurement accuracy is not enough at the present in the imaginary part of the permittivity. The imaginary parts have different values in the samples a) and b). Both were cut from the same block. For the time being we could say the imaginary part of the permittivity is between 6×10^{-4} and 6×10^{-3}, namely tanδ is between 1×10^{-4} and 1×10^{-3} at 10 GHz.

TABLE 2. Real part of complex permittivities in rock salts

Thickness / Polarization	R_p	R_s
Hall Statt 11.1mm	5.9	6.0
Hall Statt 30.1mm	5.9	6.0
Asse Mine 99.0mm	5.9	5.9

We could calculate the absorption coefficient α:

$$\alpha = \frac{\omega}{c}\sqrt{\varepsilon'}\frac{\tan\delta}{2}, \qquad (9)$$

where the notations are following, complex permittivity ε:

$$\varepsilon = \varepsilon' - j\varepsilon'' = \varepsilon'(1 - j\tan\delta), \qquad (10)$$

loss angle in the permittivity $\tan\delta$:

$$\tan\delta = \frac{\varepsilon''}{\varepsilon'}, \qquad (11)$$

complex refractive index n:

$$n = \sqrt{\varepsilon} = \sqrt{\varepsilon'}\sqrt{1 - j\tan\delta}. \qquad (12)$$

When the value of $\tan\delta$ is 1×10^{-4} or 1×10^{-3}, α becomes 0.025/m or 0.25/m, respectively, at 10 GHz. As the result we get the absorption lengths $1/\alpha$ as 40 m or 4 m. Assuming $\tan\delta$ is constant with respect to the frequency, the absorption length comes 400m or 40 m at 1 GHz.

We are planning to measure natural rock salt samples by a perturbed cavity resonator method at 9.4GHz [24]. The drawing of the cavity is shown in Fig. 5. The photograph of the apparatus is presented in Fig. 6.

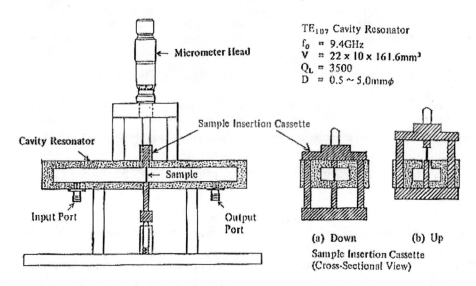

FIGURE 5. Illustration of the perturbed cavity resonator.

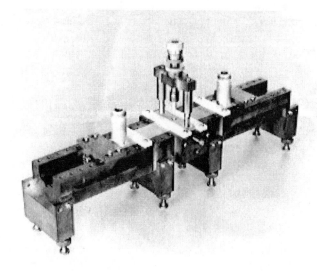

FIGURE 6. Photograph of the perturbed cavity resonator.

Using the apparatus the complex permittivity ε could be measured. The principle of the measurement in the real ε' and the imaginary ε'' of the complex permittivity is the changes in the center frequency and the width of the resonance, respectively, with and without insertion of the sample in the cavity. The thin sample is needed to the small perturbation for the resonance behavior. The imaginary ε'' depends on the change of the Q factor (the ratio of the resonance frequency to the resonance width). In order to get the value, we should take into the volume ratio of the sample to the cavity. We could cleave a synthetic rock salt sample to the size of 1 mm $\times$ 1 mm $\times$ 10.2 mm. By that method we have got the result of the synthetic rock salt as $\varepsilon = 5.8 - j3.5 \times 10^{-4}$ or $tan\delta = 0.6 \times 10^{-4}$ which is a little bit lower than the value tabulated in ref. [18]. While we have not succeeded to cleave the natural samples to the size.

We have tested a noise level of a radiometer system. The main part of the radiometer is a modified TV Broadcasting Satellite (BS) receiver, which is an amplifier-converter module with the center frequency of 10.6 GHz. The system is composed of the horn antenna having the aperture of 97 mm $\times$ 68mm, the amplifier-converter module, a 30 dB amplifier, a rectifier and a time chart recorder or an oscilloscope. The amplifier-converter module consists of a signal conversion part from the wave guide of the horn antenna to a coaxial style, 3-stage low noise HEMT amplifiers with filter at the center frequency of 10.6 GHz (band-width of 200 MHz), and frequency converter to which the local oscillator signal of 9.6 GHz is introduced from outside. After the conversion part, the signal is amplified with another 3-stage amplifier in the module. The horn antenna is directed vertical to the sky having ~30° K noise or covered with a ~300° K thermal noise material. The comparison of the signal strengths, we could get the noise level of the amplifier-converter module as low as 150° K. The noise level is well below the rock salt temperature. Due to the results we could afford to get a lot of the receivers with low cost which is the modified BS receiver used by popular TV audience.

SND is envisaged as shown in Fig. 7. Radio wave sensors of 216 are arrayed regularly in 200m repetitions inside a 1km $\times$ 1km $\times$ 1km rock salt. The attenuation length becomes longer when we use the lower frequency to be detected. However the detection threshold energy of UHE neutrinos reduces at the lower frequency. The optimization should be done to select the frequency. If the attenuation length is 400 m, the sensor separation distance less than 200 m is needed to detect the Cherenkov ring shape as well as the energy. Six radio sensors are hung on a string. The string is lowered in a well with the depth over 1km. At the total 36 wells should be bored.

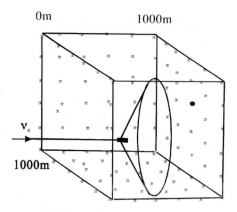

0m   1000m

ν_e

1000m

FIGURE 7. Salt Neutrino Detector underground. Excess electrons in the shower at the UHE neutrino interaction generate a coherent Cherenkov radiation with the emission angle of 66°.

Summary

Rock salt is studied as a radio wave transmission medium in the UHE cosmic neutrino detector. The radio wave would be generated by Askar'yan effect (coherent Chrenkov radiation from negative excess charges in the electromagnetic shower) in the UHE neutrino interaction in the rock salt. Some rock salt mines were investigated whether it has a possibility as a SND site or not. We found that there were potential sites to construct SND. The samples of the rock salts were gathered to measure the absorption length. As a tentative result, the absorption length of the radio wave was obtained to be between 400m and 40 m at 1 GHz.

The preliminary results of the radio wave absorption length in the rock salt show that it has a chance to be used as the medium of UHE neutrino detector when we select a rock salt mine in good transparency. We need to measure in the lower frequency and more samples as well as in the perturbed cavity resonator method to deduce the concrete conclusion. Before the site is decided, it is important to measure the absorption length in *situ*. There may be defects and impurities as well as intrusions by rocks other than rock salt. For that research it is useful to employ the ground penetrating radar, which is a well-explored technique.

Further we studied a radiometer modified from TV BS receiver, which showed the noise level of 150° K. The level is low enough to implement SND. It is advantageous if we could use the low cost receivers. Large number of antennas is needed in the configuration of the antennas spacing less than 200 m in a 1km × 1km × 1km rock salt. The attenuation in the rock salt and the radiation pattern of the coherent Cherenkov radiation require the repetition of the antennas. The radio frequency should be decided taking into account the detection energy threshold of UHE neutrinos and the attenuation length. In addition it is indispensable to study the basic process on the coherent Cherenkov radiation due to a pulsed electron beam supplied by an accelerator. By the research we could calibrate the energy of the initial electromagnetic shower by the detected radiation power, which depends on how much coherency is realized.

ACKNOWLEDGMENTS

This work was supported partly by Funds for Special Research Project at Tokyo Metropolitan University, Fiscal Year 1999. We should appreciate Profs. K. Minakata and T. Kikuchi (TMU) to support this project.

We express our gratitude to M.E. Ryouichi Ueno who discussed with and advised us about the microwave techniques. He was indispensable to carry out this study.

The research could not be possible without the assistances from and discussions with many persons because parts of the field researched were far from our specialized field. Here we express our appreciations to those who have interests and extended their favors on our research.

Prof. Tuneaki Daishido (Wasada University, Tokyo) instructed us about the low noise microwave detection and transferred us the amplifier-converter. M.E. Yoshinori Sanada (Kyoto Univ.) gave us the penetration rate of electromagnetic wave underground. Messrs. Akiharu Kitahara and Hisashi Itou (Tokyo Metropolitan Industrial Technology Research Institute) informed us low energy electron accelerators available. Messrs. Yuichi Chou and Teruhisa Hirashima (Denki Kogyo Co. Ltd.) provided us an opportunity to use a compact RF accelerator for electron beam irradiation.

The information and the samples of Austrian rock salt mines were got from Profs. Franz Mandl, Jimmy MacNaughton (Institute for High Energy Physics, Austria), Komarek (Wien University, Austria) and Dr. Doz.Weber (Ministry of Economy, Austria) and Mr. Masayuki Handa (Tobacco and Salt Museum, Tokyo).

Prof. Motoyuki Sato (Tohoku Univ.) introduced us ground-penetrating radars in a rock salt mine at Asse research mine. We are indebted for the call at the Asse mine and getting the sample to Dr. Dieter Eisenburger (Federal Institute for Geosciences and Natural Resources, BGR, Hanover, Germany), Dr. Stefan Neumaier and Mr. Eberhard Funk (Physikalisch-Technische Bundesansalt, Braunshweig, Germany). Messrs. Nobuteru Hitomi and Yoshiharu Kobayashi (Mechanical Engineering Center, KEK) cut us the rock salt samples.

We appreciate those who informed us about salt domes concerning to petroleum minings; Messrs. Shin-ichi Inoue (Abu Dhabi Oil Co., Ltd.), Hitoshi Takezaki (Japan Energy Development Co., Ltd.), Yoshihisa Tanaka, Taizo Uchimura, Toru Akutsu, Dr. Osamu Takano (Japan National Oil Corporation), and Mr. Greg Jones (Exxon, New Orleans, USA).

We are indebted to Prof. Yoshio Nakamura (Institute for Physical Geology, University of Tesax) and Mr. Marcelino Segura (United Salt Corporation, Hockley Plant) for visiting Hockley Salt Mine (United Salt Co., Houston, Texas, USA).

At the visiting of WIPP, We were taken care of by Prof. Shinjiro Mizutani (Nihon Fukushi Univ.), Dr. Hidekazu Yoshida (Nuclear Fuel Cycle Development Institute), and Dr. Erik. K. Webb (Sandia Institute, USA), Dr. Roger Nelson, Ms. Beth Bennington, and Mr. Norbert T. Rempe (Waste Isolation Pilot Plant, Carlsbad, New Mexico, Department of Energy, USA).

REFERENCES

1. Stecker, F. W., Done, C., Salamon, M. H., and Sommers, P., *Phys. Rev. Lett.* **66**, 2697-2700 (1991).
2. Barwick, S., Halzen, F., Lowder, D., Miller, T., Morse, Price, P.B. and Westphal, A., *J.Phys. G:Nucl. Part. Phys.* **18**, 225-247 (1992); Thomas K.Gaisser, Francis Halzen, Todor Stanev, *Phys. Reports* **258**, 173-236 (1995).
3. Takeda *et al..*, *Phys. Rev. Lett.* **81**, 1163-1166 (1998).
4. Greisen, K., *Phys. Rev. Lett.* **16**, 748 (1966); .Zatsepin, G.T., Kuz'min, V.A., *Zh. Eksp. Teor. Fiz., Pis' ma Red.* **4**, 114 (1966) [*Soviet Physics JETP Lett.* **4**, 78 (1966)].
5. Burdman, G., Halzen, F. and Gandhi, R., *Phys. Lett.* **B417**, 107-113 (1997); .Jain, P., Mckay, D.W., Panda, S., Ralston, J. P., *Phys. Lett.* **B484**, 267-274 (2000).
6. Gandhi, R., Quigg, C., Reno, M.H. and Sarcevic I., *Astroparticle Phys.* **5**, 81-110 (1996); *Phys. Rev.***D58**, 093009: Sigl, G. *Phys. Rev.* **D57**, 3786-3789 (1998): Kwiecinski, J., Martin, A.D., Stasto, A.M., *Phys. Rev.***D59**, 093002 (1999); Horvat, R., *Phys. Lett.* **B480**, 135-139(2000).
7. Athar H., Presentation at the seminar of high-energy theory group, Oct. 2000, TMU.
8. Fukuda Y. *et al.*, SuperK Collaboration, *Phys. Rev. Lett.* **81**, 1562(1998).
9. Askar'yan, G.A., *Zh. Eksp. Teor. Fiz.* **41**, 616-618 (1961) [*Soviet Physics JETP* **14**, 441 - 442 (1962)]; **48**, 988 - 990 (1965) [**21**, 658 - 659 (1965)].
10. Landau, L. D., Pomeranchuk, I. Ya., *Dokl. Akad. Nauk SSSR* **92**, 535 (1953); **92**, 735 (1953); Migdal, A.B., *Phys. Rev.*, **103**, 1811-1820 (1956); Klein, S., *Rev. Mod. Phys.* **71**, 1501-1538 (1999).
11. Ter-Mikaelian, M. L., *High-Energy Electromagnetic Processes in Condensed Media*, Interscience Tracts in Physics and Astronomy, Number 29, Wiley-Interscience, a Division of John Wiley & Sons, Inc., New York, (1972), ISBN 0-471-85190-6.
12. Markov, M.A. and I.M.Zheleznykh, and I.M.. *Nucl. Instrum. Methods.* **A248**, 242-251 (1986).
13. Halzen, F., Zas, E.,.Stanev, T, *Phys. Lett.* **B257**, 432-436 (1991); Zas, E., Halzen, F., Stanev, T., *Phys. Rev.* **D45**, 362-376 (1992); Alvarez-Muniz, J. and Zas, E. *Phys. Lett.***B411**, 218-224 (1997); Frichter, J.M., Ralston, J.P. and .Mckay, D.W., *Phys. Rev.***D53**, 1684-1698 (1996).
14. The most recent status of large high-energy neutrino detectors are shown in the presentation transparencies in 19th International Conference on Neutrino Physics and Astrophysics (Neutrino 2000), 2000, Sudbury, Canada at URL http://ALUMNI.LAURENTIAN.CA/www/physics/nu2000/.
15. Gorham, P., Saltzberg, D., Schoessow, Gai, P.W.,.Power, J. G., Konecny, Richard and Conde, M.E., *Phys. Rev.* **E62**, 8590-8605 (2000).
16. Halzen, F., Saltzberg, D. *Phys. Rev.Lett.*, **81**, 4305-4308 (1998).
17. J.G. Learned, G. and Pakvasa, S., *Astropart. Phys.* **3**, 267(1995); Athar, H., Jezabek, M. and Yasuda, O., *Phys. Rev.* **D62**,103007 (2000); Athar, H., *hep-ph/*9912417, 1999; Athar, H., *Nucl. Phys.* **B(Proc. Suppl.)76**, 419 (1999); Athar, H., Parente, G. and Zas, E., *hep-ph/*0006123(2000).
18. *Chronological Scientific tables* (in Japanese) edited by National Astronomical Observatory of Japan, Maruzen Co. Ltd., Tokyo, (1998) pp.486.
19. *Topography dictionary* (in Japanese) edited by Machida, T., *et al.*, Ninomiya Book Co. Ltd., Tokyo, (1981) pp.110.
20. Cook, J.C., *Geophysics*, **40**, 865-885 (1975).
21. Stanley J. L., *Handbook of World Salt Resources*, Plenum Press, New York, (1969); Michel T. H., *Salt Domes*, Gulf Publishing Company, Houston (1979).
22. Mundry, E., Thierbach, R., Sender, F and Weichart, H., *Proceedings of the Sixth International Symposium on Salt*, **Vol.I**, 585-599 (1983); Nickel, H., Sender, F., Thierbach, R. and Weichart, H., *Geophysical Prospecting* **31**, 131-148 (1983); Sato, M. and Thierbach, R. *IEEE Transactions on Geoscience and Remote Sensing*, **29**, 899-904 (1991); Eisenburger, D., *Proceedings of the 5th International Conference on Ground Penetrating Radar*, 647-659 (1994); Eisenburger, D., Gundelach, V., Sender, F., Thierbach, R., *Proceedings of the 6th International Conference on Ground Penetrating Radar*, 427-432 (1996).
23. Ueno, R. and Kamijo, Toshio, *IEICE Trans. Commun.* **E83B**, 1554-1562 (2000); Ueno, R. and Kamijo, T., *Memoirs of Graduate School of Engineering, Tokyo Metropolitan University*, 5743-5752 (1999).
24. Ueno, R. and Kamijo, T., *Memoirs of Faculty of Tech., Tokyo Metropolitan University*, 3923-3933 (1989).

ACCELERATOR MEASUREMENTS

Observation of the Askaryan Effect

D. Saltzberg[1], P.W. Gorham[2], D. Walz[3], C. Field[3], R. Iverson[3],
A. Odian[3], G. Resch[2], P. Schoessow[4], and D. Williams[1]

[1] *Department of Physics and Astronomy, University of California, Los Angeles, CA 90095*
[2] *Jet Propulsion Laboratory, Calif. Institute of Technology, Pasadena, CA, 91109*
[3] *Stanford Linear Accelerator Center, Stanford University, Stanford, CA 94309*
[4] *Argonne National Laboratory, Argonne, IL*

Abstract. We present direct experimental evidence for the charge excess in high energy particle showers and corresponding radio emission predicted nearly 40 years ago by Askaryan. We directed picosecond pulses of GeV bremsstrahlung photons at the SLAC Final Focus Test Beam into a 3.5 ton silica sand target, producing electromagnetic showers several meters long. A series of antennas spanning 0.3 to 6 GHz detected strong, sub-nanosecond radio frequency pulses produced by the showers. Measurements of the polarization, coherence, timing, field strength vs. shower depth, and field strength vs. frequency are completely consistent with predictions. We also show the emission peaks at the Cherenkov angle in sand. These measurements thus provide strong support for experiments designed to detect high energy cosmic rays such as neutrinos via coherent radio emission from their cascades in dense dielectric media.

I INTRODUCTION

During the development of a high-energy electromagnetic cascade in normal matter, Compton scattering knocks electrons from the material into the shower. In addition, positrons in the shower annihilate in flight. The combination of these processes should lead to a net 20-30% negative charge excess for the comoving compact body of particles that carry most of the shower energy. Askaryan [1] first described this effect, and noted that it should lead to strong coherent radio and microwave Cherenkov emission for showers that propagate within a dielectric. The range of wavelengths over which coherence obtains depends on the form factor of the shower bunch—wavelengths shorter than the bunch length suffer from destructive interference and coherence is lost. However, in the fully coherent regime the radiated energy scales quadratically with the net charge of the particle bunch, and hence with the incoming energy. At ultra high energies the resulting coherent radio emission may carry off a significant fraction of the total energy in the cascade.

The plausibility of Askaryan's arguments combined with more recent modeling and analysis [2–4] has led to a number of experimental searches for high energy

CP579, *Radio Detection of High Energy Particles*, edited by D. Saltzberg and P. Gorham
© 2001 American Institute of Physics 0-7354-0018-0/01/$18.00

neutrinos by exploiting the effect at energies from $\sim 10^{16}$ eV in Antarctic ice [5,6] up to 10^{20} eV or more in the lunar regolith, using large ground-based radio telescopes [7 9]. Radio frequency pulses have been observed for many years from extensive air showers [10,11]. However, it has been shown [12,13] that the dominant source of this emission is due to geomagnetic separation of charges, rather than the Askaryan effect. Thus neither the charge asymmetry nor the resulting coherent Cherenkov radiation has ever been observed.

In a previous paper [14] we described initial efforts to measure the coherent radio-frequency (RF) emission from electron bunches interacting in a solid dielectric target of 360 kg of silica sand. That study, done with low-energy electrons (15 MeV), demonstrated the presence of coherent radiation in the form of extremely short and intense microwave pulses detectable over a wide frequency range. These results, while useful for understanding the coherent RF emission processes from relativistic charged particles, could not directly test the development of a shower charge excess, as Askaryan predicted. Also, because the particles were charged, passage of the beam through any interface induced strong RF transition radiation (TR), which obscured the presence of Cherenkov radiation (CR).

We have previously reported on these measurements made at the Stanford Linear Accelerator Center (SLAC) [15]. We used high-energy photons, rather than low-energy electrons, has enabled us to clearly observe microwave Cherenkov radiation from the Askaryan effect. In this report we add a few plots: the shower profile measured at C-band, the shock-wave behavior and the intensity relative to the Cherenkov angle.

II EXPERIMENTAL SETUP

A silica sand target and antennas were placed in a gamma-ray beamline in the Final Focus Test Beam facility at SLAC in August 2000. The apparatus was placed 30 m downstream of aluminum bremsstrahlung radiators that produced a high-energy photon beam from 28.5 GeV electrons. Beyond the radiators, the electron beam was bent down, and dumped 15 m beyond our target. Typical beam currents during the experiment were $(0.2 - 1.0) \times 10^{10}$ electrons per bunch. The two radiators could be used either separately or in tandem, thereby providing 0%, 1%, 2.7% or 3.7% of a radiation length. Thus the effective shower energy induced by the photons could be varied from $(0.06 - 1.1) \times 10^{19}$ eV per bunch. The size of the photon bunch at the entrance of the silica sand target was less than several mm in all directions and is negligible compared to the instantaneous shower dimensions.

The target was a large container built largely from non-conductive materials such as wood and plastic which we filled with 3200 kg of dry silica sand. As shown in Fig. 1, the sand target was rectangular in cross section perpendicular to the beam axis, but the vertical faces on both sides were angled to facilitate transmission of radiation arriving at the Cherenkov angle (about 51° in silica sand at microwave frequencies). We avoided making sides parallel to the beam since the CR would

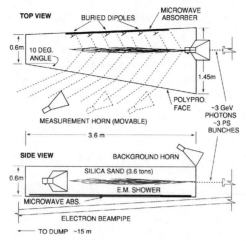

FIGURE 1. Sectional views of the target geometry.

then suffer total internal reflection at the interface. Both internal (buried) half-wave dipoles and external antennas were used for the pulse measurements. The external antennas were pyramidal "standard gain" microwave horns (1.7–2.6 GHz or 4.4–5.6 GHz). A 0.5 in. low-microwave-loss polypropylene was used for the face viewed by the external antennas.

Details of the trigger system, data acquisition, and polarization and power measurements were similar to those used in our previous experiment [14]. Briefly, time-domain sampling of the antenna voltages using high-bandwidth oscilloscopes allowed us to make direct measurements of the electric field of the pulses as a function of time, with time resolutions of 100 ps (0.2–3 GHz) and 10 ps (4–6 GHz). Because of the strength of the signals no amplification was required. Microwave absorber material was placed wherever possible to minimize reflections, and the geometry was chosen so that, where reflections could not be eliminated, they would arrive well after the expected main pulse envelope. The incident electron beam current was measured using a beam current transformer.

Because the accelerator itself uses S-band (2.9 GHz) RF for the beam generation, we took particular care to measure the background levels of RF with the electron beam on, but no photon radiators in place. A weak background RF pulse with a few mV r.m.s. in coincidence with the beam proved completely negligible compared to the signal pulses detected with the photon beam on target, typically 10-100 V peak-to-peak. In addition, we used a second S-band horn to monitor any incoming radiation (see fig. 1). These and numerous other tests such as the lack of effect from moving absorbers eliminated the possibility that stray linac RF or other RF associated with the electron beam contributed to our measurements.

We checked the static magnetic field strength in the vicinity of our sand target and found it to be comparable to ambient geomagnetic levels. This is important since stray fields present in the accelerator vault could have induced significant charge separation in showers within our target and led to other radiation mechanisms. In 0.5 Gauss, the electron gyroradius for the bulk of the shower is ≥ 1 km, implying charge separations of ≤ 1 cm over the length of a shower, well below the ~ 4 cm Molière radius.

III   RESULTS

A   Time & shower profiles

The inset to fig. 2 shows a typical pulse profile measured with one of the S-band horns aimed near shower maximum. Given that the bandwidth of the horn is 900 MHz, the ~ 1 ns pulse width indicates a bandwidth-limited signal. This behavior is seen at all frequencies observed, with the best upper limit to the intrinsic pulse width, based on the 4.4–5.6 GHz data, being less than 500 ps.

Fig. 2 also shows a set of measured peak field strengths for pulses taken at different points along the shower. Both the S-band and C-band horns were translated parallel to the shower axis, maintaining the same angle (matched to the refracted Cherenkov angle) at each point. The values are plotted at a shower position corresponding to the center of the antenna beam pattern, refracted onto the shower. The half-power beamwidth of the horn is about $\pm 10°$; thus the antenna would respond to any isotropic shower radiation over a range of $\pm 20 - 25$ cm around the plotted positions. The CR radiation pattern is expected to have a total beamwidth less than several degrees [3,16], much smaller than the antenna beam pattern.

The plotted curve shows the expected profile of the total number of particles in the shower, based on the KNG [17] approximation. Here the axes have been chosen to provide an approximate overlay of the field strengths onto the shower profile. Clearly the pulse strengths are highly correlated to the particle number. Since the excess charge is also expected to closely follow the shower profile, this result is consistent with Askaryan's hypothesis.

Since CR propagates as a conical bow shock from the shower core, the pulse wavefront traverses any line parallel to the shower axis at the speed of the shower ($\simeq c$) rather than the local group velocity c/n. Using our buried dipole array we have confirmed such behavior. The arrival times of the pulse at each dipole are shown in fig. 3 and imply a radiative bow shock at $v/c = 1.0 \pm 0.1$, inconsistent with the measured group velocity (0.6 c) in the target.

Additional observations with an external S-band horn at the top interface of the sand show evidence for total internal reflection. We observed ~ 34 dB attenuation of the pulse transmitted through the surface, which is approximately parallel to the shower axis. This latter behavior is characteristic of CR, since the Cherenkov angle is the complement of the angle of total internal reflection to vacuum.

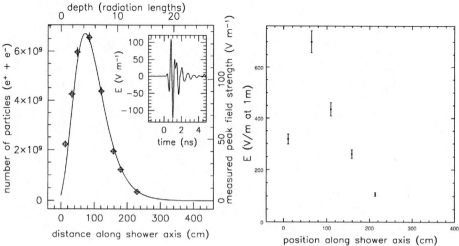

FIGURE 2. left:Expected shower profile (solid), with measured peak field strengths (diamonds) plotted normalized to the peak. Inset: typical pulse time profile. right: Profile using the C-band data.

Pulse polarization was measured with an S-band horn directed at a shower position 0.5 m past the shower maximum. Field intensities were measured with the horn rotated at angles 0, 45, 90 and 135 degrees from horizontal. Fig. 3 shows the pulse profile for both the 0° and 90° (cross-polarized) orientations of the horn. Fig. 3 also shows the degree of linear polarization and the angle of the plane of polarization, respectively. The results are completely consistent with Cherenkov radiation from a source along the shower axis. Since the position was downstream of shower maximum, late reflections from the upper surface of the sand enter the pulse profile, resulting in a loss of polarization beyond ~ 2 ns.

B Coherence

Fig. 4 shows a sequence of pulse field strengths for the external S-band horn versus the total shower energy, which was varied both by changing the beam current and the thickness of the radiators. The curve is a least-squares fit to the data, of the form $|\mathbf{E}| \propto (W_T)^\alpha$ where $|\mathbf{E}|$ is the electric field strength and W_T is the total shower energy. The fit yields $\alpha = 0.96 \pm 0.05$, consistent with complete coherence of the radiation, implying the characteristic quadratic rise in the corresponding pulse power with shower energy.

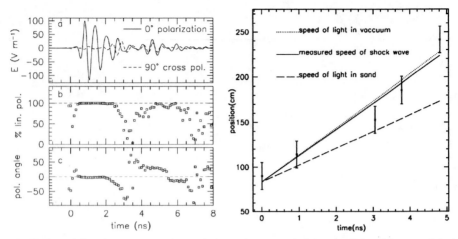

FIGURE 3. left:Polarization analysis of the pulses recorded by the S-band horns. (a) the measured field strength of the pulse, (b) the fractional linear polarization, (c) the angle of the plane of polarization, where 0° is horizontal. right: Electric field normalized to 1 m distance relative to Cherenkov angle–after accounting for refraction at interface.

C Angular dependence

The shape of the target (see fig. 1) was designed to facilitate the measurement of the shower profile. Because the antennas were so close to the shower, angular measurements could not be made by simply rotating the horn–which would then point to a different part of the shower. Instead, we took a series of measurements for which the horn axis was rotated off of the expected direction of the Cherenkov radiation from shower max (after being refracted by the sand-air interface). This required moving the horn longitudinally. As a result the field measurements needed to be corrected for differerent distances from shower max (a linear correction) as well as a small (<10%) correctoin for the changing integral track length in the antenna aperture. The angular response is shown in fig. 4. Uncertainties are due primarily to the precision of antenna positioning and interpretation of the pulse shape. The data show a peak at the expected Cherenkov angle with a width roughly consistent with the antenna response, indicating that the width of the Cherenkov "beam" is much narrower than the antenna's beam, as expected.

D Spectral dependence and absolute intensity

Fig. 4 shows the measured spectral dependence of the radiation. superimposed on a curve based on a parameterization of Monte Carlo results [3,18] which we have corrected for the differences between ice and sand and for the fact that the limited antenna apertures are only sensitive to about half of the total field strength. A parameterization is also available [19] that accounts for various near-field effects but is not straightforward to apply to our case. The error bars are conservative estimates of effects due to uncertainties in surveying, interpretation of pulse shapes, antenna response, and beam intensity. Note that Fig. 4 compares absolute field strength measurements to the predictions and the agreement is very good.

IV   DISCUSSION

We have demonstrated that the radiation we have observed is coherent and 100% linearly polarized in the plane containing the antenna and shower axis and is emitted as a shock wave at the Cherenkov angle in sand. The radiation is pulsed with time durations much shorter than the inverse bandwidth of our antennas. The strength of the pulse is strongly correlated to the size (in particle number) of the shower region that appears to produce it. All of these characteristics are consistent with the hypothesis that we have observed coherent Cherenkov radiation. Because of the strong correlation with the shower profile and the physical constraint that a shower with no net excess charge cannot radiate, we conclude that excess charge production along the shower is the source of the CR observed. These conclusions are strengthened by the agreement of the absolute field strengths with predictions.

Our observations are inconsistent with radiation from geomagnetic charge separation as observed in extensive air showers. The most striking evidence is that the plane of polarization is clearly aligned with the shower axis rather than the local geomagnetic dip angle (62°). Given the approximate east-west orientation of the shower axis, radiation from geomagnetic charge separation would have produced an electric field polarization with significant components orthogonal to what we have observed.

We note that in all cases our measurements are likely to be made in near-field conditions, and we have not attempted to correct for these effects. Recent studies [19,18] have begun to treat these issues, but are not straightforward to apply in our case. In any case, near field effects should generally *decrease* the measured field strengths relative to far field measurements.

The total energy of our cascades is as high as 10^{19} eV, but these showers consist of the superposition of many lower-energy showers. Higher-energy effects [4] would have elongated the shower and tended to increase the total tracklength of the shower, at the expense of a lower instantaneous particle number density. The net effect is that the total radiated power is expected to be approximately conserved, but the angular distribution would sharpen significantly. Scaling of our results to high energy showers should consider corrections for these effects, which are likely to increase the peak field strengths in far-field measurements.

Extrapolations to determine the energy threshold of existing experiments are possible, after correcting for the differences in material properties. For Antarctic ice

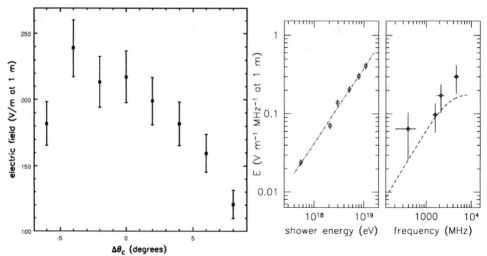

FIGURE 4. (a) Measured coherence of pulse electric field vs. shower energy, at 2.1 GHz. The curve is a least-squares fit. (b) Spectral dependence of the measured pulse field strengths. The horizontal bars indicate the antenna bandwidth, and the vertical bars the uncertainties. The curve is based on a theoretical prediction described in the text.

experiments, the use of the existing simulations [3–5] appears completely justified. For experiments observing the lunar regolith [7–9], silica sand shares many similarities with the lunar surface material, and the expected cascade energy threshold is $\sim 2 \times 10^{19}$ eV. Note that showers from a single high energy particle would be longer than the superposition of low-energy showers used in this study and while the total radiated power would be approximately conserved, the far-field angular distribution would sharpen and lead to possibly even lower thresholds.

In combination with our previous measurements of coherent TR [14], these results have established a firm experimental basis for radio-frequency detection of high energy cascades in solid media, either through interaction within a dielectric (for CR), or via passage through interfaces (for TR). Above cascade energies of $\sim 10^{16}$ eV, these processes become dominant over optical Cherenkov or fluorescence emission in the number of quanta produced [2]. Thus experiments designed to exploit this effect in the detection of ultra-high energy particles can now be pursued with even greater confidence.

V ACKNOWLEDGMENTS

We thank D. Besson, R. Rose, L. Skjerve, and M. Spencer for the loan or construction of several antennas. We thank the members of the SLAC accelerator and EF departments for invaluable assistance before, during and after the run. This research was supported in part by by DOE contract DE-FG03-91ER40662 at UCLA and the Sloan Foundation. It has been performed in part at the Jet Propulsion Laboratory, Caltech, under contract with NASA. SLAC is supported by the DOE, with work performed under contract DE-AC03-76SF00515. Argonne is supported by DOE contract W-31-109-ENG-38.

REFERENCES

1. G. Askaryan, Soviet Physics JETP **14**, 441 (1962); G. Askaryan, Soviet Physics JETP **21**, 658 (1965).

2. M. A. Markov and I. M. Zheleznykh, Nucl. Instr. Meth. A248, 242, (1986).

3. E. Zas, F. Halzen, and T. Stanev, Phys. Rev. D **45**, 362 (1992).

4. J. Alvarez–Muñiz and E. Zas, Phys. Lett. B, **411**, 218 (1997); J. Alvarez–Muñiz and E. Zas, Phys.Lett. B, **434**, 396, (1998); J. Alvarez–Muñiz, R.A. Vázquez, and E. Zas, Phys.Rev. D61 (2000), 023001.

5. G. M. Frichter, J. P. Ralston, and D. W. McKay, Phys. Rev. D. **53**, 1684 (1996).

6. D. Besson *et al.*, Proc. 26th. Intl. Cosmic Ray Conf., v. 2, 467 (1999).

7. I. Zheleznykh, Proc. 13th Intl. Conf. on Neutrino Physics and Astrophysics, World Scientific, Boston, ed. J. Schreps, p. 528 (1988); R. Dagkesamanskii, and I. Zheleznykh, JETP Lett., **50** 259 (1989).

8. T. H. Hankins, R. D. Ekers, and J. D. O'Sullivan, MNRAS **283**, 1027 (1996).

9. P. W. Gorham, K. M. Liewer, and C. J. Naudet, Proc. 26th Intl. Cosmic Ray Conf., v 2, 479, (1999); astro-ph/9906504.

10. J. V. Jelley, et al., Nuovo Cimento 46, 649, (1966).

11. D. Fegan and N. A. Porter, Nature 217, 440, (1968).

12. F. D. Kahn, and I. Lerche, Proc. Roy. Soc. A 289, 206 (1966).

13. H. R. Allan, in *Prog. in Elem. Part. & Cosmic Ray Phys.* 10, ed. J. G. Wilson & S. G. Wouthuysen (North-Holland: Amsterdam), 171, (1971).

14. P. W. Gorham, D. P. Saltzberg, P. Schoessow, *et al.*, Phys. Rev. E, **62**, 8590 (2000).

15. D. Saltzberg, P. Gorham, D. Walz *et al.*, Phys. Rev. Lett., in press, hep-ex/0011001 (2001). +

16. T. Takahashi, *et al.*, Phys. Rev. E **50**, 4041 (1994).

17. K. Kamata and J. Nishimura, 1958, Suppl. Progr. Theoret. Phys. 6, 93; K. Greisen., 1965, Prog. Cosmic Ray Phys. vol. III, J. Wilson ed., (No. Holland: Amsterdam) 1.

18. J. Alvarez–Muñiz, R.A. Vázquez, and E. Zas, Phys. Rev. D62 (2000), 063001.

19. R. V. Buniy and J. P. Ralston, astro-ph/003408 (2000).

Planning for the Generation-X Radio Cherenkov Test Beam Experiment

Alice Bean, Roman V. Buniy, and John P. Ralston

Department of Physics and Astronomy, University of Kansas, Lawrence, KS 66045

Abstract. We outline plans for future test beam experiments on coherent radio Cherenkov radiation, in particular those induced by hadronic showers at laboratory facilities. A high degree of symmetry, exploitation of exact solutions and exact features of radiation, and deliberate over-determination of observables are intrinsic to the experimental design.

I TEST-BEAM X

Coherent Radio Cherenkov radiation is a fascinating phenomenon which underlies the most sensitive known mechanism for the detection of ultra-high energy particle showers [1,2]. The process is exploited for the detection of neutrino-induced showers in cold Antarctic ice in the prototype RICE array [3]. RICE now uses more than a dozen radio receivers probing the region of neutrino energies of about 1 PeV and above. The project has recently generated a preliminary limit on ultra-high energy (UHE) neutrino fluxes, reported at this meeting [4]. Detection of showers in other radio-transparent substances such as rock-salt are also under consideration; in addition, the coherence and sensitivity of the radio regime suggest other uses which may lead to new technologies of high-energy particle detection.

Yet no experiments have been conducted to measure radio emission from hadron-induced showers. To order of magnitude, one expects that radio emission from PeV hadronic showers and electromagnetic showers are comparable, because almost all of the energy ends up at the "bottom" of the energy distribution with charge imbalances that are "order unity", namely 10-20%. But hadronic showers involve substantial uncertainties. Hadronic showers are sufficiently different from electromagnetic showers that independent tests should be done: the vast range of energies involved in a UHE shower and the intrinsic complexity of strong interactions make Monte Carlo estimates less reliable than corresponding electromagnetic shower calculations. Indeed, calculations for RICE thus far [3] have conservatively omitted the hadronic contributions.

We have developed a test-beam proposal which can be used both for hadronic or electromagnetic circumstances, and which takes into account experience gained

CP579, *Radio Detection of High Energy Particles*, edited by D. Saltzberg and P. Gorham
© 2001 American Institute of Physics 0-7354-0018-0/01/$18.00

from the previous generation of experimental work. A chronic problem has been the comparison of experiments done in near-field regime with idealized calculations sometimes limited to the asymptotically far regime. Another difficulty is the separation of different kinds of radiation, treated by textbooks as independent phenomena, which tend to mix in real experimental circumstances. Finally, the geometry of any particular experimental setup, while well-designed and clean on some criteria, may end up being very troublesome for radio-frequency optics.

We deal with the difficulties by planning from the bottom up with a high degree of symmetry and theoretical predictability. Certain configurations of beam and conductors lead to exact solutions to the electrodynamics: we engineer the experiment along these lines. We don't bother to separate out the pieces identified in textbooks so long as we get productive results in the end. Some features of the solutions are highly detailed, while others are as general as Gauss' Law: we recognize this, and take into account the limitations of theory. The experimental setup is arranged to give a high degree of internal self-calibration, so as to be hardly dependent on external standards. Indeed we plan the experiment with a high level of empirical *over-determination* of observables. Our philosophy is that when observables are over-determined by relations between the observations themselves, then the results are less subject to theory assumptions and also more amenable to systematic de-bugging.

The scheme we present is a first outline for a test-beam study exploiting the possibilities of symmetry and merging theory and experiment as much as possible step-by step: and if things go as conceived so far, would make a "pretty" experiment of high quality and intrinsic interest.

Some Helpful Experimental Predecessors: The combination of high energy physics with radio technology involves many orders of magnitude in energy and many challenges. Our work follows several efforts over the last few years aimed at verifying the effects of *radio-coherence* with laboratory test beams. We happily acknowledge the open communications and cooperative processes so important to the field. Alice Bean and Ilya Kravchenko [5] conducted informal experiments on SLAC electron beams in the mid 1990's. They saw radio pulses consistent with the total charge flowing past. Doug McKay did the theory work. Mary Anne Cummings, Paul Schoessow, and others [6] recognized the need for a hadronic test beamtest project at Fermilab. They conceived a novel concrete target arrangement, and succeeded in getting the lab's support. Several years ago Dave Besson [7] worked with Al Odian and paved the way for an electron test beam project at SLAC, conceiving a scheme with antennas buried in sand targets and box geometry to release radiation and inhibit reflections. The early efforts culminated recently in two experiments, conducted at the ANL wake-field facility and SLAC, which convincingly demonstrate that the important features of coherence have been observed in electromagnetic shower evolution [8,9]. These experiments are also discussed at this meeting by Saltzberg et al [9].

235

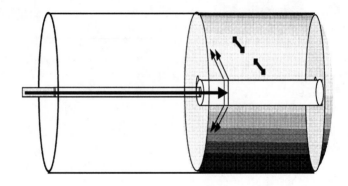

FIGURE 1. Schematic layout of the Tank. The accelerator beampipe (left side) feeds an evacuated dielectric beam pipe entering a field pre-form region with conducting boundary conditions. The length of this region of "free space" is adjustable. The tank proper (gray cylinder) contains an inner target pipe. The target pipe is illustrated empty; the tank is illustrated full. Radio receivers (black bowties) are placed at various locations inside the tank; causality is exploited. The illustration shows Cherenkov radiation (arrows) forming from the boosted Coulomb charge under free space conditions inside the target pipe.

II CONCEPTUAL OVERVIEW: THE TANK

The basic experimental set-up is dictated by the symmetry of the electrodynamics. The fields of steadily moving charges have basic symmetry along the beam direction or z-axis, and azimuthal symmetry about the beam. Shower evolution breaks the z-translational symmetry, but we retain as much z-symmetry as possible for empty target calibration purposes, explained momentarily. These observations lead to our basic design of a cylindrical tank containing a co-axial, independent hollow "target pipe". Both the tank and the target can be filled independently with dielectric fluid, or operated empty, in several combinations to be discussed. Radio antennas are placed at various locations in the tank with facilities to move any particular receiver from place to place without changing connections. We plan to use very small, very inefficient dipole antennas: the small size allows precision of placement, while fields from intense particle beams are so large that we can tolerate inefficiency. The scale of things is set by the scale of shower development: for 100 GeV showers induced by a Fermilab proton beam, we need dimensions "of order meters" of tank length and diameter. However the design "re-scales" easily, and we are also considering smaller models, on the scale of tens of centimeters, that might serve for lower energy tests or prototyping.[1]

The Tank and target-pipe arrangement allows many combinations of measurements with the same apparatus and receivers.

The combinations include the case of *Empty Target-Empty Tank*, in which the beam simply passes freely though the target pipe (evacuated for this purpose) and excites the antenna with a pulse of radio-frequency fields. The beam has been treated as a nuisance in previous experiments. We propose to use this case for calibrating the receivers. The pulse of radiation is obtained by Lorentz-transforming the Coulomb fields, or solving the wave equation directly: sometimes these fields are not recognized as "radiation", although in high energy physics the correspondence is well established, and well exploited in the "equivalent photon approximation". In more detail, the Fourier transformed electric fields,

$$\vec{E}_\omega(\vec{x}) = \int dt\, e^{i\omega t} \vec{E}(\vec{x}, t)$$

of a charge e moving with speed v, boost γ, are given by

$$\vec{E}(\omega, \vec{x}) = -i\omega \vec{A}(\omega) - \vec{\nabla}\phi(\omega);$$

$$E_\omega(\vec{x})_z = -\frac{2iQ\omega}{\epsilon v^2 \gamma^2} \exp\left(i\frac{\omega}{v}z\right) K_0\left(\frac{\omega\rho}{\gamma v}\right);$$

$$E_\omega(\vec{x})_T = \frac{2Q\omega}{\epsilon v^2 \gamma} \exp\left(i\frac{\omega}{v}z\right) K_1\left(\frac{\omega\rho}{\gamma v}\right),$$

where ρ is the transverse distance to the axis, and $\gamma^{-2} = 1 - \epsilon(v/c)^2$. The free-space fields are given by dielectric constant $\epsilon = 1$; otherwise use $\epsilon(\omega)$ appropriate to the medium. Note the argument of the Bessel function: dependence on the dimensionless ratio $\omega\rho/(\gamma v)$ causes the transverse size of the fields to scale like the boost parameter γ. When $\gamma \gg 1$ the fields may be comparatively flat out to surprisingly large distances.

Depending on frequency, the case of $\gamma \gg 1$ of relativistic beams also forces the Bessel functions to be practically evaluated at zero argument: and for $\omega \to 0$ the electric field is impeccably determined by Gauss' Law. Applying Gauss' Law in the frequency domain can be done by converting a surface integral into a time integral: for an infinite cylinder surrounding the beam, write

$$\oint \vec{E}\cdot d\vec{S} = 4\pi Q_{encl}/\epsilon;$$

$$\int d\vec{S}\cdot\vec{E} = \int_{-\infty}^{\infty} dz\, 2\pi\rho E_T = \int dt\, 2\pi\rho E_T(\rho, t)/v;$$

$$E_{\omega,T}(\rho; \omega \to 0) = \frac{2Q}{\epsilon v\rho} \exp\left(i\frac{\omega}{v}z\right).$$

We have assumed ϵ is evaluated in a region where it varies slowly. While the Bessel functions have the details, we see that the fields for $\omega\rho/(\gamma v) \ll 1$ are flat

---

[1] We thank Peter Gorham for suggesting this point.

in frequency, absolutely normalized, kinematic, and known as well as the accelerator physicists know the total charge in their beam, creating a highly accurate opportunity for calibration.

The next simplest combination is *Empty Target/Full Tank*, in which the boosted Coulomb field "refracts" into the dielectric tank medium, creating Cherenkov radiation. The radiation here is over-determined, while exhibiting all the coherence features of the subsequent combinations: *Full Target/Empty Tank*, and *Full Target/Full Tank*. Here the new feature is shower evolution. As the hadronic (or electromagnetic) shower evolves, the net electromagnetic current which produces radio emission evolves. The current will be predicted by Monte Carlo routines such as GEANT. Elsewhere there are discussions of radio emission from an evolving, coherently radiating radio source in the near or "Fresnel"and far transition regimes [10] and the asymptotically far regime [11]. The *Full Target/Empty Tank* allows investigation of shower evolution which is probed by the free-space propagating radio fields. Since the shower current is the only new thing compared to the previous case, we will be able to tell if the Monte Carlo is valid, comparing data to data. The *Full Target/Full Tank* will use buried receivers just like RICE, and is overdetermined after the previous measurements.

Obviously there are many measurements that can be made with a single, calibrated antenna which is moved about inside the tank, both transversely and longitudinally. We also plan to use ganged receivers, which can directly determine the Cherenkov angle by timing differences, and we can measure the speed of waves propagating in the medium, among other things. Before we go too far, however, we must caution that the boundary conditions can be considered with more care, which is "Part Two" of the theory planning.

More on Boundary Conditions: and Making Movies: In the discussion so far we neglected to mention the breaking of z-symmetry by the front edge of the tank, and the exit of the beam from the beampipe upstream.

The very broad transverse field profile of a relativistic beam (dependence on $\omega\rho/(\gamma v)$) makes the experimental hall a potential unwanted part of the boundary conditions. To control this, the Tank design includes a "pre-forming" hollow cylindrical conducting extension the same diameter as the tank. There is nothing in the preform but air (or maybe another, similarly cheap dielectric) and a length of dielectric evacuated beam pipe. The preform cylinder sits upstream of the tank and is designed to get the incoming fields into a known configuration before they hit the front surface of the Tank. If the pre-form region is sufficiently many wavelengths in extent, the boundary conditions are well established: we can check this by changing the length of the preform, so we don't need to rely on theory. There will also be a pulse of transition radiation from the beam's upstream emergence from the metallic beampipe: this may be controlled with an adiabatically tapered concentric absorber. The longitudinal locations of the tank, the end of the preform cylinder, and the metallic beampipe are all adjustable, creating many freedoms for consistency check.

The boundary-value problem for moving charges inside a conducting cylinder is easy: potentials are a combination of $K_0(\omega\rho/(\gamma v))$ and $I_0(\omega\rho/(\gamma v))$. Deviations from azimuthal symmetry should be small, but if any exist they are smoothly described with K_m and I_m series we also developed.

The problem of fields inside a cylinder meeting a plane boundary of dielectric at the front surface of the tank is the main complication. We solved this using the techniques for transition radiation emphasized by Bolotovsky (see his talk at this meeting) and others [12]. First, there is no particular reason to decompose the fields into those standard species of transition radiation, boosted Coulomb or Cherenkov radiation, which is sort of impossible anyway: the terminology of textbooks does not apply to real experiments. (Plus, we only want the fields whatever they are called.)

Second, the front and back boundaries of the tank are handled by Bessel series. These series have been set up by exact analytic algebra, with some care to get the solutions for forward-moving waves downstream from the tank. The sums over the series are seriously complicated: however, the finite frequency response of receivers made an otherwise infinite series converge with gratifying rapidity. Nevertheless, the operations in the frequency domain can be treacherous, because various backward and forward reflections hinge on properly deforming contours, and one must be careful.

To make sure we knew how to do this, we went the full route of making "movies", which seem to be the ultimate test of proper causal behavior of the complicated Bessel series. The movies in real space/real-time show the Coulomb field entering the Tank, reflecting from the front and "ringing" inside the Tank cavity, and exiting the Tank to re-form on the far side. The ringing, traveling wave has been observed before in the context of wake-fields [13].

In these calculations and in the whole experimental plan, causality is a great help. The complications from the Tank front surface cannot propagate through the dielectric medium faster than light speed in the medium. The leading fields of a Cherenkov pulse near the charge are very much like the free space fields. We knew this in advance, but made the calculations anyway, because they are beautiful, and worth doing well.

III SUMMARY

The self-calibrating features of empty-empty mode in our design promise measurements of high absolute accuracy. The various over-determined cases, measured in free space or medium, should help the eternal experimental difficulty of "getting things right". However certain details remain open, and make it hard to be firm on the level of accuracy achievable. One important detail is the nature of the dielectric fluid. We have considered water, because it is cheap, well-characterized and not-too-attenuating on the regime of GHz and below. Water is admittedly a complicated dielectric, and some colleagues shudder, but we have researched the dielectric properties and find hydrophobia unwarranted. Other radio-friendly fluids exist and are under consideration. Activation of the material is a consideration: this is another reason that we minimize the volume exposed to the beam with the target pipe. Activation of the antennas does not seem to be a problem. Worries raised in the workshop, that a pulse of 10^{11} particles with 100 GeV (say) might modify the dielectric properties, create a momentary plasma state, or do something unpredictable for the radio signal, have a simple answer: if that happens, it is just what we all want to know, both for Monte Carlo validation, and for RICE, where one of these days we might see a neutrino of equivalent super power make a loud radio splash.

Acknowledgments: Support by NSF and DOE, Division of High Energy Physics, and NSF Division of Polar Programs for the RICE project is gratefully acknowledged. We thank our colleagues attending the meeting for many useful discussions: in particular Boris Bolotovsky, John Learned, Al Odian and Peter Gorham were very helpful.

REFERENCES

1. G. A. Askaryan. *Sov. Phys. JETP* **14**, 441 (1962)
2. M. A. Markov and I. M. Zheleznykh. *Nucl. Instr. Meth. Phys. Res.* **A248**, 242 (1986).
3. J. P. Ralston, D. W. McKay, G. M. Frichter. *Phys. Rev.* **D53**, 1684 (1996).
4. D. Besson, these proceedings
5. A. Bean and I. Kravchenko, RICE internal communications (unpublished).
6. M. Cummings, Fermilab proposal (unpublished).
7. D. Besson, RICE internal communications (unpublished).
8. P. Gorham et al. *Phys. Rev.*, E62, 8590 (2000)
9. D. Saltzberg *et al*, hep-ex/0011001; D. Saltzberg, and these proceedings.
10. R. V. Buniy and J. P. Ralston, astro-ph/0003408.
11. E. Zas, F. Halzen, and T. Stanev. *Phys. Rev* **D 45**, 362 (1992)
12. G. M. Garibian. *Sov. Phys. JETP* **6**, 1079 (1958)
13. J. G. Power, W. Gai, and P. Schoessow *Phys. Rev.* **E 60**, 6061 (1999).

Coherent Radio Radiation of 15MeV ÷ 30GeV Electron and Photon Bunches in Thin and Thick Radiators

Sergey S. Elbakian, Edmond D. Gazazian, Karo A. Ispirian,
Rouben K. Ispirian and Khnkanos N. Sanosyan

Yerevan Physics Institute, Alikhanyan Brothers' St. 2, 375036 Yerevan, Armenia

Abstract. To understand the recent experimental results on coherent radiation in the radio wave region produced by bunches of $15MeV$ electrons [1] and $30GeV$ bremsstrahlung photons [2] as well as to carry out similar experiments with the help of $20MeV$, $50MeV$ and $4.5GeV$ electron and photon beams at Yerevan Synchrotron, the angular and spectral distributions of the radiation produced in thin and thick radiators are calculated using the theories of Cherenkov (ChR) and transition radiation (TR) and of the radiation of variable charges [3] as well as of the radiation at distances comparable with the formation zone [4]. The comparison of the obtained results with those of the Monte Carlo calculations of the type [5] and experimental results [1,2] is important for the projects and works on the detection of super high energy neutrinos with the help of radio radiation of the electromagnetic showers produced in Antarctic ice.

INTRODUCTION

The experiments [1,2] carried out at Argon and SLAC almost 40 years after the pioneering works of G. Askarian [6] have shown the importance of the accelerator investigations of the problems on the way to the high energy neutrino astrophysics. The Yerevan Physics Institute (YerPhI) physicists, who in the past have published theoretical works devoted to the charge excess and coherent radio radiation (CRR) of high energy electromagnetic showers [7,8] as well as have obtained experimental results on CRR production in wave guides (see [9]), have plans to continue studies to answer the following questions: 1)What are the real michanisms of production of CRR ? 2) What is the number of detected photons when the distance z between the detectors and radiation sources is less or of the order of the formation length L_f ? The first question concerns the role of sudden acceleration of the produced charges [10,11] and analytic calculation of CRR which usually is performed by Monte Carlo simulations [5], while the second question requires more efforts since the existing theoretical [12,13] and experimental [1] results are not sufficient.

CP579, *Radio Detection of High Energy Particles,* edited by D. Saltzberg and P. Gorham
© 2001 American Institute of Physics 0-7354-0018-0/01/$18.00

It is convenient to call the radiator thick (or thin) if its thickness $L_{rad} \geq L_f$ and $L_{rad} \gg L_{abs}$ (or $L_{rad} \ll L_f$ and $L_{rad} \ll L_{abs}$) where L_{abs} is the radiator's absorption length. As it is well known in solid radiators $L_f \simeq \lambda$ where λ is the CRR wave length. Of course, it is not reasonable to prepare thick metallic radiator because of strong absorption of the produced CRR. If the photons are detected inside the radiator or at its lateral sides and/or the radiator is sufficiently thick one can neglect the contribution of the output interface. Such assumption corresponds to the many experimental situations, in particular, to the experiments [1,2]. Such a differentiation allows one to consider separately the radiation produced at the boundary and the radiation produced in the thick radiator and make concrete conclusions in the case of thin radiators without using complicated expressions.

CRR PRODUCED IN THICK RADIATORS

Longitudinal Distribution of the Charge Excess

All the Monte Carlo simulations are carried out for sand (SiO_2) radiators with radiation length unit $X_0 = 18$cm and refraction index n = 1.55 using the modified EGS4 code EGSnrc [14] for a cut energy corresponding to the Cherenkov threshold energy as it is described in [8]. In the case of calculation of the radiation of the

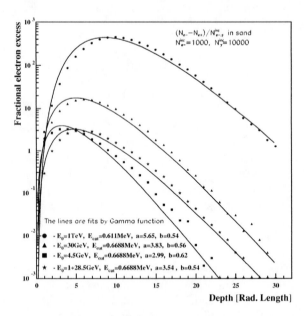

FIGURE 1. The longitudinal distribution of the charge excess. The excess points for the bremsstrahlung photons ($\star$) are multiplied by factor 100.

242

photon bunches it is assumed that an additional amorphous converter and sweeping magnet are placed before the radiator providing a pure photon Bethe-Heitler bremsstrahlung beam incident on the radiator. The thickness dependence of the charge excess $\delta N = N_{e^-} - N_{e^+}$ per incident electron for 4.5, 30 and 1000 GeV (the later energy is for comparison with [5]) electrons and 28.5 GeV bremsstrahlung beam [2] are shown in Fig. 1.

As in the case of shower curves one can approximate the excess results with the help of Gamma distributions

$$f(t) = b^a t^{a-1} exp(-bt)/\Gamma(a), \tag{1}$$

where a and b are fitting parameters given in Fig. 1, and $\Gamma(x)$ is the Gamma function. We do not show the results concerning the increase of the longitudinal and transversal beam sizes because even at lower energies [1] these sizes are small and give no essential contribution at relatively large thicknesses.

ChR Like CRR Produced in Thick Radiators

For simplicity we shall assume that the thick radiator has no absorption, and besides the above condition $L_{rad} \gg \lambda$ the radiator thickness is so large that very small quantity of charged particles come out from the radiator. Using the results and methods of the work [3], for a Gamma distribution form charge excess (1) one can derive the following frequency-angular distribution for the ChR type CRR:

$$\frac{d^2W(\nu, \theta)}{d\nu d\Omega} = \alpha(\frac{\nu}{\nu_0})^2 N_0^2 \hbar \sqrt{\varepsilon} \beta^2 \frac{sin^3\theta}{\left[1 + (\frac{\nu}{\nu_0})^2(1 - \beta\sqrt{\varepsilon}cos\theta)^2\right]^a}, \tag{2}$$

where $\nu_0 = cb/2\pi X_0(cm)$, $\beta = v/c$, and N_0 is a normalization constant determined by the charge excess distributions.

The expression (2) can be integrated over θ from 0 up to $\pi/2$ giving a lengthy expression for the frequency distribution, which for $\nu \gg \nu_0$ and Cherenkov condition $\beta\sqrt{\varepsilon} \geq 1$ fulfillment gives:

$$\frac{dW(\nu)}{d\nu} = \alpha N_0^2 \frac{2\pi\hbar\nu}{v} \left[1 - \frac{1}{\beta^2\varepsilon}\right] L_{eff}, \tag{3}$$

Note that without the effective length $L_{eff} = 5v/28\pi\nu_0$ the CRR intensity is given by the usual formula of Cherenkov intensity per cm.

For 30 GeV electrons and 28.5 GeV bremsstrahlung bunches (see Fig. 1) $N_0 = 84.54$ and 12.57; $\nu_0 = 1.53 \cdot 10^8 s^{-1}$ and $1.43 \cdot 10^8 s^{-1}$, respectively. For $E_0 = 30GeV$ electrons the angular distributions of CRR at various ν, the spectral distribution of CRR under the Cherenkov angle and the spectral distribution of CRR integrated over θ from 0 up to $\pi/2$ are shown in Fig.s 2 a), b), and c), respectively. Note that the total energy of CRR from single 30GeV electron and 28.5GeV equivalent quanta from single electron emitted under angles from 0 up to $\pi/2$ in the wave length interval $\lambda = 10 \div 20cm$ is $1.9 \cdot 10^{-23}$J and $1.19 \cdot 10^{-25}$J, respectively.

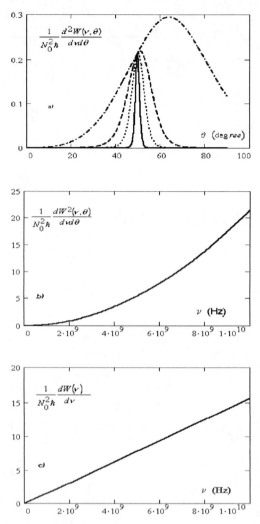

FIGURE 2. Angular distributions for $a = 4$. Solid, dotted, dashed and dash-dotted curves are for $\nu = 3 \cdot 10^9 (\times 1/9)$, $1 \cdot 10^9 (\times 1)$, $5 \cdot 10^8 (\times 4)$, and $1 \cdot 10^8 (\times 100)$, respectively. -a); Differential spectral distribution for $\theta = \theta_{Ch}$. -b); Spectral distribution integrated over all the angles. -c).

TR Produced at the Boundaries of Thick Radiators

Let N_b electron bunches, each containing N_e electrons with distances between the bunches L_b much larger than the length l_b of the bunches, $L_b \gg l_b$, enter a thick radiator without absorption from vacuum. Then the spectral distribution of the forward TR integrated over θ is given by:

$$\frac{dW_{N_e}(\lambda)}{d\lambda} = \frac{dW_1(\lambda)}{d\lambda} F_f(N_b, L_b, \lambda) F_b(N_b, L_b, \lambda) F_L(z, L_f), \qquad (4)$$

where $dW_1(\lambda)/d\lambda$ is the spectral distribution of usual TR in wave zone;

$$F_f(N_b, L_b, \lambda) = N_e[1 + N_e f(\lambda)] \qquad (5)$$

is a factor taking into account the bunch form factor $f(\lambda)$, equal to N_e^2 for $\lambda \gg l_b$;

$$F_b(N_b, L_b, \lambda) = \left[\frac{sin(\pi L_b N_b/\lambda)}{sin(\pi L_b/\lambda)} \right]^2 \qquad (6)$$

is a factor taking into account the interference between the radiation from various bunches. It is equal to N_b^2 if the detector resolution is good, $\Delta\lambda/\lambda \ll 1$, and is equal to N_b if $\Delta\lambda/\lambda \gg 1$;

$F_L(z, L_f)$ is a suppression factor depending on the distance z between the interface and the detector and on the formation length L_f. This factor is of the order of 1 in the case of thick radiators when $z \geq L_f \sim \lambda$, and will be considered in next section when $L_f \simeq \lambda\gamma^2$.

The calculated (not shown) angular dependences of TR differential and integral intensities emitted in the wave length interval $10 \div 20cm$ by a single $E = 15MeV$ and $30GeV$ electrons are similar to those given in [1] and give total energy of TR from single 30 GeV electron emitted in the angular interval $\theta = 0 \div \theta_{Ch}$ equal to 10^{-26} J. Since this TR intensity is less than the those corresponding to Cherenkov like radiation of 30 GeV electrons and is emitted under very small angles not available for the antenna, the SLAC experiment could be carried out with pure electron bunches having much larger CRR yield.

CRR PRODUCED IN THIN RADIATORS

TR Produced in Thin Radiators by Single Particles

In this case TR dominates since very small ChR comes out. For $| \varepsilon - 1 | \geq 1$ TR do not depend on ε, i.e. the TR yield is the same for metallic and dielelectric radiators [15]. The calculated differential and integral angular distributions of TR produced by a single 15MeV and 30GeV electron coming out from the metallic radiators into vacuum are slightly higher than for thick radiators considered above.

The Diffraction-Like Factor F_b

Fig. 3 shows the spectral distribution of TR taking into account the factor F_b for the beam parameters of the YerPhI 50 MeV injector in the wavelength region $\lambda \sim L_b$ ($N_e = 4.37 \cdot 10^7$, $N_b = 2860$ and $L_b = 0.105m$). In this case as expected the

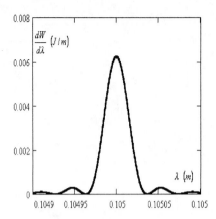

FIGURE 3. Spectral distribution of coherent TR of Yerevan Synchrotron injection beam.

coherence factor $F_f = N_e^2$, and it is assumed that the measurements are carried out in far wave zone when the formation zone suppression factor $F_L = 1$. As it is seen from Fig. 3 as a consequence of the interference of TR from N_b bunches the continuous spectrum of TR becomes a monochromatic spectrum with $\Delta\lambda$ / $\lambda \sim 1/N_b$. If the detector resolution is low, as in the case of the experiment [1], $N_b \leq F_b \leq N_b^2$. The calculations show that for the given values of the beam parameters $F_b \sim 5N_b$. Total TR energy is about $2 \cdot 10^{-7}$J which gives pulses with amplitudes about $3 \cdot 10^{-3}$.

Formation Length Suppression Factor F_L

The experimental results [1] show that at near field distances about 1m for $E_e = 15$MeV, $\lambda \sim 15$cm, the measured value $F_L \sim 1/30$. With a purpose to understand this fact and trying to construct a quantitative theory [4] we have calculated and studied the z-dependence of the flux of the Pointing vector which consists of three components due to the radiation field, particle charge, and their interference. For the simplest case of passing an interface between metal and vacuum one obtains:

$$\frac{dW^2(\lambda, \theta)}{d\omega d\theta} = \frac{2e^2}{\pi\lambda^2} \frac{sin^3\theta}{(1 - \beta^2cos^2\theta)^2} \left[1 + cos\theta - (1 - \beta cos\theta)cos[\frac{2\pi z}{\lambda}(1 - \beta cos\theta)]\right],$$

(7)

where the first, second and third terms are due to the radiation field, charge field and their interference, and only the last term depends on the distance z from the interface.

In all publications TR is investigated in the wave zone $z \gg L_f$ when after averaging in a small λ interval the last term oscillates and vanishes. We have

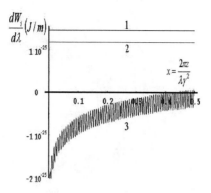

FIGURE 4. The dependence of $dW_{rad}/d\lambda$, $dW_{ch}/d\lambda$, and $dW_{int}/d\lambda$ (curves i = 1, 2, and 3, respectively) on $x = 2\pi z/\lambda\gamma^2$ for $\gamma = 30$ and $\lambda = 0.15 m$.

studied the z-dependence of these 3 components. After integrating (7) over all the angles we obtain:

$$\frac{dW_{rad}(\lambda)}{d\lambda} = \frac{4\alpha(\hbar c)}{\lambda^2}[ln\gamma + ln2 - 1/2], \tag{8}$$

$$\frac{dW_{Charge}(\lambda)}{d\lambda} = \frac{4\alpha(\hbar c)}{\lambda^2}[ln\gamma - 1/2], \tag{9}$$

$$\frac{dW_{int}(\lambda)}{d\lambda} = -\frac{4\alpha(\hbar c)}{\lambda^2}I_0, \tag{10}$$

where

$$I_0 \simeq ci(\frac{2\pi z}{\lambda}) - ci(\frac{2piz}{\lambda}\frac{1}{2\gamma^2}) - sin^2\pi z/\lambda. \tag{11}$$

Fig. 4 shows the dependence of each of three components on $x = 2\pi z/\lambda\gamma^2$.

To comprehend the z-dependence we propose to define the TR suppression factor in the form:

$$F_L(z,\lambda) = \left[\frac{dW_{rad}(\lambda)}{d\lambda} - k\frac{dW_{int}(\lambda)}{d\lambda}\right] / \left[\frac{dW_{rad}(\lambda)}{d\lambda}\right], \tag{12}$$

where the coefficient $0 < k < 1$ shows the fraction of $dW_{rad}(\lambda)/d\lambda$ which goes to $dW_{int}(\lambda)/d\lambda$. Fig. 5 shows the dependence of $F_L(z,\lambda)$ vs $x = 2\pi z/\lambda\gamma^2$. The observed [1] TR intensity suppression $F_L \sim 1/30$ can be in part be conditioned also by the effect considered in [16,17], according to which TR should be suppressed if the transversal sizes of the radiator is less than the transversal sizes of TR source $\lambda\gamma$.

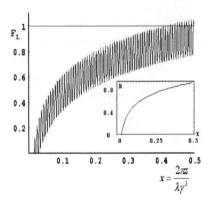

$$x = \frac{2\pi z}{\lambda \gamma^2}$$

FIGURE 5. The dependence of the suppression factor $F_L(z, \lambda)$ upon $x = 2\pi z/\lambda\gamma^2$ for $\gamma = 30$ and $k = 1$. The dependence of R upon x when $dW_{rad}/d\lambda$ and $dW_{int}/d\lambda$ are integrated over λ from 0.1 to 0.2m is shown inserted.

CONCLUSIONS

The following accelerator experiments can be carried out with

a) 15 - 50 MeV electron beams;

Since at such low energies the number and range of the particles in the region close to the maximum of the shower curves of the charge excess is small the TR intensity dominates, and it is impossible to study ChR. It is of interest to study theoretically and experimentally the dependence of the TR intensity on the distance from the radiator interface and on the radiator transversal sizes.

Since for the single electron bunches of the experiment [1] $W^{SLAC} \simeq N_e^2 (dW/d\lambda)\Delta\lambda$, while for bunch consequence of the YerPhI beam $W^{YerPhI} \simeq N_e^2(dW/d\lambda)F_b$, for the above given beam parameters and parameters of [1] one obtains $W^{YerPhI}/W^{SLAC} \simeq 2 \cdot 10^{-2}$ or taking into account that $W \simeq 4V^2/R$ where V is the measured voltage and R is the cable resistance one expects $V^{YerPhI}/V^{SLAC} \simeq 0.15$.

b) 4.5 GeV electron and photon beams fast extracted from YerPhI Synchrotron;

In this case the great number and large range of the particles near the maximum of the shower curve of the charge excess as well as the large difference between the TR and ChR radiation angles allows to study separately TR and ChR. A $3 \div 5$ radiation length thick converter can be used to enhance TR.

However, since the bunch structure of the 1-2 μs electron and photon beams fast extracted from Yerevan Synchrotron consists of about 0.7ns 400 bunches CRR measurements must be carried out at longer wave lengths, longer than 1m;

c) 30 GeV electron and bremsstrahlung beams at SLAC.

In this case also TR and ChR can be experimentally investigated separately because of the above discussed reasons. Moreover, together to the single bunch regime pulses with consequences of bunches can be used to study CRR at other wave length.

ACKNOWLEDGMENTS

The authors thank D. Saltzberg and P. Gorham for inviting one of us (K.A.I) to RADHEP-2000 and supporting this and other future works on accelerator experiments in the frames of ISTC, as well as R.O. Avakian, E.M. Laziev, G.G. Oksuzian and A. Vardanyan for interest.

REFERENCES

1. Gorham, P., Saltzberg, D., Schoessow, P., Gai, W., Power, J.G., Konecny, R., and Conde, M.E., *e-preprint* **hep-ex/0004007** (2000).
2. Saltzberg, D., Gorham, P., Walz, D., Field, C., Iverson, R., Odian, A., Resch, G., Schoessow, P., and Williams, D., *e-preprint* **hep-ex/0011001** (2000).
3. Amatuni, A.Ts., Garibian, G.M., and Elbakian, S.S., *Izvestia Akad. Nauk Arm. SSR* **16**, 101 (1963).
4. Gazazian, E.D., Ispirian, K.A., and Elbakian, S.S., *Pisma Zh. Eksp. Teor. Fiz. (to be published)*.
5. Zas, E., Halzen, F., and Stanev, T., *Phys. Rev.* **45**, 362 (1992).
6. Askarian, G.A., *Zh. Eksp. Teor. Fiz.* **41**, 317 (1961); **48**, 988 (1965).
7. Gazazian, E.D., Ispirian, K.A., and Kazarian, A.G., *Proc. of the 2-nd Intern. Symp. on Transition Radiation of High Energy Particles*, Yerevan, 1983, p. 413.
8. Avakian, R.O., Ispirian, K.A., Ispirian, R.K., and Khachatrian, V., *e-preprint* **hep-ex/0005039** (2000); *To be published in Nucl. Instr and Meth.* **B**,... (2000).
9. Laziev, E.M., and Oksuzian G.G., *see Ref. [7], p.701*.
10. Tamm, I.E., *J. Physics(Moscow)* **1**, 439 (1939).
11. Afanasiev, G.N., and Shilov, V.M., *Preprint JINR*, Dubna, E2-2000-61, 2000.
12. Bolotovski, B.V., and Serov, V.A., *Zh. Tekhn. Fiz.* **67**, 8999 (1997).
13. Verzilov, V.E., *Phys. Lett.* **A273**, 135 (2000).
14. Kawrakow I., and Rogers, D.W.O., *The EGSnrc Code System: Monte Carlo Simulation of Electron and Photon Transport*, NRCC Report PIRS-701, Canada (2000).
15. Alikhanian, A.I., Ispirian, K.A., and Oganesian, A.G., *Zh. Eksp. Teor. Fiz.* **56**, 1796 (1969).
16. Shulga, N.F., and Dobrovolski, S.N., *Pisma Zh. Eksp. Teor. Fiz.* **65**, 681 (1997).
17. Shulga, N.F., Dobrovolski, S.N., and Sishchenko, V.G., *Nucl. Instr. and Meth.* **B145**, 180 (1998).

RADAR AND OTHER TECHNIQUES

On Radar Detection of EeV Air Showers

P. W. Gorham

Jet Propulsion Laboratory, California Institute of Technology
4800 Oak Grove Drive, Pasadena, CA, 91109

Abstract. Extensive air showers produced by EeV cosmic rays leave a significant ionization trail, which may be detectable by radar reflection, by analogy to the way both meteor and lightning ionization trails are presently observed. Such methods could lead to low-cost detection and characterization of events near and above the GZK cutoff.

I INTRODUCTION

It is now clear that the primary cosmic ray spectrum extends in energy beyond the $\sim 5 \times 10^{19}$ eV Greisen-Zatsepin-Kuzmin (GZK) cutoff expected if the particles are hadronic in nature and due to sources at distances greater than about 50 Mpc, still well within the local supercluster volume, and a small fraction ($\sim 0.1\%$) of the volume of the universe to $z \simeq 3$. To answer the question of the origin of these particles demands much higher statistics than we presently have To achieve this will require either a scaling of present detection techniques to much larger collection areas (and correspondingly higher costs), or development of new cost efficient techniques that are sensitive to much larger areas.

One such possibility, that of detecting the ionization column of extensive air showers (EAS) by radar techniques, has its origin in the suggestion by Blackett & Lovell (1940) [1] that earlier detections in the 1930's of sporadic radio reflections at altitudes of 5-20 km from the troposphere and mesosphere [4,2,3,5] might be attributed to cosmic ray EAS, although at the time both the energies and fluxes of such events were only crudely known. This work appears to have led Lovell to construct a large radio antenna in Great Britain dedicated to this purpose [6]; however, for reasons that are unclear (but possibly due in part to skepticism about electron attachment lifetimes) no EAS radar work with this facility was ever reported. The idea was later re-examined in the early 1960's [7] and a proposed experiment was described [8]; however, no results from this test have appeared in the literature to date.

The present work summarizes a more detailed study [9] reported elsewhere which again investigates the viability of this approach, with the advantage of another 40 years of radar technology development that has taken place in the intervening time.

CP579, *Radio Detection of High Energy Particles,* edited by D. Saltzberg and P. Gorham

The conclusion is that, depending somewhat on the details of the lifetime of the electrons in the ionization column, the radar approach could be a very powerful technique for EeV shower detection.

II   PROPERTIES OF THE IONIZATION COLUMN

A   Electron density

Figures 1- 3 show the properties of the ionization column for horizontal air showers at various energies and altitudes. We have chosen to consider only horizontal showers here since it simplifies the analysis, and such showers are of interest for detection of penetrating primaries (such as neutrinos). The parameters of the showers have been modelled according the Nishima-Kamata-Greisen parameterization [10,11] which is reasonably acurate at these high energies.

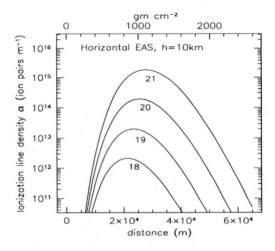

FIGURE 1. Electron ionization line density for 5 showers of energies in the 10^{18} to 10^{21} eV range. Such line densities are quite similar to those of radio meteors.

Fig. 1 shows the profile of the ionization, expressed as line density of electrons (the total number of electrons projected onto the axis of the shower). The free electrons in the ionization column form a tenuous plasma with a radio-frequency index of refraction different from the surrounding neutral atmosphere, and thus there is some reflection of any incident radio waves.

Fig. 2 shows the corresponding electron volume density, along with the plasma frequency ν_p of the ionized region:

$$\nu_p = \sqrt{(n_e e^2/\pi m_e)} \simeq 8.98 \times 10^3 \sqrt{n_e} \text{ Hz} \tag{1}$$

where n_e is the electron density in cm^{-3}. If the plasma frequency exceeds the frequency of the incident radiation, the index of refraction becomes imaginary, and total reflection of the incident radiation occurs, rather than refractive scattering.

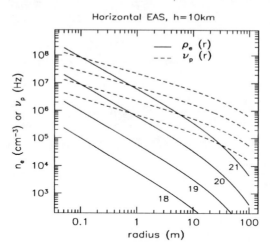

FIGURE 2. Radial dependence of the ionization density for the same showers presented in the previous figure. Also shown is the radial dependence of the plasma frequency for each case.

It is evident that the plasma frequency for this range of energies at 10 km altitude is generally below ~ 10 MHz. This would lead to strong reflections in the MF range (1-10 MHz), which may explain some of the early sporadic detection seen above. However, one drawback of such reflections is that their strength is very sensitive to geometric effects and is not easily interpreted in terms of the amount of ionization present. At higher frequencies, above the plasma frequency, the scattering takes on a much simpler form, depending only on the superposition of the fields from each electron, scattering independently with the Thomson scattering cross section

$$\sigma_T = \frac{8\pi}{3}\left(\frac{e^2}{m_e c^2}\right)^2 = 6.65 \times 10^{-29}\,\mathrm{m}^2 . \tag{2}$$

The total effective radar cross section (RCS) σ_b will then depend on the individual phase factors of each of the scattering electrons [12]:

$$\sigma_b(\mathbf{k}) = \left| \int n_e(\mathbf{r})\sqrt{\sigma_T}e^{2i\mathbf{k}\cdot\mathbf{r}}d^3\mathbf{r} \right|^2 \tag{3}$$

where $\mathbf{k}$ is the wave vector of the incident field ($k = 2\pi/\lambda$), $\mathbf{r}$ is the vector distance to the volume element at which the scattering takes place.

If we now write $\mathbf{q} = 2\mathbf{k}$ equation 3 can be seen to take the form of a Fourier transform, thus reducing the problem of estimating the effective radar cross section to that of calculating the Fourier transform (and the resulting power spectrum) of the electron number density distribution.

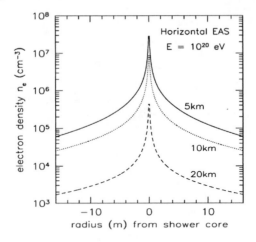

FIGURE 3. Radial profile of showers shown for the inner 40 m radii, plotted on a linear distance scale, for different shower altitudes, showing the effects of altitude for a set of horizontal showers.

A broader implication of this result can be stated as follows: *A measurement of the complex amplitude of the radar echo from an EAS in the underdense case is proportional to a measurement of one Fourier component of the shower ionization profile.* This will be true even in the case of a bistatic radar (receiver location not coincident with transmitter) although there will be an additional coefficient to account for the bistatic cross section and its angular behavior.

B   Electron lifetime

The rate of loss to recombination and attachment for atmospheric electrons is described by [13]:

$$\frac{dn_e}{dt} = N_e^0 - \alpha_e n_e^2 - \beta_e n_e \qquad (4)$$

where N_e^0, n_e are the initial and evolving electron density, respectively, α_e is the electron-ion recombination coefficient (in units of cm^3 s^{-1}), β_e and is the electron attachment rate (s^{-1}). Direct recombination of the electrons with ions occurs with low probability at the electron densities in air showers, and attachment dominates over recombination. Molecular oxygen is electronegative, forming ions with free electrons primarily through dissociative attachment, via the process

$$e^- + O_2 \rightarrow O^- + O \qquad (5)$$

where the resulting oxygen ion and atom carry off the excess kinetic energy of the electron, and the electron affinity of O_2 is of order 0.5 eV. Molecular nitrogen, in contrast, does not form attachments with electrons but can be more easily collisionally excited, and may play a role in mediating detachment processes in air.

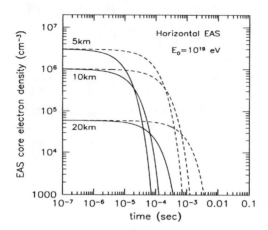

FIGURE 4. The evolution of the electron number density is shown for three 10^{19} eV EAS, at different altitudes. The solid lines correspond to the evolution for the present upper limit on the attachment rate, and the dashed lines to one-tenth of the upper limit.

Several studies of electron attachment in air using electron swarm techniques [14] have indicated a much lower attachment rate than that which is expected from using the attachment coefficient for pure molecular oxygen reduced by the molar ratio in air [15]. Measurements of the attachment coefficient are very sensitive to contamination by carbon dioxide, and CO_2 was not removed from the samples in many early measurements. Although CO_2 can be present in significant concentrations in the near sea level atmosphere, its average concentration in the atmosphere as a whole is 0.033%. Thus electron attachment rates for dry, high-altitude air must be determined with CO_2 largely removed from typical air samples.

Moruzzi & Price (1974) [15], after carefully removing all CO_2 from their samples, confirmed earlier measurements in being unable to detect any attachment in pure dry air. This result was attributed to a competing process: rapid collisional detachment from N_2 interactions, yielding an effective attachment rate that was too small too measure.

Figure 4 shows the behavior of the electron density for three different altitudes for a 10^{19} eV shower core region. Here we have plotted the electron density time evolution for both the case where the attachment rate is equal to the limit of Moruzzi & Price, and one-tenth of the limit (dashed lines). It is evident that the expected electron lifetimes are at least 10-20 μs, increasing slowly at the higher altitudes. Thus, although it is evident that atmospheric conditions may be very important in determining the success of detection of EAS by radar, the lifetime under good conditions should be adequate.[1]

III RADAR CROSS SECTION ESTIMATES

Radar return power P_r is described in terms of a model where the radiation is emitted from an antenna with peak transmitted power P_t and directivity gain $G = 4\pi\Omega_A^{-1}$, where Ω_A is the solid angle of the main beam of the antenna. The radiation is assumed to then scatter from objects in the antenna beam and be re-radiated isotropically in the frame of the scatterer, producing a R^{-4} dependence in the returned power as a function of range R. Deviations from isotropic scattering are thus absorbed into the effective radar backscatter cross-section σ_b, which can be larger or smaller than the physical cross-section of the object. In addition any real transmitting and receiving system will have less than unity efficiency, which we designate here as η. The radar equation under these conditions is [16]:

$$\frac{P_r}{P_t} = \sigma_b\eta \; \frac{G^2\lambda^2}{(4\pi)^3 R^4} \; .$$

(6)

Here we are assuming that the transmitting and receiving antennas are identical, and we are neglecting for the moment any polarization effects or losses in the medium.

Given equation 6, the problem of determining the detectability of EAS-initiated ionization columns reduces to that of determining the effective RCS σ_b for a given choice of operating radar frequency, and the noise power of the specific radar in use. The noise power is given by $P_N = kT_{sys}\Delta f$, where T_{sys} is the system noise temperature, k is Boltzmann's constant, and Δf the effective receiving bandwidth, assumed here to be matched to the transmitting bandwidth.

Combining the noise power equation above with equation 6, the signal-to-noise ratio (SNR) of the received power is

$$\frac{P_r}{P_N} = \sigma_b P_t\eta \; \frac{G^2\lambda^2}{(4\pi)^3 R^4} \; \frac{1}{kT_{sys}\Delta f} \; .$$

(7)

---

[1] We note here that this analysis has neglected to treat the 3-body attachment process which may in fact dominate the attachment lifetime if the electrons from the shower thermalize quickly. If the 3-body attachment process is dominant, as appears probable as this article goes to press, then the free electron lifetimes can be much less than for the 2-body process considered here, and may be reduced to several hundred ns. Such short lifetimes will significantly change any approach toward radar detection of air showers. It is thus important to confirm such analysis before settling on a possible experimental approach.

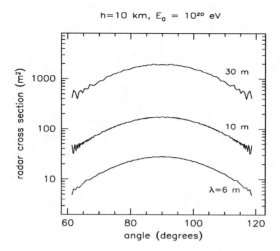

FIGURE 5. Radar cross section over a range of angles centered on normal incidence (at 90°) for a horizontal air shower with $E_0 = 10^{20}$ eV, at an altitude of 10 km, at radar wavelengths 30, 10, and 6 m. Results are based on numerically integrating the contributions of all of the individual volume elements of the shower.

Evaluating equation 7 for a nominal choice of parameters gives the SNR per received radar pulse per square meter of RCS:

$$\frac{S}{N} = 3.3 \left(\frac{\sigma_b}{1 \text{ m}^2}\right) \left(\frac{P_t}{1 \text{ kW}}\right) \left(\frac{\eta}{0.1}\right) \left(\frac{G}{10}\right)^2$$

$$\times \left(\frac{\lambda}{3 \text{ m}}\right)^2 \left(\frac{R}{10^4 \text{ m}}\right)^{-4} \left(\frac{T_{sys}}{10^3 \text{ K}}\right)^{-1} \left(\frac{\Delta t}{10 \ \mu s}\right). \tag{8}$$

We note that the reference values chosen here represent a modest radar system; in particular the peak power of 1 kW is easily attained by current standards, and the directivity $G \simeq 10$ (beam size ~ 1 sr) represents a relatively low-gain antenna. The system temperature $T_{sys} = 1000$ K is realistic for 100 MHz ($\lambda = 3$ m) however, since the brightness temperature of the sky is quite high at these frequencies.

Fig. 5 shows an example of a numerical estimate of the radar cross section for a horizontal 10^{20} eV shower at 10 km altitude, where the effects of individual electron phase factors have been accounted for [9]. It is evident that the cross sections, here peaking at more than 1000 m^2 at 10 MHz, and still several tens of m^2 at 50 MHz, are quite large. Equation 8 indicates that such showers should be detectable out to several tens of km over a reasonable range of incident angles.

IV EXAMPLE OF A POSSIBLE RADAR ARRAY

Because of the issues of antropogenic interference rejection, an EAS radar system would almost certainly have to be constructed with multiple stations configured to both transmit and receive, with pulse encoding such that each station would be sensitive to pulses produced by the other stations. Bistatic radar cross sections of EAS will vary only modestly from the monostatic RCS values treated here. Global synchronization of the stations to several tens of ns or better is routine with the Global Positioning System, particularly with the disabling of Selective Availability (SA) of the high precision GPS codes.

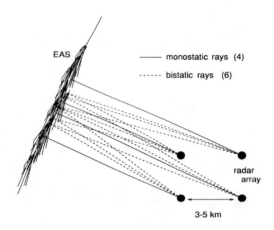

FIGURE 6. An array geometry for a possible standalone EAS radar detection system. Each of the 4 stations has both transmit & receive capability, and can distinguish the signals from the other stations as well.

To optimize the signal-to-noise ratio of the detected echoes, we wish to operate at the lowest frequency that is practical from considerations of background noise, including ionospheric noise from daytime operation. Although lower frequency operation may be possible, we choose here to consider only frequencies above about 30 MHz. In the HF-VHF regime the effective system noise temperature for a remote site (distant from urban noise sources) can be written as [16]:

$$T_{sys} = 2.9 \times 10^6 \left(\frac{f}{3 \text{ MHz}} \right)^{-2.9} \text{ K} \tag{9}$$

which yields $T_{sys} = 3600\text{K}$ at 30 MHz.

If we assume that we require an energy threshold of 10^{19} eV out to a range of 20 km, the implied RCS for the underdense case is $\sim 2 \text{ m}^2$ at this energy. The

TABLE 1. SNR budget for one station of a possible EAS standalone radar system.

Parameter	value	±dBm
Received Power	$\sigma_b \eta P_t G^2 \lambda^2 \times (4\pi)^{-3} R^{-4}$	
Peak transmit power	60 kW	77.8
Pulse duration	$10\mu s$	...
chirp bandwidth	3 MHz	...
Number of repetitions	1	0.0
Antenna gain	3	9.54
wavelength	10 m	20.0
σ_b at $E = 10^{19}$ eV	3.8 m^2	5.8
range to EAS	20 km	-172.0
xmit/rcv efficiency	0.05	-13.0
$(4\pi)^{-3}$	5.04×10^{-4}	-33.0
Received power		-104.9 dBm
Noise power	$kT_{sys}\Delta f$	
Boltmann's constant	1.38×10^{-20}mWK^{-1}Hz^{-1}	-198.6
System temperature	3650 K	35.6
effective bandwidth	100 kHz	50.0
Noise power		-113.0 dBm
Net SNR	6.4	8.0 dB

Fresnel zone length at this range is about 250 m for a shower at 10 km altitude, corresponding to about 10 g cm^{-2} along the track. Using a commercial VHF radar system, it is straightforward to achieve 60 kW peak power and pulse repetition rates of 10-50 kHz for 10 μs pulses. The sky brightness temperature will lead to a system temperature of $T_{sys} = 3600$ K at 30 MHz as noted above. Since we wish to be sensitive to as large a volume as possible, we use an antenna with a broad beam (thus corresponding to a low gain). A suitable system is a vertical monopole or discone system [17] which provides good 2π azimuthal coverage from 5° to about 50° elevation, giving an overall gain of ~ 3 with respect to an isotropic antenna. We also assume here a somewhat pessimistic overall transmit/receive efficiency of $\eta = 0.05$, and a single pulse echo (no pulse averaging) to account for possible rapid decay of the free electron population.

Using these assumptions we show the SNR budget, in dBm (decibels referenced to 1 milliwatt) in Table 1. This budget corresponds to a single monostatic or bistatic pulse measurement in the array. The resulting SNR of 6.4 provides a range resolution of 34 m at a 30 MHz operating frequency, using a chirp bandwidth of 10% (3 MHz in this case).

V   CONCLUSIONS

We have investigated the possibility of detecting the ionization column produced by extensive air showers in the 10^{18-21} eV energy range by VHF radar. We have shown that the ionization density is plausibly high enough for such methods, and that the electron lifetimes appear to also be adequate if the air column used for the detection is dry and relatively pure. In this latter respect, higher altitude showers may be more detectable. It is important for tests of this technique to bracket the possible range of electron attachment lifetimes and also to probe higher altitudes where air showers are not normally detected. For example, above 20 km, nearly all fully-developed showers (age ~ 1) are horizontal and may provide good radar targets.

We thank George Resch, David Saltzberg, Elaine Chapin, Ray Jurgens and Tanya Vinogradova for useful discussion and comments. This work was performed in part at the Jet Propulsion Laboratory, California Institute of Technology, under contract with NASA.

REFERENCES

1. Blackett, P. M. S., and Lovell, A. C. B., 1940, Proc. Royal Soc. (London) Ser. A, 177, 183.
2. Watson Watt, R. A., Bainbridge-Bell, L. H., Wilkins, A. F., and Bowen, E. G., 1936, Nature 137, 866.
3. Watson Watt, R. A., Wilkins, A. F., and Bowen, E. G., 1936, Proc. Roy. Soc. A, 161, 181.
4. Colwell, R. C., and Friend, A. W., 1936, Phys. Rev. 50, 632.
5. Appleton, E. V., and Piddington, J. H., 1937, Proc. Roy. Soc. A, 164, 467.
6. T. C. Weekes, this proceedings.
7. Suga K., 1962, Proc. Fifth Interamerican Seminar Cosmic Rays, 2, XLIX.
8. Matano, T., Nagano, M., Suga, K., and Tanahashi, G., 1968, Can. Journ. Phys. 46, S255.
9. P. W. Gorham, 2001, Astroparticle Physics, in press.
10. Kamata, K., & Nishimura, J., 1958, Suppl. Progr. Theoret. Phys. 6, 93.
11. Greisen, K., 1965, in Prog. Cosmic Ray Physics vol. III, J.G. Wilson ed., (North Holland: Amsterdam) 1.
12. Wehner, D. R., 1987, *High Resolution Radar*, (Norwood, MA: Artech House).
13. Thomas, L. 1972, Journ. Atmos. Terr. Phys. 33, 157.
14. Gallagher, J. W., Beaty, E. C., Dutton, J., and Pitchford, L. C., 1983, J. Phys. Chem. Ref. Data, 12(1), 109.
15. Moruzzi and Price (1974).
16. Skolnik, M., 1990, *Radar Handbook*, (New York: McGraw-Hill), 2nd ed.
17. Balanis, C. A., 1997, *Antenna Theory*, 2nd edition, (New York: Wiley & Sons).

Detection of Energetic Particles by a Network of HF Propagation Paths in Alaska

A.Y. Wong*

HIPAS Observatory, Fairbanks, AK and Dept of Physics and Astronomy, UCLA

*With the assistance of G. Rosenthal, J. Pau, E. Nichols of UCLA

Abstract. A network of horizontal propagation paths of HF radiation is proposed to detect ionization paths created by energetic particles near the earth's surface. Coherent phase method is sensitive to sudden changes in ionization in any portion of a long path length.

Recognizing that the flux of magnetic monopoles is low and that a large viewing region is necessary to increase the probability of observation I am proposing an electronic network using commercial cell-phone stations in the entire state of Alaska (Fig.1, 2). The reasons for performing the experiment in Alaska are: the remoteness of the Alaska region means that the level of background noise is significantly lower than that in the lower forty-eight states; the climate in Alaska does not give rise to thunderstorms in most regions, thereby reducing the amount of extraneous ionization which can confuse our detection scheme; most of the infrastructure is already present at the UCLA HIPAS Observatory (Wong, 1999;

CP579, *Radio Detection of High Energy Particles,* edited by D. Saltzberg and P. Gorham
© 2001 American Institute of Physics 0-7354-0018-0/01/$18.00

Figure 1: Location of commercial communications receiver sites in Alaska.

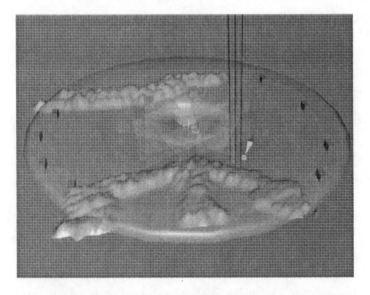

Figure 2: Artist conception of the experimental arrangement. HIPAS is in the center of the array radiating a coherent signal. The radiation pattern of each receiver is shown as an overlapping shade. The earth's magnetic field is represented by vertical magnetic field lines.

), located in the central region of Alaska as a result of our more than twenty years of performing communication experiments on the interaction of electromagnetic waves with the ionosphere (fig 3, 4).

The central idea of our concept consists of sending a coherent HF wave from a vertical antenna located at HIPAS Observatory, which propagates uniformly in all horizontal directions. Receiver sites located at various cellular stations throughout Alaska are used to record this signal in amplitude and phase. These receiver sites (fig 1) all have very accurate timing provided by T-1 lines. The phase stability is one part in 10^9. Therefore very small incremental change in phase can be detected. Receivers are located at distances from 30 km up to 500 km. All data are recorded continuously in a digital mode at the 1 MHz rate and any sudden change in the order of milliseconds or less will ensure that the data will be retained permanently and examined first on a local computer. This procedure avoids the accumulation of an overwhelming amount of uninteresting background data.

Only these pre-screened data are then sent forward in batch formats to a central telescience center using the cell station links. The data from all stations will be examined and correlated. The use of existing cell stations for data gathering and intermittent transfer greatly lowers the cost of this large-scale experiment.

The network can be calibrated by the short burst of ionization generated by a 300-Joule laser (Wuerker, 2001) at the HIPAS Observatory on dust particles. A 2.7 m rotating mirror at HIPAS can be used to focus a laser beam to impact on dust particles at a height of 100 km. The resulting electron population line density of nl $< 10^8$ cm^2 is expected to cause a detectable change in the phase of a propagating HF signal.

The propagation of electromagnetic waves through a medium containing electrons will suffer a change of phase given by the following equation (Wong, 1980):

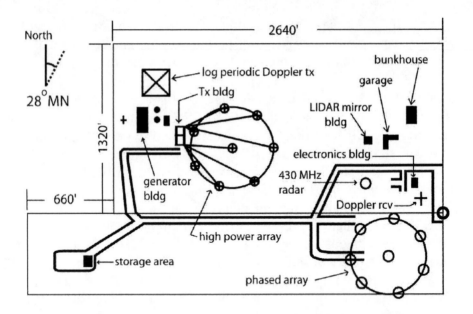

Figure 3: Antenna arrays at the HIPAS Observatory, Fairbanks, Alaska. The frequencies used range from HF to Laser regimes.

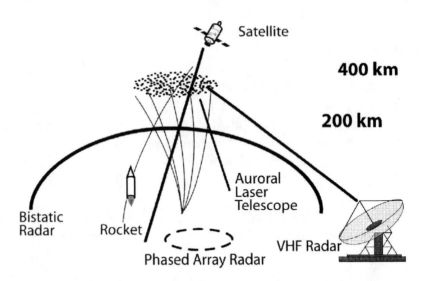

Figure 4 Schematic showing the various diagnostics used in an experiment to investigate the interaction of electromagnetic waves with the ionosphere.

$$\Delta\phi = \int_0^L \left(k_{\text{vacuum}} - k_{\text{plasma}} \right) dx .$$

Assuming constant n_e through length L, we have:

$$\Delta\phi = \frac{\omega}{c}\left[1 - \left(1 - \frac{\omega_{pe}^2}{\omega^2} \right)^{1/2} \right] L ,$$

where $k_{\text{vacuum}} = \dfrac{\omega}{c}$, and

$$\omega_{pe} = \left(\frac{4\pi n_e e^2}{m_e} \right)^{1/2} = 5.6 \times 10^4 \, n_e^{1/2} \, (\text{rad/s})$$

The minimum detectable change in phase is expected to be 0.1 degree, which corresponds to a change in the line density ($\sim n_e L$) of 6×10^3 cm^{-2}.

According to the layout of the receiving stations this detection system is expected to cover an area of approximately 500 km in radius. Anytime an energetic particle falls into this area, it will cause ionization detectable by this method. According to the theoretical work by Wick et al (2000) taking the upper flux bound of monopoles as 10^{-16} cm^{-2}sec^{-1}sr^{-1} (Amanda, Baikal and Macro) and considering a region with a radius of 500 km, an upper limit of 5 events/sec is estimated. Even if this were an optimistic estimate our detection system is designed to observe 5 events/year, which can put the flux 10^7 times lower. Wick also estimated that the ionization in the lower atmosphere will give a line density nl $\sim 10^5$ cm^2 which is well above the sensitivity limit of our detectors.

The line density might be higher if we consider the impact of the monopole on the earth's surface. The material density is higher and gives rise to secondary electron emission and transition radiation. This might result in a large electron shower in the 0-100 km height range we are detecting. We are currently performing computer modeling of such events.

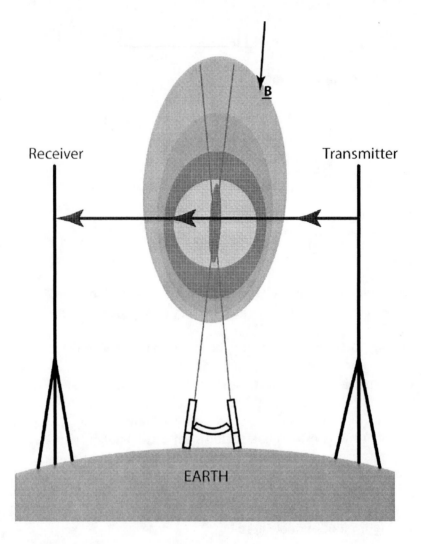

Figure 5 Schematic showing a calibration procedure using a laser to ionize dusts at a height of 100m. The electrons generated will cause phase shift in the coherent signal detected by the receiver.

In our proposed experiments the events are discriminated according to the speed of occurrence. Background events cannot in general produce such a large line density within tens of microseconds near the earth's surface.

We have also considered the duration of the electron line density at such low heights. The ionization balance equations for the concentrations N_e of the electrons and N^- of the negative ions take the form (Gurevich, 1978):

$$\frac{dN_e}{dt} = q_i + v_{ion}N_e - v_aN_e + v_dN^- - \alpha N_e(N_e + N^-)$$

$$\frac{dN^-}{dt} = v_aN_e - v_dN^- - \alpha_i N^-(N_e + N^-)$$

where q_i is the total intensity of the ionization produced by the external source, v_{ion} is the frequency of the molecule ionization by the fast electrons, v_a is the frequency of electron attachment to the molecules, and v_d is the electron detachment frequency. Under the conditions of the lower ionosphere, an important role is played by the attachment of electrons to oxygen molecules in triple collisions $v_a \cong k_1 N_{O_2} N_{O_2}$.

According to Phelps (1969), $k_1 = 1.4 \times 10^{-29} \exp(-\frac{600}{T_e}) cm^6/s$. For the electron detachment frequency, important processes in the ionosphere are photo detachment, detachment in collisions with molecules, and associative detachment:

$$v_d = v_{ph} + k_6 N'_{O_2} + k_7 N_O + k_8 N_N$$

The rate of change in the electron population is equal to the difference between rates of detachment and attachment.

With $N_e + N^- = N^+$ and v_a, $v_d \gg \alpha_I N^+$, we can express the net electron density

$$N_e \cong N^+ \frac{v_d}{v_a + v_d}$$

Depending on the excited states of molecules and the energy released from the interaction between monopoles and atmosphere and ground a rate of change of electron density between 1-100 microseconds is possible.

References

Gurevich A.V., <u>Nonlinear Phenomena in the Ionosphere, Springer-Verlag, N.Y., 1978.</u>

Wick, S., Conference paper, RADHEP – 2000, UCLA, November 16, 2000

Wong, A.Y., Nonlinear Interactions of Electromagnetic Waves with the Auroral Ionosphere, 13th Topical Conference on Radio Frequency Power in Plamas, Annapolis, MD, AIP Conference Proceedings 485, p 18-34, April, 1999

Wong, A.Y., Introduction to Experimental Plasma Physics, UCLA Publication, 1980

Wuerker, R.F., Private communication, 2001. This new Lidar facility is being put together by R. F. Wuerker.

Acknowledgment

This work is supported by SDSU foundation.

Proposed experiment to detect air showers with the Jicamarca radar system

T. Vinogradova[1], E. Chapin[1], P. Gorham[1], D.Saltzberg[2]

[1] *Jet Propulsion Laboratory, Calif. Inst. Of Technology, Pasadena, CA 91109*
[2] *Department of Physics and Astronomy, University of California, Los Angeles, CA 90095*

Abstract. When an extremely high energy particle interacts in the atmosphere, the collision induces a multiplicative cascade of charged particles, which grows exponentially until the energy per secondary degrades enough to dissipate in ionization of the surrounding air. During this process the compact cloud of energetic secondary particles travels 10-20 km through the atmosphere, leaving a column of ionization behind it. This ionized column quickly recombines, but for a period of order 0.1 ms it is highly reflective at frequencies below 100 MHz. This ionization trail, which is comparable in ionization density to that of a micro-meteor, should be clearly detectable using standard radar methods. We propose radar measurements using the facilities operated by Cornell University and the Instituto Geofisico del Peru (IGP) at the Jicamarca Radio Observatory near Lima, Peru. This facility's primary instrument is 49.92 MHz incoherent scatter radar, transmitting up to 1.5 MW of pulse power.

I RADAR DETECTION OF IONIZATION COLUMNS

A detailed description of the physics and phenomenology of radar detection of air shower ionization is available in a recent paper. [1] Here we extract the major results of this work. For radar detection of the columns that result from meteors, extensive air showers (EAS) events, or lightning, there are two regimes to consider, depending on the plasma frequency, ν_p of the ionized region:

$$\nu_p = \sqrt{(n_e e^2)/(\pi m_e)} = 8.98 \times 10^3 \sqrt{n_e} \text{ Hz}, \tag{1}$$

where n_e is the electron density in cm^{-3}. These two regimes are known in as the under- and over-dense regimes, respectively, and traditionally (in the meteor radar literature) are divided at the line density of $\alpha = 10^{14}$ m^{-1}. However, because of the significant differences in the radial distribution of the electrons in EAS ionization compared to that of meteors, we distinguish them only on the basis of the ratio of radar frequency to plasma frequency, a distinction that is also common to radar lightning measurements. Thus we will consider the over-dense portion of an ionization column to be that region where the electron density is high enough that

CP579, *Radio Detection of High Energy Particles,* edited by D. Saltzberg and P. Gorham
© 2001 American Institute of Physics 0-7354-0018-0/01/$18.00

the plasma frequency exceeds the radar frequency, and the radar cannot penetrate it and is reflected. Underdense columns are those, which have electron densities such that the local plasma frequency is below the frequency of the incoming radar, which can therefore penetrate the ionized region, and will then scatter with varying degrees of partial coherence from the volume of free electrons present.

Given an estimate for the electron density and its radial profile [1], the radar cross section of the ionization column at a given radar frequency can be parameterized as following for over- (σ_b^{od}) and under- (σ_b^{ud}) cases. This parameterization shows the result for a horizontal shower at 10 km altitude:

$$\sigma_b^{od} = 2.6 \times 10^4 (f/30 \text{ MHz})^{-1.45} (E/10^{20} \text{eV})^{0.44} (R/10 \text{ km}) \text{m}^2 \quad (2)$$

$$\sigma_b^{ud} = 33 (f/30 \text{ MHz})^{-1.84} (E/10^{20} \text{eV})^{1.9} (R/10 \text{ km}) \text{m}^2, \quad (3)$$

where is f the radar frequency, E is the shower energy and R is the range in km.

The radar returned power, P_r is described in terms of a model where the radiation is emitted from an antenna with peak transmitted power, P_t and directivity gain $G = 4\pi \Omega_A^{-1}$, where Ω_A is the solid angle of the main beam of the antenna. In terms of the effective radar backscatter cross section, σ_b, transmitting and receiving system efficiency, η, the resulting returned power [2] is

$$P_r = P_t \sigma_b \eta \frac{G^2 \lambda^2}{(4\pi)^3 R^4}. \quad (4)$$

The noise power of the radar system is given by $P_N = kT_{sys}\Delta f$, where T_{sys} is the system noise temperature, k is Boltzmann's constant and Δf is the effective receiving bandwidth. Combining the noise power equation above with equation for radar-received power one can write the signal-to-noise ratio (SNR) as

$$\frac{P_r}{P_N} = \sigma_b P_t \eta \frac{G^2 \lambda^2}{(4\pi)^3 R^4} \frac{1}{kT_{sys}\Delta f} \quad (5)$$

Given the above equation, the problem of detectability of the ionization columns, initiated by EAS reduces to evaluating of the effective radar cross section (RCS) for a given choice of operating parameters for radar system and to tuning radar system parameters to achieve the highest efficiency in experiment.

II   PROPOSED JICAMARCA EXPERIMENT

From the parameterization for the radar cross section for detecting the ionized column from the cosmic-ray showers and from the estimates of the plasma frequency of the shower cores described above, it follows that the most promising regime for investigating radar echoes from EAS is in the HF to VHF range. In addition, because of a large uncertainty in the magnitude of the echoes, a transmitter system with the highest possible pulse power should be used. For these reason we

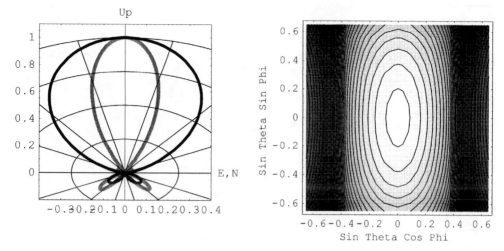

FIGURE 1. (a) Principal plane radiation patterns for Jicamarca "mattress" array. Light gray curve corresponds to the Up-East plane pattern, dark gray curve - to the Up-North. (b) Expanded view of the central region of the polar diagram. Contours are in 2 dB steps.

propose to perform the observations at the Jicamarca Radio Observatory, a facility sponsored by the National Science Foundation and operated by Cornell University near Lima, Peru.

The Observatory is the scientific facility for studying the equatorial ionosphere. It has a 1.5 MW transmitter and a main antenna with 18,432 dipoles covering an area of nearly 85,000 square meters. It is located at geographic latitude of 11.95° south and a longitude of 76.87° west (Peru). There are three additional 50 MHz "mattress" array antennas steerable to ±70° zenith angles in the east-west direction only. Each consists of 4×2 half-wave dipoles mounted a quarter wavelength above a ground screen. Two of these arrays can handle high powers. We are proposing to use the "mattress" array for detection of radar echoes from the cosmic ray showers.

The dipoles in the array are arranged in a group of four with a spacing of $\lambda/4$, where is a dipole wavelength. The spacing between 2 groups of 4 dipoles is $\lambda/2$. The principal plane patterns which are profiles along the East and North directions are shown in Figure 1a. Dark gray profile corresponds to the Up-North principal plane and the light gray profile corresponds to the Up-East plane. The half-power beam width is 0.3 rad in the North-Up plane and 0.8 rad in the East-Up plane. Figure 1b shows the polar diagram of the radar beam when oriented to transmit vertically, as predicted by numerical model. Antenna beam solid angle is 0.44 sr for the "mattress" array. The instrument parameters proposed for our observations are listed in Table 1.

Fixed parameter	Value
Transmitter pulse power	1.5 MW
Wavelength	6 m
Main beam lobe diameter	$0.3 - 0.8$ rad
Antenna gain	20.0 dB
Effective bandwidth	1 MHz
Duty cycle	6%
Tunable parameters	**Value**
Pulse duration	$10^{-6} - 10^{-3}$ s
Pulse rep. interval	0.2 m
Num, pulses	$1 - 100$

TABLE 1. Fixed and tunable parameters of the system.

In order to avoid decoding difficulties we propose using a simple square pulse with a width of 1 μs and the pulse repetition interval (PRI) of 0.2 ms, which corresponds to a one-way unambiguous sampling region up to 30 km altitude range. The resulting signal to noise ratio (SNR) can be written in the following parameterization:

$$\frac{S}{N} = 132(\frac{\sigma_b}{1\text{m}^2})(\frac{P}{10\text{kW}})(\frac{\eta}{0.1})(\frac{G}{10})^2(\frac{\lambda}{6\text{m}})^2(\frac{R}{10^4\text{m}})^{-4}(\frac{T_{sys}}{10^3\text{K}})^{-1}(\frac{\Delta t}{10\mu s}), \qquad (6)$$

where P is the pulse power, η the efficiency after losses, G the directivity gain of the antenna, T_{sys} the system thermal noise temperature, and Δt is the pulse duration. Figure 2 shows curves of the calculated RCS at 50 MHz as a function of initial cosmic ray energy for the different altitudes. The lowest through highest curves correspond to 10, 15 and 20 km altitudes, respectively. The showers are assumed to be propagating horizontally.

The other important issue for determining the efficiency of the proposed technique is the nature of the background noise. In the HF-VHF regime the effective background noise temperature for a remote site can be written as [2] $T_{\text{bkg}} = 2.9 \times 10^6 (f/3 \text{ MHz})^{-2.9} K$ which yields 830 K at 50 MHz. Considering a receiver noise in the range of 150 K, an overall system noise temperature of 1000 K is achievable at the Jicamarca site.

Using these assumptions and the discussed beam pattern above, we show an example SNR budget, in dBm (decibels referenced to 1 milliwatt) in Table 2. This is derived from numerical evaluations of the RCS for a horizontal shower of the energy 10^{19} eV, at an altitude 17 km and distance 50 km from the transmitter. These chosen parameters correspond to the beam pointing angle of 20 degrees above the horizontal plane with the maximum range of observation up to 175 km at altitudes up to 60 km.

To estimate the total number of the expected events above the given primary energy of the cosmic ray shower we use the same configuration as listed in Table 1.

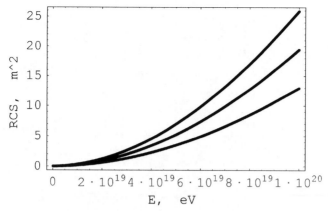

FIGURE 2. RCS for horizontally propagating showers at 10, 15 and 20 km altitudes (lowest to highest curves).

We required SNR to be more than 5.0 for the detection and estimated the effective aperture of the system based on the radiation beam profile as described above. From the energy spectrum for the ultra high-energy cosmic rays we calculated the flux of the particles as a function of energy per seconds per volume element (volume element in sr-cm^2 units). Based on the required SNR for the detection we determined the maximum range of the detection and integrated this range over projected beam area. The number of expected events for the period of observation is then the expected flux per energy bin for the time period mulitplied by the effective volume of the system (which is range-dependent as described above). Figure 3 shows the results of our estimation for 80 hours of the run time.

There are several uncertainties and possible risks for the proposed experiment. One of them is related to detailed understanding of the lifetime of the free electrons produced in the air shower ionization column; in fact radar detection of EAS events can provide independent measurements of these quantities or the limits if they cannot be detected. The other class of uncertainties is related to the ground clutter and the scattering effects. Ground-wave refraction and scattering cannot be accurately assessed until the experimental conditions are seen, and such clutter may significantly impact the SNR and usable range gate for our observations. We are going to collect data at Jicamarca during two runs of 40 hours each to demonstrate the efficacy of the technique to better estimate the lifetime of the air showers.

1. Gorham, P. W., 2000, Astropart.Phys. **15**, 177 (2001).
2. Skolnik, M.I., ed., Radar Handbook, 2nd ed., New York; McGraw-Hill, 1990.

Parameter	Value
Received power:	
Pulse power	1.5 MW
Pulse duration	1 μs
Number of repetitions	1
Antenna gain	20.0 dB
Wavelength	6 m
RCS at 10^{19} eV	0.82 m^2
Range to EAS	50 km
Loss factor	0.1
Noise power:	
System temperature	1000 K
Effective bandwidth	1 MHz
SNR	10.5

TABLE 2. SNR budget for one possible configuration of Jicamarca system.

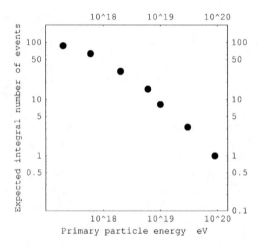

FIGURE 3. Predicted rates of the detected ultra-high energy cosmic ray showers with the Jicamarca system for a 6-days run.

Prospects of Hydroacoustic Detection of Ultra-High and Extremely High Energy Cosmic Neutrinos

L. G. Dedenko[1], Ya. S. Karlik[2], J. G. Learned[3], V. D. Svet[4], and I. M. Zheleznykh[1]

[1] Institute for Nuclear Research of Russian Academy of Sciences, Moscow, Russia
[2] Central Research Institute "Morfizpribor", St.Petersburg, Russia
[3] Department of Physics and Astronomy, University of Hawaii, USA
[4] N. N. Andreev Acoustic Institute, Moscow, Russia

Abstract. The prospects of construction of deep underwater neutrino telescopes in the world's oceans for the goals of ultra-high and super-high energy neutrino astrophysics (astronomy) using acoustic technologies are reviewed. The effective detection volume of the acoustic neutrino telescopes can be far greater than a cubic kilometer for extreme energies. In recent years, it was proposed that an existing hydroacoustic array of 2400 hydrophones in the Pacific Ocean near Kamchatka Peninsula could be used as a test base for an acoustic neutrino telescope SADCO (Sea-based Acoustic Detector of Cosmic Objects) which should be capable of detecting acoustic signals produced in water by the cosmic neutrinos with energies 10^{19-21} eV (e.g., topological defect neutrinos). We report on simulations of super-high energy electron-hadron and electron-photon cascades with the Landau-Pomeranchuk-Migdal effect taken into account. Acoustic signals emitted by neutrino-induced cascades with energies 10^{20-21} eV were calculated. The possibilities of using a converted hydroacoustic station MG-10 (MG-10M) of 132 hydrophones as a basic module for a deep water acoustic neutrino detector with the threshold detection energy 10^{15} eV in the Mediterranean Sea are analysed (with the aim of searching for neutrinos with energies 10^{15-16} eV from Active Galactic Nuclei).

I INTRODUCTION

In 1957 G. A. Askaryan suggested the use of ultrasonic and hypersonic emission from tracks of charged particles due to local heating in a medium for particle detection in dense media (Askaryan, 1957). Fifty years later, in the 21st Century, the hydroacoustic detection of ultra-high and extremely high energy neutrino interactions in the World's oceans may at last be developed to investigate problems of high energy neutrino astrophysics.

CP579, *Radio Detection of High Energy Particles*, edited by D. Saltzberg and P. Gorham
© 2001 American Institute of Physics 0-7354-0018-0/01/$18.00

Since first detection of the atmospheric neutrinos with energies $10^9 - 10^{11}$ eV in underground neutrino experiments (Reines, *et al.*, 1965; Achar, *et al.*, 1965), the target mass scale of the instruments – underground neutrino telescopes – has grown to $\sim 5 \times 10^4$ tons (Super-Kamiokande). In fact neutrino telescopes have become universal instruments for investigation of the microworld (e.g., in the search for proton decay), as well as searching for cosmic objects (e.g., solar neutrinos, supernova neutrinos, WIMPS, monopoles, etc.).

However to search for ultra-high energy (UHE, greater than 10^{15} eV) astrophysical neutrino sources, sites of the most energetic processes (accelerators) in the Universe, neutrino telescopes of effective detection volumes a cubic kilometer or more (KM3) are necessary (Learned, 1993; Learned and Mannheim, 2000). A prototype deep underwater optical neutrino telescope in Lake Baikal does operate, prototypes in the Mediterranean Sea (NESTOR, Antares) and in the deep polar ice (AMANDA in the Antarctic) are under construction now.

Neutrino telescopes with a target scale greater than 1 km^3 have also been proposed for studies of the upper boundary of the energy spectrum of cosmic neutrinos. For example, if elementary particles of maximal (Planck) masses $10^{24} - 10^{28}$ eV do exist, their interactions and decays could produce neutrinos (and other particles) of superhigh (extremely high) energies (SHE, EHE) up to $10^{24} - 10^{28}$ eV (Markov and Zheleznykh, 1980). Speculations upon the unknown source of the observed highest energy cosmic rays have provided several current models, which include significant or even dominant neutrino fluxes. For example, in the framework of some GUT models with X-particle masses of 10^{23} eV and in topological defect (TD) models, calculations of cosmic neutrino fluxes with energies up to 10^{23} eV have been performed (Sigl, 1997 and these Proceedings; Kuzmin, these Proceedings and references therein).

Several alternative methods for studying UHE and EHE neutrino interactions have been discussed and studied: deep underwater acoustics (Askaryan and Dolgoshein, 1976; Bowen, 1977; Learned, 1979), radio wave neutrino detection in cold Antarctic ice (Gusev and Zheleznykh, 1983; Markov and Zheleznykh, 1986; Boldyrev *et al.*, 1991; Ralston and McKay, 1990; Zas *et al.*, 1991, 1992; Besson; Saltzberg; Seckel; Zas–these Proceedings) and the the Radio Astronomical Method of Hadron And Neutrino Detection (Dagkesamansky and Zheleznykh, 1989; Alvarez-Muniz; Gorham; Hankins–these Proceedings). The energy thresholds for radio and acoustic detection are a few orders of magnitude greater than $10 - 50$ GeV threshold typical for the deep underwater optical detection. However the target mass scale of such alternative detectors can be a few orders of magnitude greater than that of optical detectors because the fundamental signal attenuation length in the medium is given by tens of km for the acoustic signals, compared to less than one hundred meters for light or because of the huge mass of the lunar regolith. The calculations for RAMAND (Radio Antarctic Muon And Neutrino Detector) threshold have shown (Provorov and Zheleznykh, 1995) that a 1 km^2 RAMAND in central Antarctica should be sensitive to the predicted diffuse flux of neutrinos from Active Galactic Nuclei (AGN) in the energy interval $10^{14} - 10^{16}$ eV.

II INITIAL SADCO FOCUS: TO SEARCH FOR SHE NEUTRINOS

The bipolar acoustic pulse production arises from the rapid expansion of the region of material traversed by the neutrino induced particle cascade which ionizes and heats the medium. Unfortunately it is an inefficient process, transferring only a tiny fraction (order of 10^{-9}) of the neutrino energy to acoustic radiation (largely due to the smallness of the bulk coefficient of thermal expansion). An acoustic signal is emitted by a neutrino induced cascade mainly in the direction perpendicular to the cascade axis in a rather narrow solid angle. The angular thickness of the disk is determined by the transverse size of the effective emitting region (which for high energy particle cascades in water is of the order of a few cm) divided by the cascade length (roughly ten meters), or a few milliradians. The initial spectrum peaks at a few tens of kHz.

The signal spectrum at large angles and at large distances shifts to a lower frequency region of a few kHz (Figures 1 and 2 in Dedenko *et al.*, 1997). Lower frequencies, while costing dearly in amplitude, provide potentially dramatic gains in solid angle and range over which events may be observed. If SADCO could detect 1 kHz signals produced by SHE cascades at distances of 10 to 50 km its effective detection volume could be tens to hundreds of cubic kilometers. The key to success is antenna gain.

While acoustic detection of elementary particles was suggested by G. A. Askaryan in the 1950's it was not until twenty years later that the possibility of construction of a deep ocean acoustic detector to study UHE neutrinos was widely discussed. Even then it was not pursued because at that time the threshold was thought too high and there were no plausible sources of UHE neutrinos. It was thought safer to start with optical detectors which had guaranteed cosmic ray neutrino signals in the GeV energy range. More recently with the advent of larger optical neutrino detectors, and increased interest in UHE neutrinos (as from AGNs), a deep underwater neutrino telescope SADCO with threshold energy above 5 PeV was proposed to be deployed at a depth of 3.5-4 km in the Ionean Sea near Pylos, Greece, at the site of the NESTOR optical neutrino telescope (Karaevsky *et al.*, 1993; Dedenko *et al.*, 1995). The sensitive volume of the SADCO neutrino telescope would be greater than 10^8 m^3. With the most optimistic models, dozens of events per year might be seen from the Glashow resonant neutrino interaction (6.4 PeV). But the main attraction of SADCO is its ability to search for the SHE neutrino interactions ($E \geq 10^{18}$ eV) in a huge water volume (Butkevich *et al.*, 1996). The search for TD neutrinos should thus be one of goals for SADCO if the registration volume reaches to hundreds of cubic km.

III ACOUSTIC PULSES IN WATER CAUSED BY CASCADES WITH ENERGIES 10^{20}-10^{21} EV

We have performed updated calculations of the acoustic pulse production and propagation in the ocean. Neutrinos interacting with nucleons in water produce a lepton (electron, muon, tau-lepton or scattered neutrino) and hadrons (hadron cascades). The calculations of acoustic signals emitted by electron induced cascades were carried out and the comparison with acoustic signals due to hadron cascades was made. Acoustic emission is localized in a divergent disk of a few hundred meters thickness, which is perpendicular to the cascade axis.

Figure 1 shows calculations of acoustic pulses caused by an electron-hadron cascade with energy 10^{20} eV in sea-water (14°C temperature), scaled to a reference distance of 1 meter from the cascade core. Calculations were performed for different shifts from the shower maximum along the cascade axis.

At super-high energies the Landau-Pomeranchuk-Migdal effect (decreasing cross sections for pair production and bremsstrahlung) should be taken into account. Due to this effect the longitudinal development of the electron-photon cascade in water increases by tens and even hundreds of times compared to the cascades with energies less than 10^{15} eV. At a cascade energy of 10^{20} eV the cascade length increases up to 300m, at energy 10^{21} eV up to 1-1.5 km, the density of deposited energy decreasing by nearly 100 times compared to the Bethe-Heitler showers.

Figure 2a shows acoustic pulses caused by the electron-photon cascade with energy of 10^{20} eV in sea water and 14°C temperature at a distance of 1 meter from the cascade core. Calculations were performed for different shifts from the shower

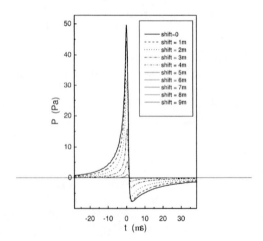

FIGURE 1. Acoustic pulses caused by the electron-hadron cascade with the energy 10^{20} eV in sea-water (14°C) temperature scaled to a reference distance of 1 meter from the cascade core. Calculations were performed for different shifts from the shower maximum along the cascade axis.

maximum along the cascade axis. Figure 2b shows pulses from electron-photon cascade with energy 10^{21} eV an equivalent distance of 1 m from the axis of cascade. The amplitude of the acoustic signal at this distance can be as much as 5 Pa. Strong acoustic emission is localized in a divergent disk of 200 m thickness, which is perpendicular to the cascade axis. The density of deposited energy in the hadron cascade would be much greater as its length is of a few tens meters and the amplitude of the acoustic pulse is approximately 50 Pa at 1 m distance from the cascade core and at 14°C, if the energy of the cascade is 10^{20} eV (Dedenko et al.,1999).

IV KAMCHATKA ARRAY AS A SADCO TEST BASE IN THE OCEAN

It is quite attractive to consider the use of already existing stationary sonar facilities, such as those placed in the Kamchatka region, as an acoustic detector of neutrinos (Karlik et al., 1997; Dedenko et al., 1997). This sonar has a large plane phased-array, with 2400 hydrophones. The array is installed on the sea shelf and connected with on-shore equipment by cable. The sector of view is 120°. The angular resolution in the horizontal plane is 0.8° in each of 150 (virtual) parallel fan-shaped beams. The vertical angular width is 7°. The gain of this array is 2500 at 1400 Hz.

Evaluations of the effective detection volume of the Kamchatka Array were performed for three cascade energies (10^{20} eV, 3.16×10^{20} eV, 10^{21} eV), real ocean conditions and for different probabilities of false alarm (false impulse signal) have been made. It is seen from the Table 1 that in summer time and under calm wind conditions, the Kamchatka array can have a significant detection volume (tens of cubic km) for searching for acoustic signals from 10^{21} eV cascades. On the basis of the Kamchatka array, we hope to develop a special measurement system for search for electron-hadron cascades induced by super-high energy (topological defects) cosmic neutrinos with $E > 10^{20} - 10^{21}$ eV and higher energies in water volumes of tens cubic kilometres and more.

Summer, wind velocity	False alarm probability		
	once per day	once per week	once per month
2 m/sec, calm	11/-/-	9.5/-/-	6.6/-/-
5.5 m/sec	3.9/5.3/81	2.2/4/-	1.3/2.2/-
10.2 m/sec	0.98/1.3/24	0.73/1.4/-	0.56/0.98/-

Table 1: Acoustic detection volumes in cubic kilometres for cascade energies10^{20} eV/3.16×10^{20} eV/10^{21} eV. ("-" denotes not calculated).

Acoustic signals emitted over a large solid angle in the ocean generally are complex when observed from some distance due to the guiding of sound around the depth of sound velocity minimum (the SOFAR zone). The small solid angle of

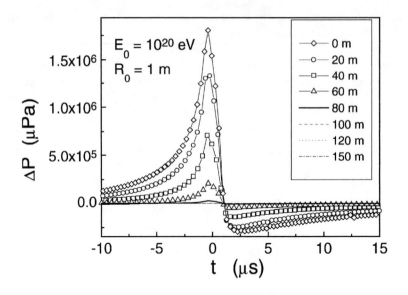

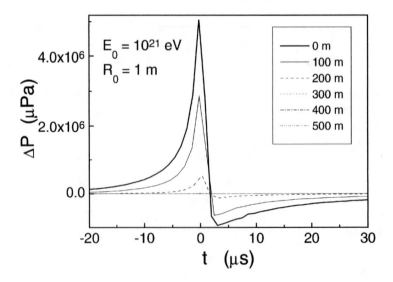

FIGURE 2. a) Acoustic signal from electron-photon cascade with the energy 10^{20} eV in sea water at temperature 14°C calculated for different shifts along the cascade axis. b) For 10^{21} eV.

peak emission from neutrino events may thus be employed to advantage in discriminating against background. Despite the fact that the frequency range of the existing sonar is not optimal (we would like higher bandwidth) the very large detection volume can compensate for the frequency deficit and greatly increases the detection probability of extremely energetic particles.

From the practical standpoint, the economic benefit of such an exploration is evident, because this sonar is in operation. Our suggested variant of acoustic detection of elementary particles can thus be very suitable for preliminary search experiments. Moreover this permits the development of different methodologies and algorithms for the detection and recognition of neutrino events. It is expedient to construct a special acoustical transducer suspended from a ship for simulation of neutrino impulse signals with different space-time parameters and calibrate the system. Little else is needed in terms of hardware.

V OPPORTUNITY FOR HYDROACOUSTIC MONITORING WITH AN OPTIMAL SADCO FREQUENCY BAND OF 10–20 KHZ

We have demonstrated, at least by computer simulation, the possibility to arrange a hydroacoustic detector of $10^{20} - 10^{21}$ eV neutrinos with a detection volume of some cubic kilometers using the Kamchatka array, when there is an underwater sound channel. However the frequency range of this array $(1.0 - 1.5$ kHz) is not optimal because of low signal/noise input ratio and impulsive noise from other sources of sound. But there is an opportunity for creation of acoustic detectors of even greater volume, using optimal frequencies (10-20 kHz) for signals which are generated by neutrino induced cascades with energies of about 10^{18-19} eV.

Without binding to any concrete site of deployment of the acoustic detector, it is possible to estimate the necessary parameters and the potential of such a system and to estimate the approximate volume needed for average hydro-acoustic conditions. The main sources of noise in this frequency band, taking into account that the receiving array is removed from a sea surface by hundreds meters and works in conditions of the underwater sound channel (USCH), are the noises of nearby navigation systems. With a high degree of probability (>96%) it is possible to consider that for the chosen frequency range the level of the this navigational noise is about $(1 - 2) \times 10^{-4}$ Pa. If the array amplification (concentration of acoustic array) is about 1500 -2000, that is close to the value considered by Karlik (Karlik *et al.*, 1997). After beamforming, the level of a noise will be lowered to $(2.5 - 5.0) \times 10^{-6}$ Pa in any angular beam in a horizontal plane and the angular resolution would be about $(1.5 - 2.0)°$.

The neutrino-induced cascades with energy 10^{18-19} eV in the specified frequency range can generate acoustic pulses with pressure of 35 μPa over distances of 8- 10 km. For the given level of noise the output signal/noise ratio at the specified array amplification will be more than 10 dB, which guarantees the probability of detection greater than 95%. Even if considering that the entire detection zone is concentrated in USCH of the width of 100–200 m (zone of the channel) the rough estimated volumes of the acoustic detector can be from 15 to 60 km^3, which is a rather large size. Note that approximately such USCHs exist in the Mediterranean Sea from May through November at depths of an array deployment from 200 to 500 meters. It goes without saying that the creation of such a hydroacoustic array represents a complex technological and economic problem, though soluble.

However it is expedient to consider a shorter and less expensive approach. At the end of the 1990s, in all leading countries there was a change of hydroacoustic underwater technologies, which for whatever reasons became ineffective. A typical example is the American Navy System SOSUS, which is now partially removed or used in civil applications. There is a similar situation in Russia. In particular, we can consider, for example, the hydroacoustic system MG-10M, which was used by the USSR Navy before and now is withdrawn from operation. The interest in this station (or to its receiving array) is due to the fact this array has an amplification about 1700 at an average frequency of 15 kHz, i.e. approximately what is necessary for optimal neutrino detection. On the other hand, the receiving array of this system is not so large — it is a cylinder with a diameter of 1.6 meters and height of 1 meter. The array contains 132 hydroacoustic sensors directed in a vertical plane on a cylindrical surface. The mass of the array is about 1200 kg and it can work at depths up to 500 meters. The frequency band is (4–25) kHz, and its sensitivity is about 170 $\mu V/Pa$. The specified array of MG-10M practically completely satisfies the required parameters of the acoustic neutrino detector mentioned above.

Certainly, such array requires the development of a special platform for its installation on the bottom and a modern fiber optic digital network to transfer signals from elementary sensors to coastal equipment. However taking into account the existing level of development of underwater telecommunications cable systems capable of transferring hundreds of gigabits up to 400 km without regeneration, making such lines is a standard technological task. The advantages of this approach are that the technology of such arrays has been repeatedly confirmed. In particular the lifetime of such a system is more than 25-30 years. On the other hand, the necessary expenses for modernization of this array are not large compared to design and construction of a new hydroacoustic array.

The MG-10M array could be a basic module of the neutrino telescope with a "low" energy detection threshold of $(1-5) \times 10^{15}$ eV for searching for AGN neutrinos, in particular, electron anti-neutrinos with energies close to 6.4×10^{15} eV owing to the Glashow resonance reaction $\bar{\nu}_e + e^- \rightarrow W^- \rightarrow$ hadrons.

VI   CONCLUSIONS

We have outlined a program for study of acoustic detection of neutrinos in the oceans employing existing sonar arrays with great amplification ($\sim$1500), e.g. the huge Kamchatka hydroacoustic array (frequencies less than ~1.5 kHz) and the existing and available compact hydroacoustic system MG-10M (an optimal average frequency ~15 kHz). First it will permit development of detection techniques in the real ocean, something not heretofore available to neutrino physicists. Moreover, the existing equipment should permit a useful physics exploration for neutrinos of extreme energies, not far above energies of cosmic rays already observed by extensive air shower arrays.

These investigations are supported by the RFBR-INTAS grant IR-97-2184 and the RFBR grant 00-15-96632. One of the authors (I.Zh.) gratefully acknowledges the RADHEP-2000 Organizing Committee for hospitality and support.

VII REFERENCES

Achar, C. V. et al., *Phys. Lett.*, 18, 196 (1965).

Askaryan, G, A., *Atomic Energy*, 3, 152, (1957).

Askaryan, G. A., Dolgoshein, B. A., 1976 DUMAND Summer Workshop, Hawaii.

Boldyrev, I. N. et al., *Proc. 3rd Int. Workshop on Neutrino Telescopes*, ed. Milla Baldo Ceolin, Venezia, 337 (1991).

Bowen, T., *Conference Papers, 15th ICRC*, Plovdiv, 6, 277 (1977).

Butkevich, A. V. et al., *Proc. Int. School Particles and Cosmology (1995)*, Baksan, World Scientific, 306 (1996).

Dedenko, L. G. et al., *Proc. 24th ICRC*, Rome, 1, 797 (1995).

Dedenko, l. G. et al., *Proc. 25th ICRC*, Durban, 7, 89 (1997).

Dedenko L. G. et al., *Izvestiya RAN, ser. fiz.*, (in Russian), 63, 589, (1999).

Gusev, G. A. and Zheleznykh, I. M., *Pisma Zh. Exp. Teor. Fiz.*, 38, 505 (1983).

Karaevsky, S. Kh. et al., *Proc. 23rd ICRC*, Calgary, 4, 550 (1993).

Karlik, Ya. S. et al., in Proc. XXXII-nd Rencontres de Moriond, ed. Girand-Herand, J. Tran Tranh Van, Edition Frontiers, France, 283 (1997).

Learned, J. G., *Phys. Rev. D*,19, 3293 (1979).

Learned, J. G., *Nucl. Phys. B (Proc. Suppl.)*, 33 A,B, 77 (1993).

Learned, J. G. and Mannheim, K., Ann. Rev. Nucl. Part. Sci., 50, 679 (2000).

Markov, M. A. and Zheleznykh, I. M., in *Proc 1979 Dumand Summer Workshop at Khabarovsk and Lake Baikal*, ed. J.Learned, p.177 (1980).

Markov, M. A., Zheleznykh, I.M., *Nucl. Inst. Methods*, A 248, 242 (1986).

Provorov, A. L. and Zheleznykh, I. M., *Astropart. Phys.*, 4, 55 (1995).

Ralston, J. P. and McKay, D. W., *Nucl. Phys. B (Proc. Suppl.)*, 14A, 356 (1990).

Reines, F. et al., *Phys. Rev. Lett.*, 15, 429 (1965).

Sigl, G., *Phys. Lett. B*, 392, 129 (1997).

Zas, E., Halzen, F., and Stanev, T., *Phys. Lett. B*, 257, 432 (1991); *Phys. Rev. D*, 45, 362 (1992).

ZeV Air Showers:
The View from Auger

Enrique Zas

Departamento de Física de Partículas,
Universidade de Santiago de Compostela, E-15706 Santiago, Spain.
zas@fpaxp1.usc.es

Abstract. In this article I briefly discuss the characteristics of the Auger observatories paying particular attention to the role of inclined showers, both in the search for high energy neutrino interactions deep in the atmosphere and as an alternative tool for the study of cosmic rays, particulartly their composition.

I   INTRODUCTION

The detection of high energy showers with the radio technique has been shown to have a high potential for astroparticle physics thoroughout this conference. One of the advantages of radio detection is that provided an appropriate wavelength can be chosen, exceeding the shower dimensions, the emission from all the shower particles becomes coherent. When the emission from all particles is coherent the emitted power should scale with the square of the primary energy. The technique thus becomes most advantageous for the detection of the highest energy particles. The detection of air showers with the radio technique was started in the 1950's [1] but it experienced many difficulties and other methods took the leading role in cosmic ray detection, at first arrays of particle detectors and Čerenkov telescopes and more recently air fluorescence detectors.

One of the most intriguing questions in Astroparticle Physics concerns precisely the origin and nature of the highest energy cosmic rays. The existence of events with energy above 10^{20} eV has been known since the 1960's [2] soon after Volcano Ranch, the first large air shower array experiment, started operation. Since then they have been slowly but steadily detected by different experiments as illustrated in Fig. 1. The observation of high energy cosmic rays has been recently reviewed by Nagano and Watson [3] who have shown that there is very good agreement between different experiments including the low and high energy regions of the spectrum. By now over 17 published events above 10^{20} eV [4] and preliminary new events from HiRes [5] are enough to convince the last skeptics about the non observation of the Greisen-Zatsepin-Kuz'min (GZK) cutoff, expected because of proton interactions

CP579, *Radio Detection of High Energy Particles,* edited by D. Saltzberg and P. Gorham
© 2001 American Institute of Physics 0-7354-0018-0/01/$18.00

with cosmic microwave photons [6]. The data suggest that the spectrum continues but little is known about the nature of the arriving particles. A large effort is being made in the 2000's to understand these particles. A new generation of large aperture experiments has started with HiRes already in operation [5], the Auger observatory in construction and with plans for using new techniques such as detection from satellites [7], with radar [8] and with radiotelescopes pointing to the moon [9] (See Fig. 1).

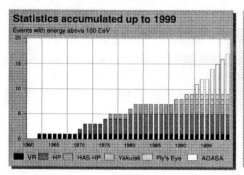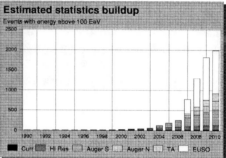

FIGURE 1. Left: Events with $E > 10^{20}$ eV detected by different experiments: Volcano Ranch (VR), Haverah Park (HP), Horizontal Air Showers in Haverah Park (HAS-HP), Yakutsk, Fly's Eye and AGASA. Right: Data buildup after 2000 from: HiRes, South and North Auger observatories (Auger S,N), Telescope Array (TA) and EUSO.

The Auger project is the last approved large aperture experiment to explore the high energy tail of the cosmic ray spectrum, those particles with energies exceeding 10^{19} eV [10]. The observatory has been shown to provide quite large acceptance for the detection of inclined showers induced by high energy neutrinos [11]. Recent analysis of inclined shower data from the Haverah Park array [12,13] has shown that it is possible to enhance the acceptance of air shower arays and to study with them the nature of the cosmic ray particles themselves [13]. In this article, after briefly addressing shower development, I will discuss the main characteristics of the Auger Observatories. The role of inclined showers for neutrino detection and composition measurements will be stressed, making reference to a model to describe the muon densties at ground level induced by inclined cosmic rays. Recent conclusions about composition at high energies using vertical and inclined showers will be reviewed.

II THE AUGER OBSERVATORIES

As a high energy cosmic ray enters and interacts in the atmosphere it gives rise to different generations of secondary particles that through succesive interactions constitute the extensive atmospheric shower. By the time the shower reaches ground level the number of particles in the shower front, mostly photons, electrons and positrons can exceed 10^{12} for the highest energy cosmic rays. As the shower penetrates to further depths the number of photons and that of electrons and positrons follow a characteristic behavior not too far from a gaussian which reaches a maximum between 1000 and 2000 meters above sea level for vertical showers above the EeV energy scale. Muons arise mainly in the decays of charged pions produced in hadronic interactions. Unlike electrons, muons do not shower and are practically only subject to minimum ionization losses. Their depth development increases following the pions but hardly decreases after reaching its maximum. Typically only muons with energies above the GeV scale reach ground level because of decay and energy loss. At ground level the photons which are most abundant have an average energy of order 1 MeV. Electrons and muons are typically ultrarelativistic with respective average energies of order 5 MeV and 1 GeV.

In the plane transverse to shower axis the shower front has a particle density which decreases as the separation from shower axis (r) increases due to multiple elastic scattering. Photons are more numerous than electrons but the lateral distributions are similar, decreasing as $r^{-\alpha}$. Although most particles are contained within the Molière radius (of order 100 m), for showers above 1 EeV the particle density remains significant even when r exceeds one km. The muons have a significantly flatter lateral distribution. Although they are outnumbered by electrons, the muon density can dominate at large r (in the km scale). The shower front develops a characteristic curvature depending on the position of shower maximum. The thickness of the shower front is mostly governed by the different delays that the particles accumulate as they deviate from the shower axis. Higher energy particles tend to deviate less and thus arrive earlier.

Extensive air showers have similar distributions whatever the nature of the initial particle. The establishment of composition is one of the toughest challenges in the detection of high energy cosmic rays. Simulations reveal some differences that can be used for this purpose. One of them is the total number of muons in a shower relative to electrons. Showers induced by photons mainly cascade into electrons and photons and only ocasionally a photon photoproduces mostly pions, channelling part of its energy into a hadronic shubshower. As a result showers induced by photons typically have of order ten times fewer muons than showers induced by protons of the same energy. If the cosmic rays are hadrons, the number of muons serves as a discriminator between heavy and light nuclei, the former having a somewhat higher muon content.

The Auger project is conceived as two 3,000 km^2 twin observatories in the northern and southern hemispheres, situated at mid latitudes. Each observatory is a hybrid experiment combining the two only succesful techniques for the study of EeV cosmic rays up to now, namely an array of particle detectors and a fluorescent detector (see Fig. 2). The *Engineering Array* is now being constructed in *Pampa Amarilla* in Mendoza, Argentina. It is a fraction of the Southern Auger array covering only 55 km^2 and which should be finished during the year 2001. The first physics results could be coming very soon. The Northern observatory is planned to be sited in Utah, U.S.A.

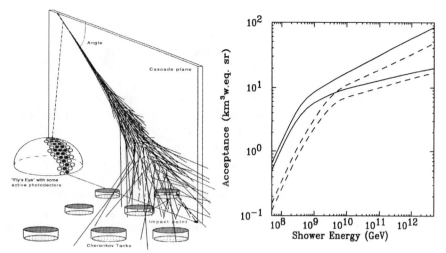

FIGURE 2. Left: Shower detected by tanks and by a Fluorescent light detector or "eye". Right: The full (dashed) line is the acceptance of the Auger Array for detecting the electrons and photons produced in hadronic (electromagnetic) showers of $\theta > 60°$. Lower curves refer to showers wih axis falling on the array.

The ground array uses cylindrical water Čerenkov detectors of 10 m² surface area and 1.2 m of height, each instrumented with three photodetectors. They are going to be arranged in a hexagonal grid separated 1.5 km from each other and extending over a surface area of 3,000 km². Two such tanks have been running in coincidence with the AGASA array. When electrons and photons reach the tank they are typically absorbed and they give a Čerenkov light signal which is proportional to the total energy carried by them. On the other hand most muons travel through the whole tank and give a light signal that is proportional to their track length within the tank. The arrival directions of the incident cosmic rays is determined from the arrival times of the shower front, typically with one degree accuracy. Each of the tanks is powered with a solar cell and the data are transmitted from the detectors to a central station by conventional wireless communication technology. Final triggering will be made at the central station.

The relative contributions of each particle species to the tank signals are comparable for a proton initiated shower because the average energies carried by photons, electrons and muons partly compensate the diferences in particle density. As we approach shower core both the relative number of muons and the time spread of the signal become smaller. At distances of order 1.5 km however the signals in the tanks can spread over times of order 2 μs. At large distances to shower axis the relative signal induced by muons becomes larger than that of electrons and photons. As the signal will be digitized with few nanosecond timing, the identification of large Čerenkov spikes from the individual muons will allow their separation, particularly well away from the shower core.

The particle density and the arrival times serve for the determination of the shower energy and its direction. The ground array has a large aperture, a 100% duty cycle and a uniform right ascension exposure for the study of anisotropy. The shower energy determination is usually determined by the particle density at a given distance to shower axis. Although this distance is chosen to minimize the effects of fluctuations and different primaries, the results are unavoidably dependent to some extent on both the nature of the arriving particle and the interaction model used for shower simulation.

As the shower develops in the atmosphere, nitrogen is excited and emits fluorescence photons in proportion to the number of ionizing particles, a few photons per meter of ionizing track length. By measuring the light signal and its arrival time from different positions along the shower depth development it is possible to detect and study very high energy cosmic rays. The fluorescence technique requires mirrors with imaging capabilities and covering a sufficient field of view to capture the depth development of the shower. The arrival direction of the cosmic ray particle is determined by the geometry and timing of the arriving light. This method has been very succesful at detecting EeV cosmic rays in dark nights.

The fluorescence detector in the Southern observatory will consist on four "eyes", three near the perimeter of the surface array and one roughly in the central part, at locations which are slightly elevated with respect to the rest of the detector. These eyes are located to monitor the atmosphere on top of the ground array. In the current design each eye is based on Schmidt optics and consists of mirror modules each with a 1.7 m diaphragm preceeding a 7 m diameter mirror and limiting the field of view of each mirror to approximately a $30° \times 30°$ fraction of the sky [14]. Each eye views 30° upwards from the horizon and combines several of these mirror modules in the azimuthal directions. The central eye requires 12 modules to cover 360° in azimuth and the other three eyes to be sited on the perimeter of the ground array only require 6 or 7 to cover 180° or 210° in azimuth [15]. The focal plane of each mirror is instrumented with a camera, an array of 20×22 optical modules, each of viewing aproximately $1.5° \times 1.5°$.

The detection of 10^{20} eV showers with the fluorescence technique necessarily requires the collection of light which is produced over 20 km away and thus subject to significant dispersion and attenuation in the atmosphere. Moreover the conversion of light to shower signal is affected by uncertainties in the geometrical reconstruction of the shower direction. Much of the geometrical uncertainty in the reconstruction is eliminated if the showers are detected from two or more eyes, that is from at least two different locations. This is the stereo viewing which constitutes one of the important advantages of HiRes [5]. In the Auger observatory many of the showers will be viewed in stereo and even by three eyes.

The fluorescence detector measures the particle production as a function of depth into the atmosphere and it is therefore a calorimetric energy determination. The uncertainty in the energy determination is therefore reduced with respect to a particle array which is measuring the particle content at a particular point in shower development and thus is subject to fluctuations between different showers. The ability to follow the depth development of the shower is also an important advantage because shower maximum can be determined directly. Such measurements of depth of maximum are most important for the establishment of primary composition.

The Auger observatory will be the first hybrid detector combining the fluorescence technique with a ground array for the detection of EeV cosmic rays. The combination of the two techniques is an improtant step because besides adding all the features of the two techniques, it will serve for cross calibration. The angular resolution of the fluorescence technique improves when used in combination with the ground array, because much of the geometrical uncertainty in the reconstruction of the shower profile is eliminated when the impact point of the shower axis is determined by the ground array. The power for composition of using both the method of establishing the depth of maximum and that of measuring muon content will help to eliminate part of the ambiguity associated to the interdependence between composition and interaction models.

III   INCLINED SHOWERS AND COMPOSITION

The fact that high energy particles exceeding the GZK cutoff have been observed allows one to make a strong case for the existence of high energy neutrinos. Although the origin and nature of these particles is unknown it is difficult to conceive the observed flux of any particle species without the existence of a flux of high energy neutrinos. The detection of inclined showers has been for long known to be a possible way to detect very high energy neutrinos interacting in the atmosphere [16]. When a neutrino interaction happens deep into the atmosphere, the showe can reach its maximum very close to ground level in spite of being close to horizontal. Such a shower would look much like an ordinary vertical shower with high electron and photon content and a front curvature corresponding to shower maximum near ground level. The Auger observatory will be sensitive to high energy neutrinos. Its acceptance for the electromagnetic component of deep and inclined showers induced by neutrinos exceeds 10 km^{3}sr of water equivalent [11] (See Fig. 2).

The original motivation for the study of inclined showers induced by cosmic rays was to understand the background of cosmic ray signals to neutrino detection, but these showers have proved to be of great interest on their own. The study of cosmic ray showers by particle arrays has been mostly restricted to relatively vertical showers, typically resticting zenith angles to less than 45°. The particle densities in such such showers keep the characteristic circular symmetry allowing an easy reconstruction of the event energy by measuring it at a given distance to the shower axis. The acceptance $\mathcal{A}$ of a ground array of area A for cosmic rays depends on θ_{max}, the maximum zenith angle that the array can detect:

$$\mathcal{A} = \int A \cos\theta \, d(\sin\theta) \, d\phi = \pi A[1 - \cos^2\theta_{max}] \tag{1}$$

If only zenith angles below $\theta_{max} = 45°$ are analysed with the Auger observatory, its acceptance would be 4,500 km^{2}sr. The acceptance of the observatory will double if the analysis of showers can extend to zenith angles less than 90°. It has recently become quite clear that inclined showers produced by cosmic rays can be analysed at least with arrays of water Čerenkov tanks [17,12,13]. Moreover the analysis of these showers has also shown to have a remarkable potential for the study of primary composition [13].

Much development has been possible by separately modelling the distortion of the muon density patterns in inclined showers under the influence of the Earth's magnetic field [17]. The showers can first be studied in the absence of a magnetic field where two important facts emerge for inclined showers: a) Most of the muons are produced in a well defined region of shower development which is quite distant from the ground and b) the lateral deviation of a muon is inversely correlated with its energy. Most of the characteristics of the muon densities in inclined showers are governed by the distance and depth travelled by the muons which is of order 4 km for vertical showers, becomes 16 km at 60° and continues to rise as the zenith angle rises to exceed 300 km for a completely horizontal shower. This distance determines the minimum energy needed for a muon to reach ground level and thus fixes the average energy of these muons that can be of a few hudred GeV.

In the absence of a magnetic field we can assume that all muons are produced at a given altitude d with a fixed transverse momentum $p_\perp$ that is uniquely reponsible for the muon deviation from shower axis. In the plane transverse to shower axis at ground level (transverse plane) the muon deviation, $\bar{r}$, is inversely related to muon momentum p. The density pattern has full circular symmetry. When the magnetic field effects are considered the muons deviate a further distance δx in the perpendicular direction to the magnetic field projected onto the transverse plane $\vec{B}_\perp$, given by:

$$\delta x = \frac{e|B_\perp|d^2}{2p} = \frac{0.15|B_\perp|d}{p_\perp}\,\bar{r} = \alpha\,\bar{r}, \qquad (2)$$

where in the last equation $B_\perp$ is to be expressed in Tesla, d in m and $p_\perp$ in GeV.

Eq. 2 is telling us that all positive (negative) muons that in the absence of a magnetic field would fall in a circle of radius $\bar{r}$ around shower axis, are translated a distance δx to the right (left) of the $\vec{B}_\perp$ direction. As the muon deviations are small compared to d they can be added as vectors in the transverse plane and the muon density pattern is a relatively simple transform of the circularly symmetry pattern. The dimensionless parameter α measures the relative effect of the translation. For small zenith angles d is relatively small and $\alpha << 1$ so that the magnetic effects are also small, and results into slight elliptical shape of the isodensity curves. For high zeniths however $\alpha > 1$ the magnetic translation exceeds the deviation the muons have due to their $p_\perp$ and *shadow* regions with no muons appear as confirmed by simulation. For an approximate $p_\perp \sim 200$ MeV and $B_\perp = 40\ \mu$T this happens when d exceeds a distance of order 30 km, that is for zeniths above $\sim 70°$. The muon patterns in the transverse plane can be projected onto the ground plane to compare with data as well as standard simulation programs. Realistic density patterns are obtained if these ideas are modified accounting for the energy distributions of the muons at a given $\bar{r}$. In the simulations it is the average muon energy which is inversely related to $\bar{r}$.

For each zenith angle the shape of the lateral distribution of the muons does not change for showers of energy spaning over four orders of magnitude. Different primary particles and interaction models also have similar distribution functions in shape. As a result one only needs the total number of muons to describe a shower of given zenith. This normalization scales with the proton energy E as:

$$N = N_{ref} \, E^{\beta} \tag{3}$$

where β and N_{ref} are slightly model dependent constants [12].

The inclined shower data obtained in the Haverah Park array was analysed with the help of the model described above. The Haverah Park detector was a 12 km^2 air shower array using 1.2 m deep water Čerenkov tanks that was running from 1974 until 1987 in Northern England which has been described elsewhere [18]. It is the array that has been made closest to the ground array of the Auger observatory because it also consisted on water Čerenkov tanks Particular care was taken to account for new corrections to the tank signals that arise when horizontal events are detected. These include light that falls directly into the photoubes, enhanced delta ray signal because muons are more energetic, catastrophic energy losses for the muons and a signal due to electromagnetic particles from muon decay.

The event rate as a function of zenith angle has been simulated using the modelled muon distributions. The qualitative behaviour of the registered rate is well described in the simulation and the normalization is also shown to agree with data to better than 30% using the measured cosmic ray spectrum for vertical incidence, assuming proton primaries and using the QGSM model [12]. More impressive are the results of fits of the models for muon densities to the observed particle densities sampled by the different detectors on an event by event basis. This result demonstrates that the acceptance of these detectors can be extended to practically all zenith angles. The analysis of the nearly 10,000 events recorded with zenith angles above 60° is a complex process that involves a sequence of time fits to get the arrival directions and density fits to obtain the energy under the assumption that the primaries are protons. The curvature of the shower front is considered and care is taken to account for the correlations between the shower energy and the impact point of the shower.

The analysed data are subject to a set of quality cuts: the shower is contained in the detector (distance to core less than 2 km), the χ^2 probability of the event is greater than 1% and the downward error in the reconstructed energy is less than 50%. These cuts ensure that the events are correctly reconstructed and exclude all events detected above 80°. Examples of reconstructed events compared to predictions are illustrated in Fig. 3. Two new events with energy exceeding 10^{20} eV have been revealed. The results have been compared to a simulation that reproduces the same fitting procedure and cuts using the cosmic ray spectrum deduced from vertical air shower measurements in reference [3]. The agreement between the integral

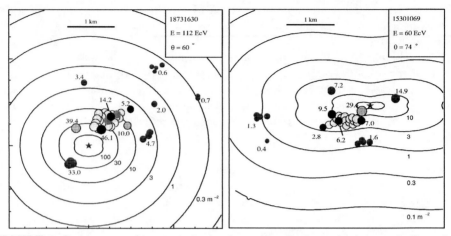

FIGURE 3. Density maps of two events in the plane perpendicular to the shower axis. Recorded muon densities are shown as circles with radius proportional to the logarithm of the density. The detector areas are indicated by shading; the area increases from white to black as 1, 2.3, 9, 13, 34 m^2. The position of the best-fit core is indicated by a star. Selected densities are also marked. The y-axis is aligned with the component of the magnetic field perpendicular to the shower axis.

rate above 10^{19} eV measured and that obtained with simulation is striking when the QGSJET model is used for the interactions. Sibyll leads to a slight underestimate [13].

The universality of the muon lateral distribution function is very powerful and once the equivalent proton energy is determined for all events, corresponding energies can be easily obtained for different assumptions about primary composition. In the case the incoming particles are iron nuclei (photons), the primary energy can be calculated multiplying the equivalent proton energy by a factor which is ~ 0.7 (6) for 10^{19} eV and varies slowly as the primary energy raises. As a result when a photon primary spectrum is assumed, the simulated rate seriously underestimates the observed data by a factor between 10 and 20. A fairly robust bound on the photon composition at ultra high energies can be established assuming a two component proton photon scenario. The photon component of the integral spectrum above 10^{19} eV (4×10^{19} eV) must be less than 41% (65%) at the 95% confidence level. Details of the analysis are presented in [13] and will be expanded elsewhere.

IV SUMMARY

The Auger detector will soon give an important contribution to the observation of the high energy tail of the cosmic ray spectrum. Its acceptance is close to 9,000 km^2sr when considering the inclined air showers. Its hybrid character will be of great value to cross calibrate the two classes of detectors and for establishing the composition. The combined analysis of vertical and horizontal showers will also set important limits to composition at high energies. The combination of two different techniques for composition studies will be of great importance in reducing the uncertainties associated to the hadronic interaction models.

ACKNOWLEDGEMENTS

The author thanks the organizers of such a pleasant conference bringing together people from so many different fields, an also thanks D. Saltzberg for helpful comments after carefully reading the manuscript. This work was supported in part by the European Science Foundation (Neutrino Astrophysics Network N. 86), by the CICYT (AEN99-0589-C02-02) and by Xunta de Galicia (PGIDT00PXI20615PR).

REFERENCES

1. T.J. Weekes in these Proceedings.
2. J. Linsley, *Phys. Rev. Lett.*, **10** (1963) 146.
3. M. Nagano and A.A. Watson, *Rev. Mod. Phys.* 72 (2000) 689.
4. E. Zas in *Proc. of the Int. Workshop on Observing Ultra High Eenergy Cosmic Rays from Space and Earth*, (2000) Metepec, Puebla, Mexico, to be published by AIP.
5. P. Sokolsky in these Proceedings.
6. K. Greisen; *Phys. Rev. Lett.*, **16** (1966) 748. G.T. Zatsepin and V.A. Kuz'min, *JETP Lett.*, **4** (1966) 78.
7. O. Catalano and L. Scarsi in *Proc. of the Int. Workshop on Observing Ultra High Eenergy Cosmic Rays from Space and Earth*, (2000) Metepec, Puebla, Mexico, to be published by AIP.
8. P. Gorham in these Proceedings.
9. J. Alvarez-Muñiz in these Proceedings.
10. *The Pierre Auger Project Design Report.* By Auger Collaboration. FERMILAB-PUB-96-024, Jan 1996. 252pp.
11. J. Capelle, J.W. Cronin, G. Parente, and E. Zas, *Astropart. Phys.* **8** (1998) 321.
12. M. Ave, J.A. Hinton, R.A. Vázquez, A.A. Watson and E. Zas, *Astropart. Phys.* **14** (2000) 109.
13. M. Ave, J.A. Hinton, R.A. Vázquez, A.A. Watson and E. Zas, *Phys. Rev. Lett.* **85**, (2000) 2244.
14. H. Bluemer *et al.* in *Proc. of the XXV ICRC*, Salt Lake City 1999, Vol 5 p. 345.
15. B. Dawson, Auger Technical Note GAP-099-024.
16. V.S. Berezinsky and G.T. Zatsepin, *Yad. Fiz.* **10** (1969) 1228. [*Sov. J. Nucl. Phys* **10** (1969) 696].
17. M. Ave, R.A. Vázquez, and E. Zas, *Astropart. Phys.* 14 (2000) 91.
18. R.M. Tennent, *Proc Phys Soc* **92** (1967) 622. M.A. Lawrence, R.J.O. Reid, and A.A. Watson, *J Phys G* **17** (1991) 733.
19. J.R.T. de Mello Neto, GAP note 1998-020.

The View From HiRes

P. Sokolsky

High Energy Astrophysics Institute
University of Utah
Salt Lake City, Utah

Abstract. The status and results from the HiRes atmospheric fluorescence experiment are discussed. In particular, results on composition from the HiRes Prototype and CASA/MIA hybrid experiment and preliminary results from the HiRes monocular data on the highest energy cosmic rays are presented. Evidence for the cosmic ray spectrum continuing past the Greisen-Zatsepin-Kuzmin cutoff is evaluated and the impact of stereo data on this question is discussed

INTRODUCTION

Penzias and Wilson's unexpected discovery of the universal microwave background in 1965 is now the centerpiece of our understanding of the evolution of the universe and the Big Bang scenario[1]. It also has deep implications for our understanding of the origins of the highest energy cosmic rays. In 1966, Greisen realized that ultrahigh energy cosmic ray protons would be affected by interactions with this radiation. Greisen predicted that, if cosmic ray sources were distant enough from us and if their flux extended past 10^{20} eV (100 EeV), then they would interact inelastically with the black body radiation[2]. This interaction has a threshold given by the onset for single pion photoproduction. He predicted that a smooth power law spectrum of cosmic rays would be cut off by this onset of inelastic interactions near 60 EeV if sources were uniformly distributed throughout the universe. This effect was independently predicted by Zatsepin and Kuzmin. The search for this so-called Greisen-Zatsepin-Kuzmin 'GZK' cut-off has been one of the primary goals of ultra-high-energy (UHE) cosmic ray physicists ever since. The lack of observation of a cut-off could imply that UHE cosmic ray sources are relatively close to us, on the order of 50 Mpc or less. If that is the case, then significant anisotropies towards the source directions should be observed. Of course, the degree of anisotropy depends on the magnitude of extragalactic magnetic fields and the composition of the cosmic rays themselves. Nevertheless, if local AGN's are the sources, then one would expect that the highest energy cosmic rays will align in their directions.

The Fly's Eye and the AGASA experiments observed events in 1991 and 1993 respectively that appear to be clearly beyond the predicted cut-off. Since then, significant additional data from AGASA and from the new High Resolution Fly's Eye (HiRes) experiment have become available and appear to confirm the original findings. No significant astronomical alignments of these events with nearby

CP579, *Radio Detection of High Energy Particles,* edited by D. Saltzberg and P. Gorham
© 2001 American Institute of Physics 0-7354-0018-0/01/$18.00

astrophysical objects has been found. This has stimulated a great deal of interest and all kinds of hypotheses ranging from decays of topological defects to violation of special relativity have been proposed. Much more data is needed to clarify the situation. In this talk, I will discuss the results from the original Fly's Eye and the Hires/CASA/MIA hybrid prototype and present preliminary results from the HiRes monocular data. HiRes stereo data is just becoming available. At this point, it is primarily useful in checking the monocular reconstruction and Monte Carlo energy resolution results but will be critical in determining the composition of cosmic rays at the highest energies.

The Atmospheric Fluorescence Technique

The atmospheric fluorescence technique as we presently understand it was developed by G. Cassiday, E. Loh, and colleagues at the University of Utah, based on a prototype detector built by K. Greisen at Cornell University[3]. The longitudinal development of the EAS in the atmosphere is measured directly by this technique. As charged particles in an EAS cross the atmosphere, they excite nitrogen fluorescence. Unlike Cherenkov radiation, this UV light is isotropic and hence EAS can be viewed from all directions by appropriate detectors. The fluorescent photon yield is about four photons per meter per ionizing particle and is approximately independent of altitude (up to 10 km). The light is emitted in several bands from 300 to 400 nm. The atmosphere has negligible absorption and has a horizontal scattering length of 10 to 15 km at these wavelengths. A single Fly's Eye detector subdivides the hemisphere of the sky into pixels defined by photomultiplier tube apertures. As a shower develops through the atmosphere, a pixel whose field of view intersects the shower position will detect signals whose amplitude, after correction for geometrical and atmospheric effects, are proportional to the number of ionizing particles passing through that part of the atmosphere. The moving shower defines a "track" through the atmosphere which in turn results in a "track" of pixels in the detector. The integral of the reconstructed shower profile is directly proportional to the primary energy. The method is essentially calorimetric, measuring the total energy deposition in the atmosphere by means of the amount of fluorescence produced. This technique does not require a complex Monte Carlo calculation to determine the energy scale. Instead, the energy scale is given by the N_2 fluorescence efficiency for ionizing particles which has been measured by a number of groups [4].

The University of Utah group built the first Fly's Eye in 1982. A second partial eye, called Fly's Eye II was built 3.5 km from the first eye in 1985. An EAS can then be detected by both eyes simultaneously. This stereo data set had much more precision and redundancy than the monocular eye measurements. The Fly's Eye experiment took data from 1982 to 1992. While the instantaneous aperture of a fluorescence detector is very large (about about 1000 km^2str for the monocular Fly's Eye and 400 km^2str for the stereo case) the effective on time of fluorescence detectors is limited to approximately ten percent by the requirement of dark nights and good weather.

Results From the Fly's Eye Experiment

Stereo and Monocular Results on the Cosmic Ray Spectrum

The Fly's Eye data[5] is divided into poorer resolution monocular data and better resolution but lower statistical significance stereo data. The Fly's Eye stereo spectrum exhibits significant structure (at the 5 sigma level of statistical significance) between 1 and 10 EeV (see Fig 1). Of particular interest is a steepening near 3×10^{17}, now dubbed the "second knee", to which we will return later.

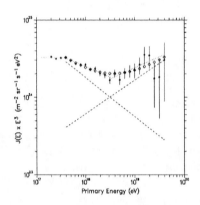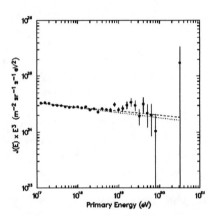

FIGURE 1. The Fly's Eye spectrum. Left hand figure is the stereo spectrum. The data is shown as solid dots. Dashed lines indicate hypothetical two component spectrum whose sum is indicated by the dotted line. Right hand figure is the monocular spectrum. The data is shown as solid dots. The dashed line is the best single power law fit to the whole spectrum. The dotted line is the best fit to the data below 10^{19} eV.

The monocular Fly's Eye data has significantly greater statistics than the stereo data. Although a flattening of the spectral index is evident above 10 EeV, the pronounced dip structure near 3 EeV observed in the stereo data is absent. The discrepancy arises because the monocular data has poorer energy resolution. If the monocular energy resolution is used with the stereo spectrum, the spectrum obtained is consistent with the observed monocular data.

While the Fly's Eye monocular spectrum generally conforms to expectations for a GZK cutoff (a pile-up of events below the cut-off would generate the observed flattening of the spectrum just below60 EeV), a single exceptional event was observed with energy well beyond it. This event with an estimated energy of 320 EeV was detected on October 15, 1991. In all respects, this event appears to be well reconstructed and its longitudinal shower profile is shown in Figure 2. The position of shower maximum in the atmosphere for the best fit at 13 km is consistent with expectations for either a proton or heavy nucleus induced shower.

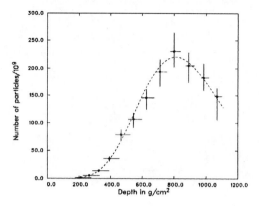

FIGURE 2. The longitudinal profile of the highest energy event observed by the monocular Fly's Eye. Vertical error bars are dominated by photostatistics. Note the horizontal error bars are correlated since the depth of shower in the atmosphere is completely determined by the overall geometry of the event.

Composition of Cosmic Rays from 10^{17} to 10^{19} eV

Indirect information on cosmic ray composition can be extracted from the distribution of EAS shower maxima in the atmosphere. Heavy nuclei such as iron will interact higher in the atmosphere with an interaction length of 15 gm/cm^2 and their shower maxima are expected to be near 600 gm/cm^2 at .1 EeV. Proton primaries will interact with a mean free path near 70 gm/cm^2 and have shower maxima near 700 gm/cm^2 and larger shower by shower fluctuations in atmospheric depth. The average depth of shower maximum increases logarithmically with primary energy . The slope of the average shower X_{max} versus primary energy plot is called the elongation rate. Different nuclei are expected to have similar elongation rates but have different intercepts on an X_{max} versus energy plot. It is possible to look for consistency with a light, predominantly protonic composition, or a heavy composition composed of CNO and Fe group nuclei. In particular, data can be compared to the two extreme possibilities of a pure proton and a pure Fe CR flux. Expected elongation rates for individual nuclei are 50 to 55 gm/cm^2 per decade of energy and are very insensitive to the details of hadronic interaction models. A cosmic ray flux that changes from a predominantly heavy to a light composition would be reflected in a measured elongation rate larger than expected for a constant composition. Figure 3 shows the stereo Fly's Eye data and the expected positions of X_{max} for Fe and p incident particles. It is apparent that the composition appears to be getting lighter in the decade between 1 and 10 EeV. The change from a heavy to a light composition occurs in the same energy interval where significant change in the spectral slope is observed. A straightforward interpretation of this is the appearance of a light, hard spectrum extragalactic cosmic ray flux which begins to dominate over the heavy softer galactic component

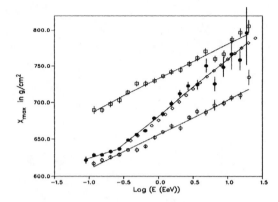

FIGURE 3. The stereo Fly's Eye elongation rate. Average shower maximum is shown in solid circles. Open circles indicate predictions for a pure iron flux while open boxes represent a pure proton flux. Open diamonds correspond to predictions of the two component power law model show in Fig. 1.

THE HIGH RESOLUTION FLY'S EYE DETECTOR

The High Resolution Fly's Eye Detector (HiRes)[6] is a collaboration between the University of Adelaide, Columbia University, the University of New Mexico, Rutgers, Montana State University, the University of Tokyo, and the University of Utah. This experiment will increase sensitivity to cosmic rays above 100 EeV by at least an order of magnitude with respect to the Fly's Eye results. It is designed to study the cosmic ray energy flux above 3 EeV and measure its composition and anisotropy with significantly improved resolution. Construction was completed in the summer of 1999. The two sites, the original Fly's Eye I site (now called HiRes I) and a new HiRes II site are 12.5 km distant. The HiRes detector gets its increased aperture and improved resolution through decreasing the pixel size from 5.5 degrees by 5.5 degrees (the tube aperture of the Fly's Eye detector) to one degree by one degree and by increasing the mirror area from 1.7 to 3.5 m^2. The resultant increased signal to noise ratio makes showers at 30 km impact parameter distance easy to detect. The improved sampling along the track allows even short tracks to be well reconstructed. In addition, all events will be seen in stereo so that the measurement will always be overconstrained. One major advantage of this approach is that events will have two independent measurements of their energy and longitudinal shower profile. This makes it possible to measure the detector resolution function in X_{max} and energy directly from the data. Knowledge of this resolution function is critical in making sure that an apparently continuing spectrum is not due to spill down effect produced by the detector resolution. At the highest energies, the aperture is 13,000 km^2str.

Results from the HiRes Prototype

An earlier prototype version of the HiRes experiment ran for three years overlooking the CASA/MIA muon and electron ground array detectors[7] 3.5 km away. In this experiment, cosmic ray EAS were detected in both the fluorescence and ground array modes simultaneously. The energy range is restricted to .1 to 1 EeV by the size of the ground array. Results confirm that the elongation rate is inconsistent with a constant composition and is in agreement with a gradual change from a heavy to a light composition (see Figure 4). The muon number density at 600 m as a function of energy is also measured. While the predictions for a pure Fe composition are somewhat below the measured muon densities, indicating a possible problem with the hadronic model, the slope of this energy dependence is consistent with a change from a heavy to a light composition (see Figure 5). As in the case of the Xmax elongation rate, the slope of the muon densitiy versus energy plot is more reliably predicted than the absolute magnitude of the density. The shape of the cosmic ray spectrum measured in this hybrid experiment is shown if Figure 6. While statistics are limited, this higher resolution data shows the beginning of the same "second knee" structure seen in the earlier Fly's Eye experiment. Fig. 6 also shows the combined results of a number of experiments from below the first "knee" to 1 EeV. There is significant evidence that the "second knee" represents a spectral index change (.3) very similar to the first.

The same data set can be used to derive the average shower profile for energies near .1 EeV. In this case, individual showers are normalized to same energy and X_{max} position in the atmosphere, converted to shower size versus shower age, and averaged. The results, shown in Fig. 7 show excellent agreement with the Gaisser-Hillas parametrization. Fluctuations about the mean are consistent with photostatistics.

Preliminary Monocular HiRes Results

HiRes I was completed in 1997 and has been accumulating data since then. Events seen by only one detector can be reconstructed by fitting to the distribution of time of arrival of light at each phototube. This monocular timing fit only works well for tracks approaching 20° in projected angular length. Because HiRes I has a limited vertical field of view (2-17° above the horizon) most events are significantly shorter. Monte Carlo studies show that the width of EAS showers in the atmosphere, expressed in gm/cm^2 is quite constant, independent of energy, composition or model. This has been confirmed by the HiRes prototype studies discussed above. To improve resolution we do a combined fit to the timing and to the central value of the predicted shower width. Monte Carlo studies show that the resolution improves as the energy increases with small non-gaussian tails. We expect a 20% energy resolution near 100 EeV. The monocular aperture of HiRes I is very similar to the stereo aperture and approaches 10,000 km^{2}str near 100 EeV.

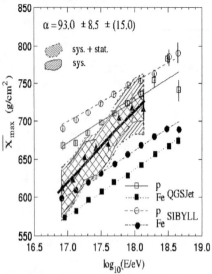

FIGURE 4. HiRes Prototype/MIA elongation rate. Dark solid line and cross hatched area indicate the central value of the data and systematic unertainty bounds. Lines and open and closed circles and squares indicate predictions for pure proton and pure Fe flux using two different hadronic models.

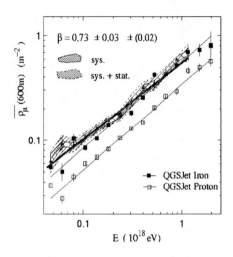

FIGURE 5. HiRes Prototype/MIA muon density at 600m as a funtion of energy. Dark solid line and cross hatched area indicate the central value of the data and systematic uncertainty bounds. Solid squares are predictions for pure Iron. Open squares are predictions for pure protons. Both predictions assume the QGSJet hadronic model. Note that theoretical predictions underestimate the number of muons.

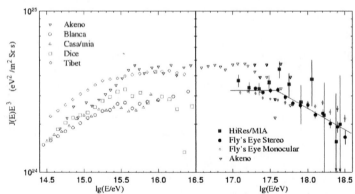

FIGURE 6. Combined world data on the cosmic ray spectrum from below the "knee" to the "second knee". Note magnitude of spectral beak for "second knee" is similar to the first "knee" of the specrum

Preliminary results on the spectrum are shown in Figure 8. The solid line is a fit to the original stereo Fly's Eye spectrum, smoothly extrapolated to the highest energies. The data indicates a number of events nominally above 100 EeV. This data is analyzed using a standard desert atmospheric model and does not yet have corrections for the night to night variation in the atmospheric transmission. It is clear, however, that

302

systematic atmospheric effects will not cause all of the events to move below 100 EeV. In particular, the highest energy event with an energy of 270 EeV, observed at an impact parameter of 30 km, would have its energy shifted down to 140 EeV if the atmosphere became as clear as it possibly could be - i.e. if it was a pure molecular atmosphere for that night. This is still significantly above the universal GZK prediction of 60 EeV.

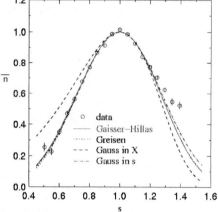

FIGURE 7. Average shower profile measure by the HiRes prototype. The Gaisser-Hillas function is an excellent representation of the data. Deviations of the last few data points are due to systematic effects and are not significant.

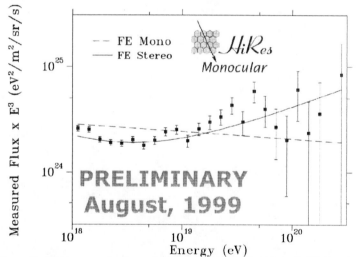

FIGURE 8. Preliminary results from the HiRes I monocular data. Solid line indicates flux extrapolated from Fly's Eye stereo spectrum. Dashed line indicates flux extrapolated from fit to Fly's Eye monocular data.

Several hundred stereo events have now been reconstructed. These events can be used to test the monocular reconstruction method. At least below 10^{19} eV where sufficient statistics exists, results indicate that the monocular energy resolution is about 25%, quite consistent with Monte Carlo calculations. It is also clear that the trend towards improving resolution with energy predicted by the Monte Carlo is occurring in the data. As more stereo statistics are accumulated, the question of non-gaussian tails on the resolution function can be addressed in more detail. However, there is no evidence from the current data that these tails or larger than those exhibited by the Monte Carlo.

REFERENCES

1. Penzias, A.A., and Wilson, R.W., *Ap.J* ,**142,** 419 (1965).
2. Greisen K., *Phys. Rev. Lett.* **16,**748 (1966).
3 Sokolsky, P. , Sommers, P. ,and Dawson, B. R. ,*Physics Reports* **217**, No. 5,(1992).
4. Kakimoto, F., et al., *Nucl. Inst. and Meth.*, (1996).
5. Bird, D. J., et al., *Phys. Rev. Lett.* **71**, 3401 (1993); *Ap.J.* **424**, 491 (1994);*Ap. J.*,**441,**151(1995).
6. Jui, C.C.H., "Results from the High Resolution Fly'e Eye Experiment" in *26th International Cosmic Ray Conference-1999*, edited by B.L. Dingus et al., AIP Conference Proceedings 516, New York: American Institute of Physics, 1999, pp.370-383.
7. Abu-Zayyad, T., et al.,*Phys. Rev. Lett.***84**, 4276-4279 (2000).

10^{20}eV Cosmic Ray and Particle Physics with IceCube

J. Alvarez-Muñiz[1] and F. Halzen

Physics Department, University of Wisconsin, Madison, WI 53706, USA

Abstract. We show that a kilometer-scale neutrino observatory, though optimized for detecting neutrinos of TeV to PeV energy, can reveal the science associated with the enigmatic super-EeV radiation in the Universe. Speculations regarding its origin include heavy relics from the early Universe, particle interactions associated with the Greisen cutoff, and topological defects which are remnant cosmic structures associated with phase transitions in grand unified gauge theories. We show that it is a misconception that new instruments optimized to EeV energy can exclusively do this important science. Because kilometer-scale neutrino telescopes such as IceCube can reject the atmospheric neutrino background by identifying the very high energy of the signal events, they have sensitivity over the full solid angle, including the horizon where most of the signal is concentrated. This is critical because upgoing neutrino-induced muons, considered in previous calculations, are absorbed by the Earth. Previous calculations have underestimated the event rates of IceCube for EeV signals by over one order of magnitude.

I INTRODUCTION

It is nothing less but exhilarating to contemplate future neutrino detectors reaching effective volumes of 10^{13} tons and effective areas of 10^6 km^2 by exploiting totally novel detection methods such as radio, acoustic, atmospheric fluorescence and horizontal air shower techniques. This is at a time when we are operating a single neutrino telescope of only 10^3–10^5 m^2 effective area, depending on the science [1]. Its extension to a kilometer-scale neutrino observatory, IceCube, is still at the proposal stage. Neutrino detectors can be classified in four categories, which are delineated by the energy for which the instruments have been optimized:

1. **MeV** detectors: for studying the sun and detecting supernovae,

2. **GeV–TeV** DUMAND-class telescopes: possibly, the first instruments to look beyond the sun, but, more importantly, with sufficiently low threshold to demonstrate the novel techniques that use ice and water as a Cherenkov medium by detecting atmospheric neutrinos. AMANDA is the first in this category, others will be commissioned in Mediterranean waters,

---

[1] Current Address: Bartol Research Institute, University of Delaware, Newark, DE 19716

3. **TeV–PeV** kilometer-scale observatories such as IceCube [2]. These represent, as far as we know, the best-buy for opening up the field of high energy neutrino astronomy,

4. **EeV** detectors specializing in answering the mystifying questions raised by the existence of 100 EeV cosmic rays, and the apparent absence of a Greisen cutoff.

Although optimized in different energy bands, the missions of these instruments can overlap. In this talk we discuss the potential of IceCube to do the science envisaged for the projects of interest to this meeting, such as the radio observatories, the Auger air shower array and the space-based atmospheric fluorescence detector OWL. These discussions are important for two reasons: i) for exploring the full potential of a detector, possibly beyond the specific goals it was designed for, and ii) in order to avoid compromising the performance of an instrument by concentrating on science better done by others. While i) is obvious, ii) is important and often controversial. For instance, should one consider the study of oscillating atmospheric neutrinos, superbly performed with Superkamiokande-type detectors, when optimizing the performance a high energy neutrino telescope?

II   ICECUBE

As far as astronomy beyond the GeV signals of EGRET is concerned, the case for using neutrinos as messengers is compelling [3]. Of all high-energy particles, only weakly interacting neutrinos can directly convey astronomical information from the edge of the universe and from deep inside the most cataclysmic high-energy processes. Copiously produced in high-energy collisions, travelling at the velocity of light, and undeflected by magnetic fields, neutrinos meet the basic requirements for astronomy. Their unique advantage arises from a fundamental property: they are affected only by the weakest of nature's forces (but for gravity) and are therefore essentially unabsorbed as they travel cosmological distances between their origin and us.

The first suggestions that kilometer-size neutrino telescopes were required to do the science originated with early estimates of the flux produced by the highest energy cosmic rays interacting with microwave photons. With time, and after consideration of the diverse scientific missions of astroparticle physics with high energy neutrinos, we have confirmed that the science does require construction of a kilometer-scale neutrino detector, and the challenge has therefore been one of technology. The only demonstrated solution is to use a "natural" detector consisting of a thousand billion liters (a teraliter) of instrumented natural water or ice. After commissioning and operating the Antarctic Muon and Neutrino Detector Array (AMANDA), the AMANDA collaboration is ready to meet this challenge and has proposed to construct IceCube, a one-cubic-kilometer international high-energy neutrino observatory in the clear deep ice below the South Pole Station.

IceCube will be an array of 4,800 optical modules within a cubic kilometer of clear ice. Frozen into holes 2.4 kilometers deep, to be drilled by hot water, the uppermost optical modules will lie 1,400 m below the surface. Simulations anchored to AMANDA data show that the direction of muon tracks can be reconstructed to 0.5 degrees above 1 PeV. IceCube will be capable of identifying neutrino type, or flavor, by mapping showers of Cherenkov light from electron and from tau neutrinos. Most important, it will measure neutrino energy. Energy resolution is critical, because there should be very little background from atmospheric neutrinos at energies above 100 TeV.

III   EeV SCIENCE

In this talk we concentrate on super-EeV science such as topological defects, super-heavy relics and neutrinos associated with the Greisen cutoff. Their detection is usually not considered as a high priority in designing the architecture of neutrino telescopes.

It has been realized for some time that topological defects are unlikely to be the origin of the structure in the present Universe [4]. Therefore the direct observation of their decay products, in the form of cosmic rays or high energy neutrinos, becomes the only way to search for these remnant structures from grand unified phase transitions [5]. This search represents an example of fundamental particle physics that can only be done with cosmic beams. We here point out that a kilometer-scale neutrino observatory, such as IceCube, has excellent discovery potential for topological defects. The instrument can identify the characteristic signatures in the energy and zenith angle distribution of the signal events. Our conclusions for topological defects extend to other physics associated with $10^{20} \sim 10^{24}$ eV energies.

To benchmark the performance of IceCube relative to OWL [6], chosen as an example of an instrument optimized to $\sim 10^2$ EeV energy, we use the following theorized sources of super-EeV neutrinos:

- generic topological defects with grand-unified mass scale M_X of order 10^{15} GeV and a particle decay spectrum consistent with all present observational constraints [7],

- superheavy relics [8,9], which we normalize to the Z-burst scenario [10] where the observed cosmic rays with $\sim 10^{20}$ eV energy, and above, are locally produced by the interaction of super-energetic neutrinos with the cosmic neutrino background,

- neutrinos produced by superheavy relics which themselves decay into the highest energy cosmic rays [11], and

- the flux of neutrinos produced in the interactions of cosmic rays with the microwave background [12]. This flux, which originally inspired the concept of a kilometer-scale neutrino detector, is mostly shown for comparison.

TABLE 1. Comparison of neutrino event rates for three representative neutrino fluxes for OWL [6] and IceCube [13].

	Volume	Eff. Area	Threshold
OWL	10^{13} ton	10^6 km^2	3×10^{19} eV
IceCube	10^9 ton	1 km^2	10^{15} eV*

	Events per Year		
	TD	Z_{burst}	$p\gamma_{2.7}$
OWL ν_e	16	9	5
IceCube ν_μ	11	30	1.5

*actual threshold ~100 GeV; requiring >1 PeV eliminates atmospheric ν background.

Our results are summarized in Table 1 where we compare the event rates for IceCube, discussed later, with those for OWL calculated in reference [6]. The conclusion is clear, while effective volume and area for OWL apparently exceed those of IceCube by many orders of magnitude, the events rates are comparable. This is a consequence of the duty cycle, reduced efficiency, and higher threshold of the OWL detector.

Cognoscenti will notice that the event rates claimed for IceCube are roughly two orders of magnitude larger than those found in the literature for a generic detector with 1 km^2 effective area. The reason for this is simple. Unlike first-generation neutrino telescopes, IceCube can measure energy, and can therefore separate interesting high energy events from the background of lower energy atmospheric neutrinos by energy measurement. The instrument can identify high energy neutrinos over 4π solid angle, and not just in the lower hemisphere where they are identified by their penetration of the Earth, as is the case with AMANDA. This is of primary importance here because neutrinos produced, for instance by the decay of topological defects, have energies large enough to be efficiently absorbed by the Earth. The observed events are dominated by neutrinos interacting in the ice or atmosphere *above* the detector and near the horizon where the atmosphere alone represent a target density for converting neutrinos of 36 kg/cm^2. This event rate typically dominates the one for up-going neutrinos by an order of magnitude. We will show that the zenith angle distribution of neutrinos associated with EeV signals form a striking signature for their extremely high energy origin.

IV   NEUTRINO EVENTS

We calculate the neutrino event rates by convoluting the $\nu_\mu + \bar{\nu}_\mu$ flux from the different sources considered in this talk, with the probability of detecting a muon produced in a muon-neutrino interaction in the Earth or atmosphere:

$$N_{\text{events}} = 2\pi \ A_{\text{eff}} \ T \int \int \frac{dN_\nu}{dE_\nu}(E_\nu) P_{\nu \to \mu}(E_\nu, E_\mu(\text{thresh}), \cos\theta_{\text{zenith}}) \ dE_\nu \ d\cos\theta_{\text{zenith}}$$

$$(1)$$

where T is the observation time and θ_{zenith} the zenith angle. We assume an effective telescope area of $A_{\text{eff}} = 1$ km^2, a conservative assumption for the very high energy neutrinos considered here. It is important to notice that the probability $(P_{\nu \to \mu})$ of detecting a muon with energy above a certain energy threshold $E_\mu(\text{threshold})$, produced in a muon-neutrino interaction, depends on the angle of incidence of the neutrinos. This is because the distance traveled by a muon cannot exceed the column density of matter available for neutrino interaction, a condition not satisfied by very high energy neutrinos produced in the atmosphere. They are absorbed by the Earth and only produce neutrinos in the ice above, or in the atmosphere or Earth near the horizon. The event rates in which the muon arrives at the detector with an energy above $E_\mu(\text{threshold}) = 1$ PeV, where the atmospheric neutrino background is negligible, are summarized in Table 2. We discuss them in more detail by introducing Figs. 1–5.

TABLE 2. Neutrino event rates (per year per km^2 in 2π sr) in which the produced muon arrives at the detector with an energy above $E_\mu(\text{threshold})=1$ PeV. Different neutrino sources have been considered. The topological defect models (TD) correspond to highest injection rates Q_0 (ergs cm^{-3} s^{-1}) allowed in Fig. 2 of [7]. Also shown is the number of events from p-γ_{CMB} interactions in which protons are propagated up to a maximum redshift $z_{\text{max}} = 2.2$ [12] and the number of neutrinos from the Waxman and Bahcall limit on the diffuse flux from optically thin sources [14]. The number of atmospheric background events above 1 PeV is also shown. The second column corresponds to downward going neutrinos (in 2π sr). The third column gives the number of upward going events (in 2π sr). We have taken into account absorption in the Earth according to reference [15]. IceCube will detect the sum of the event rates given in the last two columns.

Model	$N_{\nu_\mu + \bar{\nu}_\mu}$ (downgoing)	$N_{\nu_\mu + \bar{\nu}_\mu}$ (upgoing)
TD, $M_X = 10^{14}$ GeV, $Q_0 = 6.31 \times 10^{-35}$, p=1	11	1
TD, $M_X = 10^{14}$ GeV, $Q_0 = 6.31 \times 10^{-35}$, p=2	3	0.3
TD, $M_X = 10^{15}$ GeV, $Q_0 = 1.58 \times 10^{-34}$, p=1	9	1
TD, $M_X = 10^{15}$ GeV, $Q_0 = 1.12 \times 10^{-34}$, p=2	2	0.2
Superheavy Relics Gelmini *et al.* [8]	30	1.5×10^{-7}
Superheavy Relics Berezinsky *et al.* [11]	2	0.2
Superheavy Relics Birkel *et al.* [9]	1.5	0.3
p-γ_{CMB} ($z_{\text{max}} = 2.2$) [12]	1.5	1.2×10^{-2}
W-B limit $2 \times 10^{-8} \ E^{-2}$ (cm^2 s sr GeV)$^{-1}$	8.5	2
Atmospheric background	2.4×10^{-2}	1.3×10^{-2}

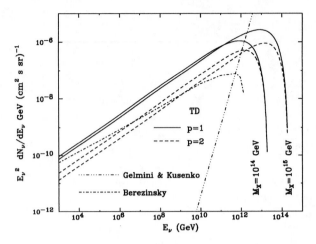

FIGURE 1. Maximal predictions of $\nu_\mu + \bar{\nu}_\mu$ fluxes from topological defect models by Protheroe and Stanev (p=1,2). Also shown is the $\nu_\mu + \bar{\nu}_\mu$ from superheavy relic particles by Gelmini and Kusenko and the flux by Berezinsky *et al.*.

Figure 1 shows the $\nu_\mu + \bar{\nu}_\mu$ fluxes used in the calculations. We first calculate the event rates corresponding to the largest flux from topological defects [7] allowed by constraints imposed by the measured diffuse γ-ray background in the vicinity of 100 MeV. The corresponding proton flux has been normalized to the observed cosmic ray spectrum at 3×10^{20} eV; see Fig. 2 of reference [7]. Models with $p < 1$ appear to be ruled out [16] and hence they are not considered in the calculation. As an example of neutrino production by superheavy relic particles, we consider the model of Gelmini and Kusenko [8]. In Figs. 2 and 3 we show the event rates as a function of neutrino energy. We assume a muon energy threshold of 1 PeV. We also show in both plots the event rate corresponding to the Waxman and Bahcall "bound" [14]. This bound represents the maximal flux from astrophysical, optically thin sources, in which neutrinos are produced in p-p or p-γ collisions. The atmospheric neutrino events are not shown since they are negligible above the muon energy threshold we are using. The area under the curves in both figures is equal to the number of events for each source. In Fig. 4 we plot the event rates in which the produced muon arrives at the detector with an energy greater than E_μ(threshold). In Fig. 5 we finally present the angular distribution of the neutrino events for the different sources. The characteristic shape of the distribution reflects the opacity of the Earth to high energy neutrinos, typically above ~100 TeV. The limited column density of matter in the atmosphere essentially reduces the rate of downgoing neutrinos to interactions in the 1.5 km of ice above the detector. The events are therefore concentrated near the horizontal direction corresponding to zenith angles close to 90°. The neutrinos predicted by the model of Gelmini and

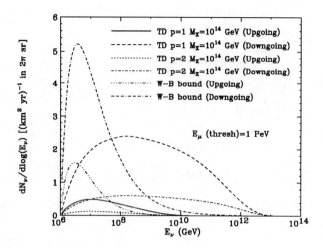

FIGURE 2. Differential $\nu_\mu + \bar{\nu}_\mu$ event rates in IceCube from the topological defect fluxes in Fig.1. The muon threshold is E_μ(threshold)=1 PeV. We have separated the contribution from upgoing and downgoing events to stress the different behavior with energy. The event rate expected from the Waxman and Bahcall bound (see text) is also shown for illustrative purposes. The rate due to atmospheric neutrinos is negligible (see Table 2) and hence it is not plotted.

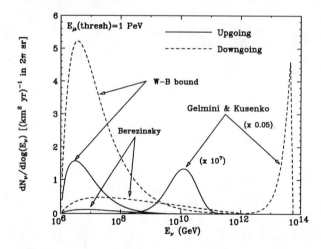

FIGURE 3. Differential $\nu_\mu + \bar{\nu}_\mu$ event rates in IceCube from super-heavy relic particles. We have separated the contribution from upgoing and downgoing events to stress the different behavior with energy. The muon threshold is E_μ(threshold)=1 PeV. The event rate due to atmospheric neutrinos as well as the one expected from the Waxman and Bahcall bound (see text) is shown for illustrative purposes. The rate due to atmospheric neutrinos is negligible (see Table 2) and hence it is not plotted.

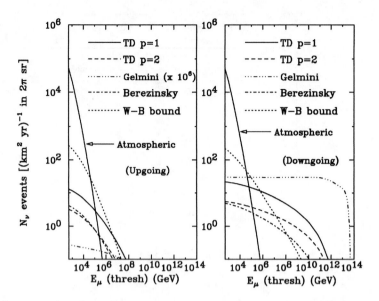

FIGURE 4. $\nu_\mu + \bar{\nu}_\mu$ event rates in IceCube from the fluxes in Fig.1. The plot shows the number of events in which the produced muon arrives at the detector with an energy above E_μ(thresh). Atmospheric neutrino events and the event rate expected from the Waxman and Bahcall upper bound (see text) are also plotted. The topological defect (TD) models shown (p=1 and p=2) correspond to $M_X = 10^{14}$ GeV. Upgoing and downgoing events are shown separately.

Kusenko are so energetic that they are absorbed, even in the horizontal direction as can be seen in Fig. 5.

Energy measurement is critical for achieving the sensitivity of the detector claimed. For muons, the energy resolution of IceCube is anticipated to be 25% in the logarithm of the energy, possibly better. The detector is definitely able to determine energy to better than an order of magnitude, sufficient for the separation of EeV signals from atmospheric neutrinos with energies below 100 TeV. Notice that one should also be able to identify electromagnetic showers initiated by electron and tau-neutrinos. The energy response for showers is linear, and expected to be better than 20%. Such EeV events will be gold-plated, unfortunately their fluxes are expected to be even lower. For instance for the first TD model in Table 2 (p=1, $M_X = 10^{14}$ GeV and $Q_0 = 6.31 \times 10^{-35}$ ergs cm^{-3} s^{-1}), we expect ~ 1 contained shower per year per km^2 above 1 PeV initiated in charged current interactions of $\nu_e + \bar{\nu}_e$. The corresponding number for the Gelmini and Kusenko flux is ~ 4 yr^{-1} km^{-2}.

One should also worry about the fact that a very high energy muon may enter the detector with reduced energy because of energy losses. It can, in principle, become indistinguishable from a minimum ionizing muon of atmospheric origin [17]. We

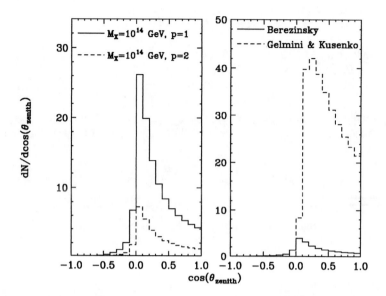

FIGURE 5. Zenith angle distribution of the $\nu_\mu + \bar{\nu}_\mu$ events in IceCube in which the produced muon arrives at the detector with energy above 1 PeV. Left: Topological defect models. Right: Superheavy relics. $\cos(\theta_{\text{zenith}}) = -1$ corresponds to vertical upgoing neutrinos, $\cos(\theta_{\text{zenith}}) = 0$ to horizontal neutrinos and $\cos(\theta_{\text{zenith}}) = 1$ to vertical downgoing neutrinos. The detector is located at a depth of 1.8 km in the ice.

have accounted for the ionization as well as catastrophic muon energy losses which are incorporated in the calculation of the range of the muon. In the PeV regime region this energy reduction is roughly one order of magnitude, it should be less for the higher energies considered here.

In conclusion, if the fluxes predicted by our sample of models for neutrino production in the super-EeV region are representative, they should be revealed by the IceCube observatory operated over several years.

ACKNOWLEDGEMENTS

We thank J.J. Blanco-Pillado for making available to us his code to obtain the neutrino fluxes from topological defects and E. Zas for helpful discussions. This research was supported in part by the US Department of Energy under grant DE-FG02-95ER40896 and in part by the University of Wisconsin Research Committee with funds granted by the Wisconsin Alumni Research Foundation. The research activities of J.A. at Bartol Research Institute are funded in part by NASA grant NAG5-7009.

REFERENCES

1. A. Andreś et al. (The AMANDA Collaboraton), talk presented at the 9th International Conference On Neutrino Physics And Astrophysics (Neutrino 2000), Sudbury, Canada, June 2000, astro-ph/0009242; R. Wischnewski et al. (The AMANDA collaboration), Nucl. Phys. Proc. Suppl. **85**, 141-145 (2000).
2. The IceCube NSF proposal *http://pheno.physics.wisc.edu/icecube/*
3. T.K. Gaisser, F. Halzen and T. Stanev, Phys. Rep. **258**, 173 (1995), and references therein.
4. See Proceedings of the First International Workshop on Particle Physics and the Early Universe, Ambleside, England, 1997, Ed. L. Roszkowski, pp. 403-432.
5. P. Bhattacharjee, C.T. Hill and D.N. Schramm, Phys. Rev. Lett. **69**, 567 (1992).
6. D. Cline and F.W. Stecker, contributed to OWL/AW Neutrino Workshop on Observing Ultrahigh Energy Neutrinos, Los Angeles, California, Nov. 1999, astro-ph/0003459.
7. R.J. Protheroe and T. Stanev, Phys. Rev. Lett. **77**, 3708 (1996) and Erratum **78**, 3420 (1997).
8. G. Gelmini and A. Kusenko, Phys. Rev. Lett. **84**, 1378 (2000).
9. M. Birkel and S. Sarkar, Astropart. Phys. **9**, 297 (1998).
10. T. Weiler, Astropart. Phys. **11**, 303 (1999); D. Fargion, B. Mele and A. Salis, Astrophys. J. **517**, 725 (1999).
11. V. Berezinsky, M. Kachelreiß and A. Vilenkin, Phys. Rev. Lett. **79**, 4302 (1997).
12. F.W. Stecker et al., Phys. Rev. Lett., **66**, 2697 (1991).
13. J. Alvarez-Muñiz and F. Halzen, Phys. Rev. D **63**, 037302 (2001).
14. E. Waxman and J. Bahcall, Phys. Rev. D **59**, 023002 (1998); E. Waxman and J. Bahcall, hep-ph/9902383.
15. R. Gandhi, C. Quigg, M.H. Reno, I. Sarcevic, Phys. Rev. D **58**, 093009 (1998).
16. G. Sigl, S. Lee, P. Bhattacharjee and S. Yoshida, Phys. Rev. D **59**, 043504 (1998).
17. T.K. Gaisser, talk given at OECD Megascience Forum Workshop, Taormina, Italy, May 1997, astro-ph/9704061.

APPENDIX

Poems by Gurgen Askaryan

Translated from Russian by Karo Ispirian

I am leaving this world for you
Choosing a rest for me
Far from vanity, from shame and passion.
The life is a fair,
God is the owner of the world
In which the genius is a target.

* * *

Ask forgiveness of the animals
Whom you killed, hurt and ate,
Ask pardon and help them still you are alive,
and it is not late.
You see, the life is awful
For them and you.
Terrible is the epoch
Never harm other destiny.

* * *

I decided to prolong the agony
To listen the Mozart harmony.

* * *

It is dear to me
what is disgusting for the others.
My diseased sister,
discomfort at home,
slipshod dressing,
and the desperate smell of my cats,
their wisdom and grace
calling exhausted to conceive.

Thy eyes are empty mirrors
Reflecting anyone who approaches you,
Their depth lied so long to me,
Nevertheless, I dream to be sunken in them.

The prescription of love apparently has not passed,
And I am bewitched by the play of reminiscences;
How frequently the face of the fate is combined with you
And how they diverge in my deceptive consciousness.

*   *   *

I don't like flowers, especially in a man's hands,
Their such appearance defames the taste of the owner
I don't like flowers, especially in coffins
They render business coloring to the tragedy.
I don't like when somebody picks flowers
To admire their agony
And to throw next day as a trash
And take a shower to escape the stench

*   *   *

Mummy, mummy, your phloxes are again bloomed,
Alas, no mother, no those phloxes, no me.

List of Participants

Alvarez-Muñiz, Jaime
Arisaka, Katsushi
Badran, H. M.
Barwick, Steve
Bean, Alice
Besson, David Z.
Blanovsky, Anatoly
Bolotovskii, Boris
Buniy, Roman V.
Chiba, Masami
Cummings, Mary Anne
Dagkesamanskii, Rustam D.
Duda, Gintaras
Gelmini, Graciela
Gerhardt, Lisa
Gorham, Peter
Halzen, Francis
Hankins, Tim
Hundertmark, Stephan
Ispirian, Karo
Kuiper, Thomas
Kusenko, Alexander
Kuzmin, Vadim
Learned, John
Liewer, Kurt
McKay, Doug

Naudet, Charles J.
Nemati, Bijan
Odian, Allen
Ostrowski, Michal
Postma, Marieke
Ralston, John P.
Razzaque, Soebur
Rosner, Jonathan L.
Saltzberg, David
Schlein, Peter
Seckel, David
Sigl, Guenter
Sinha, Kalpana *(in absentia)*
Sokolsky, Pierre
Takahashi, Yoshi
Teshima, Masahiro
Urbina, Javier
Varieschi, Gabriele
Vinogradova, Tatiana
Weekes, Trevor C.
Weiler, Tom
Wick, Stuart
Williams, Dawn R.
Wong, Alfred Y.
Zas, Enrique
Zheleznykh, Igor

Author Index

A

Alvarez-Muñiz, J., 117, 128, 305

B

Baishya, R., 98
Bean, A., 234
Besson, D. Z., 139, 157
Bolotovskii, B. M., 14
Buniy, R. V., 147, 234

C

Chapin, E., 271
Chiba, M., 204

D

Dagkesamanskii, R. D., 189
Datta, P., 98
Dedenko, L. G., 277

E

Ekers, R. D., 168
Elbakian, S. S., 241

F

Field, C., 225

G

Gazazian, E. D., 111, 241
Gorham, P. W., 177, 225, 253, 271

H

Halzen, F., 305
Hankins, T. H., 168
Husain, A., 204

I

Ikeda, M., 204
Inuzuka, M., 204
Ispirian, K. A., 111, 241, 317
Ispirian, R. K., 241
Iverson, R., 225

K

Kamijo, T., 204
Karlik, Y. S., 277
Kawaki, M., 204
Kephart, T. W., 43
Kuzmin, V. A., 23

L

Learned, J. G., 277
Liewer, K. M., 177

M

McKay, D. W., 139

N

Naudet, C. J., 177

O

Odian, A., 225
O'Sullivan, J. D., 168

321